AF807287

Information Systems Engineering and Management

Volume 54

Editorial Board

Abdelkader Hameurlain, Université Toulouse III Paul Sabatier, Toulouse, France

Ali Idri, ENSIAS, Mohammed V University, Rabat, Morocco

Ashok Vaseashta, International Clean Water Institute, Manassas, VA, USA

Ashwani Kumar Dubey, Amity University, Noida, India

Carlos Montenegro, Francisco José de Caldas District University, Bogota, Colombia

Claude Laporte, University of Quebec, Québec, QC, Canada

Fernando Moreira, Portucalense University, Berlin, Germany

Francisco Peñalvo, University of Salamanca, Salamanca, Spain

Gintautas Dzemyda, Vilnius University, Vilnius, Lithuania

Jezreel Mejia-Miranda, CIMAT - Center for Mathematical Research, Zacatecas, Mexico

Jon Hall, The Open University, Milton Keynes, UK

Mário Piattini, University of Castilla-La Mancha, Albacete, Spain

Maristela Holanda, University of Brasilia, Brasilia, Brazil

Mincong Tang, Beijing Jiaotong University, Beijing, China

Mirjana Ivanovíc, Department of Mathematics and Informatics, University of Novi Sad, Novi Sad, Serbia

Mirna Muñoz, CIMAT Center for Mathematical Research, Progreso, Mexico

Rajeev Kanth, University of Turku, Turku, Finland

Sajid Anwar, Institute of Management Sciences, Peshawar, Pakistan

Tutut Herawan, Faculty of Computer Science and Information Technology, University of Malaya, Kuala Lumpur, Malaysia

Valentina Colla, TeCIP Institute, Scuola Superiore Sant'Anna, Pisa, Italy

Vladan Devedzic, University of Belgrade, Belgrade, Serbia

Series Editor

Álvaro Rocha, ISEG, University of Lisbon, Lisbon, Portugal

The book series "Information Systems Engineering and Management" (ISEM) publishes innovative and original works in the various areas of planning, development, implementation, and management of information systems and technologies by enterprises, citizens, and society for the improvement of the socio-economic environment.

The series is multidisciplinary, focusing on technological, organizational, and social domains of information systems engineering and management. Manuscripts published in this book series focus on relevant problems and research in the planning, analysis, design, implementation, exploration, and management of all types of information systems and technologies. The series contains monographs, lecture notes, edited volumes, pedagogical and technical books as well as proceedings volumes.

Some topics/keywords to be considered in the ISEM book series are, but not limited to: Information Systems Planning; Information Systems Development; Exploration of Information Systems; Management of Information Systems; Blockchain Technology; Cloud Computing; Artificial Intelligence (AI) and Machine Learning; Big Data Analytics; Multimedia Systems; Computer Networks, Mobility and Pervasive Systems; IT Security, Ethics and Privacy; Cybersecurity; Digital Platforms and Services; Requirements Engineering; Software Engineering; Process and Knowledge Engineering; Security and Privacy Engineering, Autonomous Robotics; Human-Computer Interaction; Marketing and Information; Tourism and Information; Finance and Value; Decisions and Risk; Innovation and Projects; Strategy and People.

Indexed by Google Scholar. All books published in the series are submitted for consideration in the Web of Science.

For book or proceedings proposals please contact Alvaro Rocha (amrrocha@gmail.com).

Kali Charan Rath · Alvaro Rocha ·
Siba Sankar Mahapatra · Anoop Kumar Sood ·
Amaresh Kumar
Editors

Engineering, Technology and Management

Practices for Sustainable Manufacturing Industry, Volume 1

Springer

Editors
Kali Charan Rath
Department of Mechanical Engineering
Gandhi Institute of Engineering
and Technology University
Gunupur, Odisha, India

Siba Sankar Mahapatra
Department of Mechanical Engineering
NIT Rourkela
Rourkela, Odisha, India

Amaresh Kumar
Production and Industrial Engineering
National Institute of Technology
Jamshedpur (NIT)
Jamshedpur, Jharkhand, India

Alvaro Rocha
School of Economics and Management
University of Lisbon, ISEG
Lisbon, Portugal

Anoop Kumar Sood
Department of Mechanical
and Manufacturing Engineering
National Institute of Advanced
Manufacturing Technology (NIAMT)
Ranchi, Jharkhand, India

ISSN 3004-958X ISSN 3004-9598 (electronic)
Information Systems Engineering and Management
ISBN 978-3-031-93653-1 ISBN 978-3-031-93654-8 (eBook)
https://doi.org/10.1007/978-3-031-93654-8

© The Editor(s) (if applicable) and The Author(s), under exclusive license to Springer Nature Switzerland AG 2025

This work is subject to copyright. All rights are solely and exclusively licensed by the Publisher, whether the whole or part of the material is concerned, specifically the rights of translation, reprinting, reuse of illustrations, recitation, broadcasting, reproduction on microfilms or in any other physical way, and transmission or information storage and retrieval, electronic adaptation, computer software, or by similar or dissimilar methodology now known or hereafter developed.
The use of general descriptive names, registered names, trademarks, service marks, etc. in this publication does not imply, even in the absence of a specific statement, that such names are exempt from the relevant protective laws and regulations and therefore free for general use.
The publisher, the authors and the editors are safe to assume that the advice and information in this book are believed to be true and accurate at the date of publication. Neither the publisher nor the authors or the editors give a warranty, expressed or implied, with respect to the material contained herein or for any errors or omissions that may have been made. The publisher remains neutral with regard to jurisdictional claims in published maps and institutional affiliations.

This Springer imprint is published by the registered company Springer Nature Switzerland AG
The registered company address is: Gewerbestrasse 11, 6330 Cham, Switzerland

If disposing of this product, please recycle the paper.

Contents

Chapter 1
Advances Sustainability in Manufacturing: Exploring Green Additive Manufacturing and Environmental Impact Mitigation

Koyilada Benarji, Jinoop Arackal Narayanan, Shashikant Das, and Sujit Mishra

Abstract Additive Manufacturing (AM) holds significant promise for transforming manufacturing industries towards sustainability. This paper explores the principles, advancements, and challenges in sustainable additive manufacturing, contrasting it with the environmental impact of traditional manufacturing processes. Principles such as reducing environmental impact, energy efficiency, material efficiency, and economic viability are discussed, emphasizing the role of AM in promoting eco-friendly production practices. The study explores progress in environmentally friendly additive manufacturing (AM) technologies, focusing on eco-conscious materials, energy-conserving processes, and techniques for minimizing waste, highlighting the industry's commitment to sustainability. Direct energy deposition (DED) uses a laser to create a melt pool for either powder or wire feed, enabling high deposition rates and allowing multi-material builds. It is ideal for fabricating large components and repairing damaged parts. Powder Bed Fusion (PBF) employs selective melting of metal powder layers using either a laser or electron beam, achieving high precision suitable for complex geometries and fine features. While DED is favoured for large structures, PBF is preferred in applications requiring lightweight, intricate parts with minimal post-processing. The inclusion of Life Cycle Assessment (LCA) into Additive Manufacturing (AM) and its effects on evaluating environmental

K. Benarji (✉)
Shri Vishnu Engineering College for Women, Andhra Pradesh, India
e-mail: drkbenarjime@svecw.edu.in

J. A. Narayanan
Teesside University, Middlesbrough, Tees Valley, UK
e-mail: J.ArackalNarayanan@tees.ac.uk

S. Das · S. Mishra
Centurion University of Technology, R. Sitapur, Odisha, India
e-mail: Shashi350me@gmail.com

S. Mishra
e-mail: SujitMishra@cutm.ac.in

© The Author(s), under exclusive license to Springer Nature Switzerland AG 2025
K. C. Rath et al. (eds.), *Engineering, Technology and Management*, Information Systems Engineering and Management 54, https://doi.org/10.1007/978-3-031-93654-8_1

consequences and economic viability. The research also delves into the obstacles and prospective developments in the field of sustainable AM. By addressing these challenges and embracing future directions, AM can play a vital role in achieving a circular economy and reducing environmental impact in manufacturing industries.

Keywords Additive manufacturing · Sustainability · Energy efficiency · Environmental impact · Life cycle assessment · Green technologies · Circular economy

Preface
This chapter addresses the significant role that sustainable practices play in transforming manufacturing processes, specifically through additive manufacturing (AM). The focus on green manufacturing technologies highlights AM's capacity to mitigate environmental impacts associated with traditional manufacturing. This chapter explores innovations in AM, like the development of eco-friendly materials and energy-efficient methodologies, as well as Life Cycle Assessment (LCA) to evaluate sustainability comprehensively. By investigating the economic and environmental advantages of AM, this work aims to demonstrate its pivotal role in advancing a circular economy within industrial sectors.

1.1 Introduction

Additive Manufacturing (AM), has gained recognition as a promising approach that creates objects through sequential layering. This method offers notable benefits, including improved material utilization and decreased waste production, when contrasted to traditional subtractive production techniques [1, 2]. Sustainable Additive Manufacturing (AM) takes this a step further by optimizing energy and material usage, making the process more eco-friendly and economically viable [3, 28]. The ecosystem for sustainable additive manufacturing is shown in Fig. 1.1. By adopting AM, industries can significantly reduce material waste and manufacturing costs while offering substantial environmental benefits. This approach focuses on producing goods closer to where they are needed, reducing transportation needs and greenhouse gas emissions. Moreover, Additive manufacturing allows for the utilization of diverse materials, including those that are recyclable and biodegradable. This capability promotes innovation in eco-friendly production methods and helps support efforts to protect the environment. Through design optimization and the use of multi-materials, AM extends product lifespan and reduces the consumption of critical raw materials, aligning with sustainability goals [4]. Sustainable AM seeks to improve the environmental footprint of manufacturing by reducing energy consumption, thereby achieving greener production methods. By combining Life Cycle Assessment (LCA) with Life Cycle Cost (LCC) methodologies, AM aims to strike a balance between

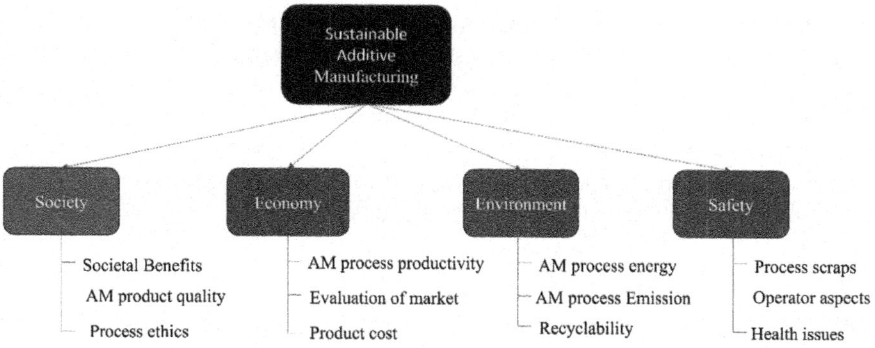

Fig. 1.1 An ecosystem for sustainable additive manufacturing

economic viability and environmental sustainability, contributing to the achievement of the Sustainable Development Goals (SDGs) [5]. The implementation of AM within supply chains encourages distributed production, which leads to decreased transportation-related emissions and fosters localized manufacturing [6, 7].

Additionally, AM supports the circular economy by recycling materials into new products, reducing environmental impact and improving profitability. This transformative technology has revolutionized industries like aerospace and automotive by efficiently producing complex parts, thereby contributing to environmental sustainability. In this chapter, we will explore the various dimensions of green additive manufacturing, including its technological foundations, environmental benefits, and practical applications across industries. By examining the intersection of AM and sustainability, we aim to provide insights into the transformative potential of green additive manufacturing and its role in shaping a more sustainable future for manufacturing industries worldwide.

The objective of this research endeavour is to examine the foundational principles, technological advancements, and inherent challenges associated with sustainable additive manufacturing (AM), while also conducting a comparative analysis of its ecological footprint in relation to conventional manufacturing methodologies. The research aims to demonstrate how AM can support environmentally friendly production by focusing on aspects such as minimizing environmental harm, improving energy and material efficiency, and ensuring economic viability. This research entails a comprehensive examination of environmentally sustainable additive manufacturing (AM) technologies, which encompasses the utilization of eco-friendly materials, energy-conserving methodologies, and approaches aimed at minimizing waste generation. Additionally, it investigates the implementation of Life Cycle Assessment (LCA) within the realm of AM to assess ecological ramifications and economic viability. Additionally, the study identifies the challenges faced by the industry in implementing sustainable AM practices and explores potential future directions for enhancing sustainability in manufacturing. This research presents a distinctive and thorough examination of the sustainability dimensions associated with additive manufacturing. It encompasses a discussion of eco-friendly technologies and sustainable materials while simultaneously assessing strategies for energy efficiency and waste minimization. The incorporation of Life Cycle Assessment (LCA) establishes a rigorous framework for assessing the environmental and economic repercussions of additive manufacturing. This research provides novel perspectives on the ability of AM to significantly enhance the innovation of a circular economy and mitigate the ecological footprint of the manufacturing industry.

1.2 Environmental Impact of Traditional Manufacturing

Traditional manufacturing processes, especially prevalent in sectors like aircraft manufacturing, involve removing large pieces of material from a workpiece, resulting in significant waste and emissions [4]. These processes consume substantial energy and materials, leading to high CO_2 emissions and environmental degradation. The dependence on petroleum-based oils and lubricants within conventional manufacturing processes, in conjunction with the utilization of aqueous cutting fluids, presents significant environmental challenges attributable to their toxic characteristics and the generation of carbon dioxide emissions during their production phase [8].

Moreover, the conveyance of unprocessed materials and completed goods within conventional manufacturing systems significantly exacerbates atmospheric pollution and is predominantly dependent on fossil fuel resources. Processes like metal injection moulding release harmful by-products, impacting air and water quality [7]. These practices contribute to direct emissions from activities like mining and indirect emissions through electricity use, exacerbating environmental challenges such as metal scarcity [4]. Efforts to alleviate the ecological ramifications of conventional manufacturing processes encompass the integration of minimum quantity lubrication (MQL) and cryogenic refrigerants such as liquid nitrogen (LN_2) and carbon dioxide (CO_2). The implementation of MQL considerably diminishes the reliance on traditional coolants, thereby fostering sustainable machining methodologies and curtailing environmental contamination [8, 9]. Technological advancements such as high-pressure jet-assisted machining and cryogenic machining are designed to diminish energy consumption and material wastage, thereby aligning with objectives of sustainability.

Traditional manufacturing's reliance on non-renewable energy sources and inefficient material usage contrasts with sustainable additive manufacturing methods. Additive manufacturing significantly reduces waste generation by building parts layer by layer, minimizing material usage. Metal additive manufacturing (MAM) can cut waste by up to 90% compared to traditional methods, aiming to use materials only where necessary. Moreover, MAM reduces environmental impact by over 60% for industrial machinery and aeronautical sectors, primarily through reduced material consumption and efficient resource utilization [10, 11].

1.2.1 Case Studies Illustrating Environmental Challenges

Case Study 1: Air Pollution from Industrial Facilities.

Description: A production facility situated within a highly populated region discharges significant quantities of atmospheric contaminants, such as sulfur dioxide (SO_2) and nitrogen oxides (NOx), resulting in respiratory health issues and ecological deterioration.

Impact: Increased healthcare costs, reduced air quality, and ecosystem damage.

Solutions: Installation of pollution control technologies, adoption of cleaner production methods, and regulatory compliance.

Case Study 2: Waste Generation in Textile Manufacturing.

Description: A textile factory generates large volumes of wastewater contaminated with dyes, chemicals, and heavy metals during fabric dyeing and finishing processes.

Impact: Water pollution, aquatic ecosystem degradation, and public health risks.

Solutions: Implementation of water recycling and treatment systems, adoption of eco-friendly dyes and chemicals, and adherence to environmental regulations.

Case Study 3: Carbon Footprint of Automobile Manufacturing.

Description: An automotive assembly plant consumes substantial energy and resources in manufacturing vehicles, contributing to a high carbon footprint.

Impact: Increased GHG emissions, climate change effects, and energy inefficiency.

Solutions: Implementation of energy-efficient technologies, use of lightweight materials, and adoption of renewable energy sources.

Case Study 4: Deforestation for Paper Production.

Description: Logging activities for paper production result in deforestation of vital forest ecosystems, leading to habitat loss and biodiversity decline.

Impact: Loss of wildlife habitat, soil erosion, and disruption of ecosystem services.

Solutions: Sustainable forestry practices, promotion of recycled paper, and conservation initiatives.

These case studies highlight the diverse environmental challenges associated with traditional manufacturing processes and underscore the urgent need for sustainable alternatives. By understanding the environmental impacts of conventional manufacturing, stakeholders can work towards implementing greener practices and fostering a more sustainable industrial ecosystem.

1.3 Principles of Sustainable Manufacturing

1.3.1 Reduce Environmental Impact

The concept of sustainable manufacturing seeks to mitigate adverse environmental impacts through the reduction of waste and emissions, as well as the careful selection of materials and processes that are less detrimental to the ecosystem. For instance, biochar, derived from plant waste, can be used to make plastics stronger and more heat-resistant while reducing resource usage and pollution. Biochar, coming from renewable sources and locking carbon away, plays a key role in creating environmentally friendly materials. The process of making these new plastics with biochar, called in situ polymerization, efficiently incorporates biochar into the plastic, using less energy compared to traditional methods, thus making the process greener [12]. Additionally, reducing energy use is crucial for lowering environmental harm in additive manufacturing. By making machines more efficient, optimizing resource and energy use, and enhancing the quality and efficiency of the manufacturing process, we can significantly lessen the adverse impacts on the environment [13]. The development of goods characterized by extended lifespans mitigates the necessity for regular substitutions, thereby fostering practices of recycling and reuse and facilitating a circular economy wherein materials are utilized for extended durations. This methodology emphasizes the utilization of renewable energy resources to diminish the carbon footprint and safeguard the well-being and safety of laborers and the surrounding community [3]. Additive Manufacturing (AM) can significantly reduce material consumption and CO_2 emissions by enabling the precise redesign and manufacturing of products, leading to less waste and lower energy use. By producing goods near the point of consumption, Additive manufacturing (AM) additionally aids in mitigating the ecological ramifications linked to extensive transportation distances, thereby fostering more sustainable supply chain practices. AM supports the recycling of unused raw materials, minimizing production waste and promoting a circular economy by reusing materials that would otherwise contribute to industrial waste. This phenomenon may yield commodities characterized by extended longevity, thereby diminishing the ecological impact linked to the fabrication of novel merchandise [14].

1.3.2 Energy Efficiency

Sustainable manufacturing prioritizes using energy as efficiently as possible, incorporating renewable energy sources to lessen the carbon footprint. Additive manufacturing, which often uses significant energy, especially for large metal parts, can be made more environmentally friendly by enhancing machine efficiency and optimizing material use. Research indicates that using less energy in additive manufacturing can significantly lower its environmental impact, making the process both cost-effective and faster [13]. Sustainable additive manufacturing (AM) systems are engineered to be less energy-demanding by refining the production methodology, utilizing solely the requisite quantity of materials, and minimizing waste generation. AM technologies further facilitate the fabrication of lightweight products, leading to energy conservation during the operational stage of the product, particularly in industries such as automotive and aerospace where weight minimization is of paramount importance. Moreover, localized production through AM diminishes energy expenditures linked to transportation and logistics, thereby augmenting overall energy efficiency by positioning manufacturing closer to the consumption point [3].

1.3.3 Material Efficiency

This principle involves selecting materials that are environmentally friendly, recyclable, and require less energy to produce, thereby reducing the overall environmental impact of manufacturing. Integrating biochar into polyamide 11, a biobased polymer, through in situ polymerization enhances the composite's mechanical and thermal properties without additional processing steps, utilizing agricultural or forestry waste and improving material performance [12]. Using materials wisely in additive manufacturing involves ensuring precise material use, minimizing waste, and saving resources. By monitoring material usage and recycling materials that might otherwise be discarded, additive manufacturing becomes more environmentally friendly and cost-effective [13]. Additive Manufacturing (AM) technologies facilitate the reclamation of materials, including metallic powders and polymer filaments, which can be re-utilized in novel manufacturing methodologies, thereby augmenting material efficacy and promoting the principles of a circular economy. Designing products optimized for AM further improves material efficiency by creating lightweight and functionally optimized components, reducing material use without compromising product integrity. This process inherently produces less waste compared to traditional manufacturing methods [3].

1.3.4 Economic Viability

Sustainable manufacturing ensures processes are economically viable, balancing cost with benefits to society and the environment. Additive manufacturing can save money by using materials and energy efficiently, which lowers costs and increases product value. Research shows that reducing energy use and optimizing material use can lessen environmental harm while enhancing the economic potential of additive manufacturing [13]. Sustainable additive manufacturing (AM) has the capacity to generate limited quantities of tailored products with greater economic efficiency compared to conventional mass production methods, as it obviates the necessity for tooling and molds, diminishes transition costs, and facilitates direct fabrication from three-dimensional computer-aided design (CAD) models. This approach speeds up modifications and customizations, saving material and reducing waste. The use of digital files for designs, which can be quickly shared and modified, reduces the time and cost associated with design changes, enhancing economic efficiency. The material-efficient process of AM reuses waste material, such as powder or resin, not utilized during manufacturing, significantly reducing material costs and enhancing economic viability [3]. Utilizing sustainable materials like biochar promotes renewable resources and reduces waste by integrating agricultural or forestry by-products into new materials. The process of in situ polymerization, where biochar is mixed with polymers directly during creation, ensures a more environmentally friendly manufacturing approach by reducing energy-intensive steps and enhancing material compatibility. These biochar-enhanced materials can produce complex objects with high precision and minimal waste, aligning with sustainable design principles by optimizing material usage and reducing production waste [12].

1.3.5 Material Selection for Eco-friendly Production

Material selection is crucial for achieving eco-friendly production in additive manufacturing. Choosing sustainable materials with minimal environmental impact promotes green manufacturing practices. Recycled and bio-based materials reduce reliance on virgin resources and mitigate waste generation. Selecting materials with lower carbon footprints and eco-label certifications contributes to sustainability efforts. Prioritizing environmentally friendly materials allows manufacturers to create products with reduced environmental impact and support circular economy principles. Materials that create less waste and are energy-efficient, as well as those that can be recycled or sourced sustainably, support the circular economy by allowing materials to be reused or repurposed instead of discarded. Choosing materials with lower environmental impacts, such as those requiring less energy to produce or having a lower carbon footprint, ensures the manufacturing process is as green as possible [3]. Materials utilized in Additive Friction Stir Deposition (AFSD) exhibit a higher degree of sustainability relative to alternative methodologies such as Direct Metal Laser Sintering (DMLS), owing to their reduced energy consumption and the ability to reclaim material chips, thereby enhancing sustainable production paradigms [15, 16].

1.3.6 Energy-Efficient Manufacturing Processes

Energy efficiency is key to green manufacturing, aiming to minimize energy consumption and optimize resource utilization. In additive manufacturing, energy-efficient processes help reduce greenhouse gas emissions and decrease manufacturing costs. Techniques such as process optimization, parameter tuning, and equipment efficiency improvements enhance energy efficiency in AM operations. The incorporation of renewable energy sources, such as wind and solar power, within manufacturing facilities significantly diminishes the environmental impact. The adoption of energy-efficient methodologies enables manufacturers to realize financial savings, improve their competitive edge, and fulfil sustainability objectives. AFSD exhibits superior energy efficiency compared to DMLS, attributed to its reduced energy requirements for material production and the absence of a necessity for inert gas, thereby further curtailing energy consumption [15].

Through the incorporation of sustainable design methodologies, the selection of environmentally benign materials, and the application of energy-efficient production techniques, manufacturers possess the capacity to mitigate ecological repercussions, foster innovation, enhance product excellence, and satisfy the increasing consumer demand for sustainable alternatives within the marketplace.

1.4 Advancements in Eco-Friendly Additive Manufacturing Technologies

Additive manufacturing (AM) is advancing towards enhanced resource efficiency, mitigating material waste through the incremental construction of components from three-dimensional models, thereby facilitating the transition from a linear to a circular economic model. Recent advancements in AM techniques have enabled the fabrication of components that exhibit quality comparable to those produced by conventional manufacturing processes, thereby broadening the industrial applications of AM via expedited tooling and hybrid manufacturing methodologies. The incorporation of environmental life cycle assessment (ELCA), life cycle costing (LCC), and social life cycle assessment (SLCA) instruments within AM procedures critically evaluates sustainability performance, with the objective of identifying environmentally sustainable alternatives that do not incur prohibitive costs. Nevertheless, obstacles persist in the incorporation of technical feasibility within sustainability evaluations for AM, underscoring the necessity for further scholarly inquiry.

1.4.1 Sustainable Materials for Additive Manufacturing

Sustainable materials are pivotal in advancing green AM. Researchers are exploring alternative materials that minimize environmental impact without compromising performance. This encompasses bio-sourced polymers derived from renewable materials, including corn starch or algal biomass, repurposed plastics and metals sourced from post-consumer or industrial refuse, and biodegradable substances engineered to undergo natural decomposition upon the conclusion of their operational lifespan.

The development of sustainable material feedstocks and filaments compatible with existing AM processes expands the range of eco-friendly options available. Additive Manufacturing (AM) technologies facilitate the manipulation of a diverse array of materials, encompassing metals, polymers and ceramics, which are selected for the fabrication of lightweight yet robust components, thereby enhancing the sustainability of the manufacturing process. Recycled plastics, particularly polypropylene (PP) integrated with tire waste, exemplify novel methodologies for the development of sustainable composites within the realm of AM [5]. Biopolymers, originating from natural resources including flora and microorganisms, play a critical role in sustainable additive manufacturing (AM) owing to their biocompatibility, biodegradability, and non-toxic nature, thereby rendering them exceptionally suitable for biomedical applications such as drug delivery systems and wound healing. Polymers like polylactic acid (PLA) are biodegradable and derived from sustainable sources, making them environmentally friendly options for various industries [17].

1.4.2 Energy-Efficient AM Processes

Energy efficiency is crucial in green AM, focusing on reducing system energy consumption to minimize environmental impact. Optimizing AM process parameters, such as adjusting energy usage and improving processing phases, enhances energy efficiency. Smart AM machinery uses intelligent algorithms to find optimal process parameters, reducing waste and energy consumption. Techniques such as Direct Energy Deposition (DED) and Powder Bed Fusion (PBF) may be enhanced for energy efficiency through the meticulous regulation of energy input to facilitate the melting of materials. Wire Arc Additive Manufacturing (WAAM) and Electron Beam Melting (EBM) represent more energy-efficient alternatives in comparison to laser-based methodologies [13]. Innovations such as nozzle optimization, adaptive layering techniques, and intelligent heating strategies further optimize energy usage while maintaining print quality. Additionally, advances in machine learning enable predictive modeling and real-time optimization of AM parameters, enhancing energy efficiency and process reliability [3].The mostly used laser additive manufacturing processes are discussed below.

1.4.2.1 Directed Energy Deposition

In Directed Energy Deposition (DED), a laser generates a melt pool on the substrate surface or an already deposited layer, where the raw material is fed under an inert gas environment to reduce the risk of oxidation. Figure 4.1 illustrates the conventional schematic of a DED system. The DED process offers several advantages over alternative metal additive manufacturing techniques, including the capacity to fabricate large structures, achieve high deposition rates, produce fully dense components, repair damaged elements, and incorporate multiple materials within a singular build

[18]. The DED process may be classified as powder-fed deposition (PFD) or wire-fed deposition (WFD), depending on the material form employed in the construction of the component.

The Powder Feed Delivery (PFD) system consists of a powder and gas delivery mechanism. The powder delivery apparatus conveys powder to the deposition site via a carrier gas, specifically helium and argon. The argon gas serves as a shielding agent for reactive substances. It is recognized as the most practical and widely employed method within the Directed Energy Deposition (DED) process. When the laser emitted from the coaxial nozzle traverses the powder cloud, it transforms into a molten pool that is subsequently deposited onto the substrate or a previously deposited layer.

In the Wire Feed Deposition (WFD) system, a molten pool is formed and deposited onto the substrate or a previously deposited layer as the laser induces melting of the feed metal wire from the nozzle, as illustrated in Fig. 1.3. The WFD process is capable of achieving a deposition rate of 100%, which surpasses that of the Powder Feed Deposition (PFD) process as shown in Fig. 1.2. The entirety of the system operates within an inert atmosphere to mitigate the risk of oxidation. The parameters governing the WFD process (Fig. 1.3) are critical determinants of the resultant material properties. Figure 1.4 illustrates the common process parameters for DED. This process is capable of fabricating large volumes and repairing damaged components, which may serve as either substrates or build rates. It provides a superior deposition rate, produces fully dense components, and facilitates multi-material integration within a single build (Functional Graded Materials) when compared to the Powder Bed Fusion (PBF) process [19, 20].

The mechanical characteristics of components fabricated via Direct Energy Deposition (DED) can be modulated by manipulating the microstructural attributes

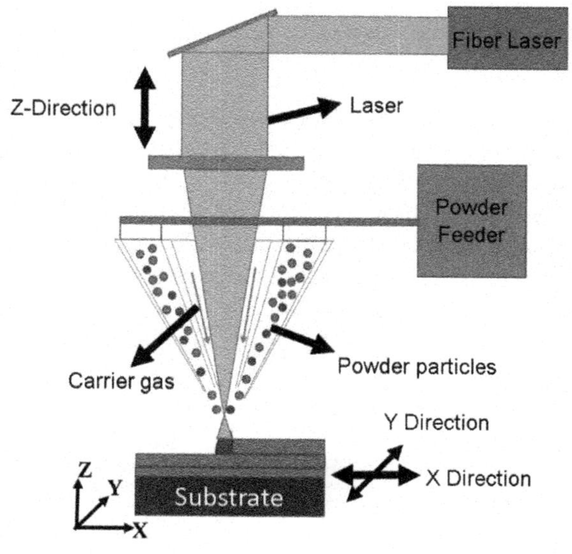

Fig. 1.2 Typical schematic of PFD system

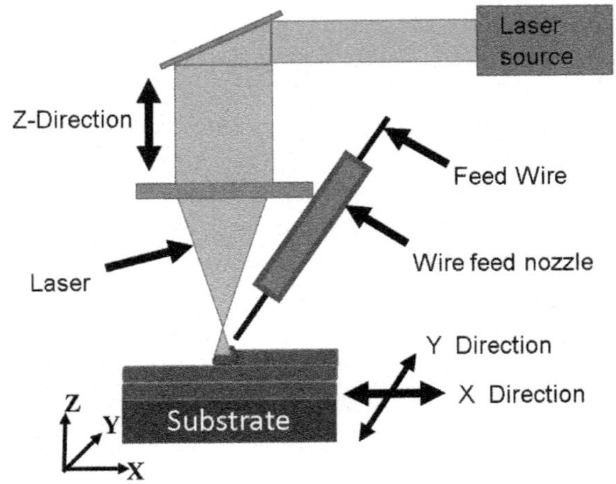

Fig. 1.3 Schematic diagram of WFD system

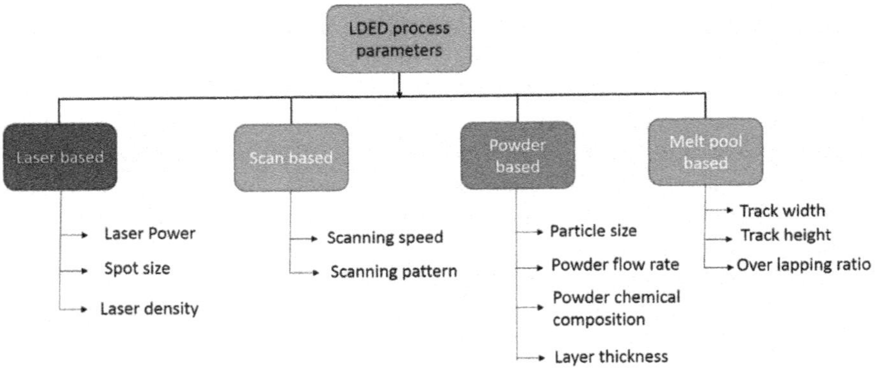

Fig. 1.4 Process parameters of DED system

through variations in chemical composition and thermal gradients. The DED technique has been effectively employed for the processing of a diverse array of materials, including iron-based alloys [21, 22], nickel-based alloys [23, 24], and cobalt-based alloys [25, 26], among others, for a multitude of engineering applications. Moreover, DED presents a viable alternative for the fabrication of near net-shaped structures, thereby minimizing material wastage and conserving costly metals such as nickel, chromium, and molybdenum.

1.4.2.2 Powder Bed Fusion Process

Powder bed fusion (PBF) constitutes a prominent category of additive manufacturing technologies employed for the production of intricate geometries characterized by high resolution and hollow cooling channels [27]. Among the most prevalent PBF methodologies are selective laser sintering (SLS) and selective laser melting (SLM). The latter is frequently referred to as direct metal laser sintering (DMLS). In this process, the laser selectively sinters each cross-sectional layer sequentially in accordance with the sliced three-dimensional data until the desired geometry is achieved. The unsintered powder serves as mechanical support throughout the operational procedure [28, 29].

The principal constituents of Powder Bed Fusion (PBF) encompass the powder supply piston, sintering piston, feeding roller, laser apparatus, control system, powder bed, and inert gas supply mechanism, as illustrated in Fig. 1.5. The morphology of the powder, its packing characteristics, and the distribution of particle sizes exert a profound influence on the porosity of the resultant components. The utilization of fine spherical powders characterized by a minimal presence of satellites and an average particle size distribution ranging from 10 to 45 μm is essential in the PBF process to enhance operational efficiency. The process is conducted utilizing a singular material per build, rendering it unsuitable for applications requiring multi-material configurations. The entirety of the process occurs within an inert gas atmosphere, specifically Argon and Nitrogen, which mitigates the risk of undesirable reactions between gaseous elements and the powder material. For instance, the reactivity of aluminum in the presence of atmospheric oxygen leads to the formation of an oxide layer, which adversely affects the mechanical properties of the component. PBF is influenced by intricate physical and chemical phenomena, encompassing heat and mass transfer as well as the absorption of light energy. The PBF process experiences elevated cooling rates, ranging from 10^3 to 10^8 K/s, attributable to the diminutive size of the melt pool (in microns) and the limited interaction duration between the laser beam and powder particles, facilitated by high scanning velocities.

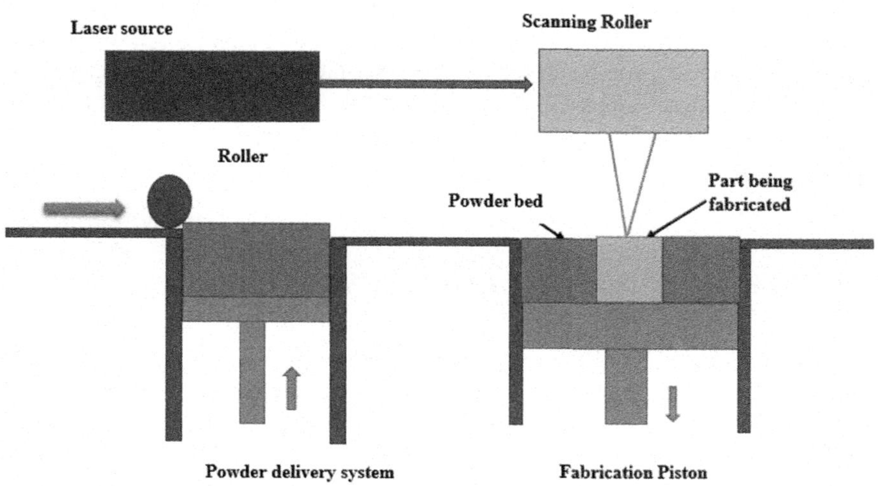

Fig. 1.5 Schematic diagram of powder bed fusion

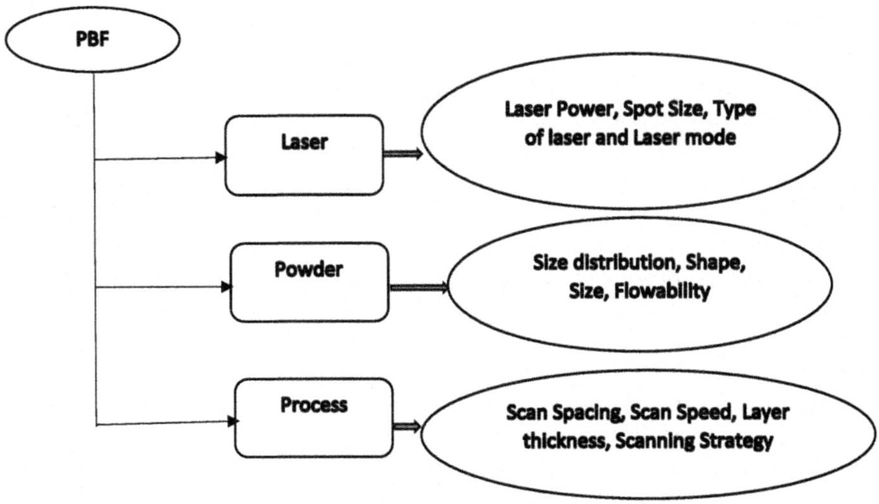

Fig. 1.6 Process parameters involved in SLM process

The unique microstructure and improvement in mechanical properties can be acquired due to the high cooling rates [30]. Apart from this, it improves chemical homogeneity and dislocation density, and modifies the microstructure with grain refinement. Moreover, the output product of PBF can be directly used in applications where it is required with minimum post-processing steps. Selecting appropriate input process parameters are much more important during PBF process and categorized based on the laser, powder, and process as shown in Fig. 4.5. PBF can process materials like metals, plastics, and ceramics but most research has been focused on metallic components and their alloys. Titanium, steel, aluminum, copper, and nickel-based alloys and their composites are successfully fabricated by PBF, which exhibits superior performance over these alloys fabricated by the conventional methods [31–33]. However, PBF works on limited powder materials, hence further research has been focused on promoting additive technology into the different material fields.

1.4.3 Waste Reduction and Recycling Strategies

Waste reduction and recycling are critical in green AM, addressing material waste generated during printing and post-processing. Strategies to minimize waste include optimizing support structures, reducing material overages, and implementing closed-loop recycling systems. Advances in powder bed fusion technologies enable in-situ powder recycling, where excess powder is reclaimed and reused, minimizing material waste [3]. Recycling methodologies for polymers encompass mechanical, chemical, and thermal techniques. Mechanical recycling converts post-consumer plastics into granules suitable for the fabrication of new products, whereas chemical recycling dismantles plastics into their fundamental chemical constituents for subsequent reuse. Thermal recycling facilitates the decomposition of polymers into monomers or alternative chemicals intended for the creation of new products or as fuel [32]. The utilization of recycled polymers such as PLA (Poly Lactic Acid), ABS (Acrylonitrile Butadiene Styrene), and PET (polyethylene terephthalate) within additive manufacturing (AM) demonstrates considerable potential for promoting sustainability by repurposing waste materials. The application of lifecycle assessments for 3D printed products can elucidate methods for energy conservation and waste reduction, with certain investigations indicating that the fabrication of filament from recycled materials necessitates less energy than the production of virgin materials. The integration of Wire Arc Additive Manufacturing (WAAM) enables an enhanced material utilization ratio, thereby optimizing material efficiency and minimizing waste generation [5].

The recent innovations in environmentally conscious additive manufacturing highlight the sector's dedication to sustainable practices. Through the utilization of eco-friendly materials, energy-conserving methodologies, and strategies aimed at minimizing waste, producers can alleviate their ecological footprint, lower production expenses, and satisfy the increasing consumer demand for sustainable products within the international market.

1.5 Life Cycle Assessment (LCA) in Additive Manufacturing

- **Understanding LCA**: Life Cycle Assessment (LCA) constitutes a systematic approach employed to assess the ecological ramifications associated with a product, spanning from the extraction of raw materials to the disposal at the end of its life cycle. In the context of additive manufacturing (AM), this entails a thorough investigation of the environmental consequences related to the incremental fabrication of objects through a layer-by-layer process.
- **Environmental Benefits**: AM can significantly reduce environmental impact, particularly in the industrial machinery and aeronautical sectors, by minimizing material waste. This reduction can exceed 60%, leading to substantial environmental savings.
- **Economic and Environmental Impact**: Although additive manufacturing (AM) has the potential to reduce expenses associated with aeronautical components by a notable 33.2%, it may lead to an escalation in costs for industrial machinery components by a substantial 79.3% owing to elevated machinery and material expenses. Notwithstanding this cost discrepancy, AM presents considerable ecological advantages across both industries.

- **Sector-Specific Impacts**: The influence of additive manufacturing (AM) differs across various sectors. For example, the aeronautical industry experiences considerable reductions in fuel consumption and CO2 emissions, thereby emphasizing the necessity of acknowledging the distinct attributes inherent to each sector [11].
- **LCA in Additive Manufacturing**: Life Cycle Assessment (LCA) systematically examines the ecological ramifications of additive manufacturing (AM) systems by scrutinizing energy and resource utilization across the entire product life cycle, encompassing stages from raw material extraction to final disposal. This methodology facilitates the identification of significant environmental impact determinants, including energy consumption and carbon dioxide emissions, thereby informing environmentally conscious design practices [11].

1.5.1 Life Cycle Stages of AM Products

- **Raw Material Extraction**: The procedure encompasses the extraction and refinement of fundamental substances utilized in additive manufacturing feedstocks, including polymers, metals, and ceramics.
- **Manufacturing Process**: Includes energy consumption, waste generation, and emissions during AM printing, post-processing, and finishing operations.
- **Product Use**: Examines the energy utilization and atmospheric emissions throughout the operational duration of additive manufacturing products, encompassing both maintenance procedures and user engagement activities.
- **End-of-Life Management**: Entails disposal, recycling, or reuse of AM products, including material recovery and waste treatment processes [3].

LCA Methodologies for Assessing Environmental Impact

- **Definition and Scope**: Setting clear goals and defining the boundaries of the assessment.
- **Inventory Analysis**: Collecting data on inputs (materials and energy) and outputs (emissions and waste).
- **Impact Assessment**: Evaluating the environmental impacts across the product's life cycle.
- **Interpretation:** Reviewing and analyzing the results to draw conclusions and make recommendations for reducing environmental impact [3].

1.5.2 Integration of LCA and Economic Analysis

Researchers constructed analytical models to evaluate the environmental and economic ramifications of additive manufacturing (AM), juxtaposing centralized, decentralized, and hybrid supply chain configurations. Decentralized production was identified as the most environmentally sustainable alternative. A separate investigation concentrated on the production of metallic components through AM, integrating life cycle assessment (LCA) and life cycle cost (LCC) metrics, underscoring the advantages of outsourcing post-processing activities to diminish expenditures. An extensive evaluation compared the efficacy of Direct Energy Deposition (DED) with traditional manufacturing methods for the refurbishment of glass bottle molds, thereby illustrating the sustainability attributes of AM.

By conducting LCAs, stakeholders can identify opportunities for improvement, optimize processes, and promote environmentally friendly practices in AM.

1.5.2.1 Case Studies Applying LCA to Additive Manufacturing Processes

Case Study 1: Environmental Impact of 3D-Printed Polymer Products.

Description: A Life Cycle Assessment (LCA) study was conducted to evaluate the ecological footprint of 3D-printed polymer products in relation to traditional manufacturing methodologies.

Findings: The study revealed substantial decreases in energy usage, material wastage, and greenhouse gas emissions linked to additive manufacturing (AM) when juxtaposed with conventional production techniques.

Implications: The research elucidated the ecological advantages of AM for particular applications and underscored potential avenues for enhancements in AM methodologies and materials.

Case Study 2: Life Cycle Assessment of Metal Additive Manufacturing.

Description: Evaluated the environmental impact of metal additive manufacturing methods, including selective laser melting (SLM) and electron beam melting (EBM), across their entire life cycle.

Findings: Investigated the consumption of resources, energy utilization, and emissions across the various stages of the additive manufacturing life cycle, underscoring the critical role of process optimization and material selection in mitigating environmental impacts.

Implications: Provided insights to guide decision-making processes concerning the selection of additive manufacturing techniques, material applications, and strategies for end-of-life management aimed at minimizing ecological footprints and advancing sustainability.

Case Study 3: Comparative Life Cycle Assessment of Additive and Subtractive Manufacturing Techniques.

Description: Compare the environmental performance of additive manufacturing (AM) and subtractive manufacturing (SM) processes for producing a specific component.

Findings: Assessed resource consumption, energy use, and emissions associated with both manufacturing methods, considering factors such as material efficiency, energy consumption, and waste generation.

Implications: Provided insights into the environmental trade-offs between AM and SM processes, guiding manufacturers in selecting the most sustainable manufacturing approach for their specific applications.

These empirical investigations illustrate the utilization of Life Cycle Assessment (LCA) methodologies to assess the ecological ramifications of additive manufacturing techniques and to facilitate informed sustainable decision-making. By conducting LCAs, stakeholders can identify opportunities for improvement, optimize processes, and advance the adoption of environmentally friendly practices in additive manufacturing.

1.6 Case Studies in Green Additive Manufacturing

Case studies in green additive manufacturing

Green additive manufacturing (AM) practices have gained traction across industries, with companies implementing sustainable initiatives to minimize environmental impact and promote responsible manufacturing. This section presents case studies highlighting success stories, sustainable applications, and lessons learned from green AM practices:

Success Stories of Companies Implementing Green AM Practices:

Case Study 1: General Electric (GE) Aviation.

Description: GE Aviation, a top aerospace manufacturer, has adopted eco-friendly additive manufacturing practices to cut waste, boost efficiency, and make aircraft component production more sustainable.

Initiatives: GE Aviation employs additive manufacturing methodologies to fabricate lightweight, fuel-efficient aircraft components, including turbine blades and fuel nozzles, by leveraging advanced materials and sophisticated design optimization strategies.

Results: By leveraging AM, GE Aviation has achieved significant reductions in material waste, energy consumption, and production lead times, leading to cost savings and environmental benefits.

Case Study 2: Adidas.

Description: Adidas, a global sportswear manufacturer, has integrated green AM practices into its footwear production processes to enhance sustainability and innovation.

Initiatives: Adidas utilizes 3D printing technology to produce custom-fit midsoles for its performance footwear lines, such as the Adidas Futurecraft series. These midsoles are designed to optimize cushioning, support, and energy return while minimizing material waste.

Results: By implementing AM, Adidas has achieved greater customization, improved product performance, and reduced environmental impact, reinforcing its commitment to sustainable manufacturing practices.

Examples of Sustainable AM Applications Across Industries:

Case Study 3: Automotive Industry.

Description: Automakers such as BMW and Volkswagen are leveraging additive manufacturing to produce lightweight, fuel-efficient vehicle components, including engine parts, interior trim, and structural components.

Initiatives: AM technologies enable automakers to optimize part designs, consolidate assemblies, and reduce material usage while maintaining structural integrity and performance.

Results: By incorporating AM into their manufacturing processes, automotive companies have realized benefits such as reduced vehicle weight, improved fuel efficiency, and enhanced sustainability across the product lifecycle.

Case Study 4: Medical Industry.

Description: Medical device manufacturers, including Medtronic and Stryker, are harnessing additive manufacturing to produce patient-specific implants, surgical instruments, and prosthetics with enhanced functionality and biocompatibility.

Initiatives: AM enables the customization of medical devices to match individual patient anatomy, improving treatment outcomes and patient satisfaction while minimizing material waste and surgical time.

Results: Through the adoption of AM, medical device manufacturers have achieved advancements in patient care, reduced surgical complications, and optimized resource utilization in healthcare facilities.

Lessons learned and best practices from case studies

Lesson 1: Design for Additive Manufacturing (DfAM): Successful green AM initiatives emphasize design optimization for additive manufacturing, leveraging topology optimization, lattice structures, and generative design techniques to minimize material usage and enhance performance.

Lesson 2: Material Selection and Certification: Companies implementing green AM practices prioritize the use of sustainable materials, including recycled polymers, bio-based resins, and metal powders sourced from responsible suppliers with eco-certifications.

Lesson 3: Collaboration and Knowledge Sharing: Collaboration among stakeholders, including manufacturers, researchers, and regulatory agencies, is essential for driving innovation and advancing sustainable AM applications. Knowledge sharing through industry consortia, conferences, and collaborative research initiatives facilitates the exchange of best practices and lessons learned in green AM.

These case studies illustrate the diverse applications and benefits of green additive manufacturing practices across industries. By highlighting success stories, sustainable applications, and lessons learned, companies can inspire further adoption of green AM practices and contribute to a more sustainable future for manufacturing.

1.7 Challenges and Future Directions in Additive Manufacturing

Challenges:

- **Environmental Impact**: High energy consumption and material waste, especially in producing complex metal parts, pose significant environmental challenges.
- **Sustainability Evaluation**: Lack of a comprehensive sustainability evaluation system that considers all aspects of additive manufacturing, including production time, product quality, and post-processing.
- **Assessment Methods**: Current methods for assessing the environmental impact of AM processes may not fully capture its unique aspects, leading to incomplete analyses.
- **Awareness and Education**: Educating engineers and designers about the potential and benefits of additive manufacturing is crucial for its widespread adoption and understanding.
- **Technological Limitations**: Performance limitations, such as the inability to incorporate microelectronics directly into components, hinder the technology's full exploitation.
- **Technology Selection**: The wide variety of AM technologies makes it challenging to select the most appropriate one for specific applications, complicating integration into existing manufacturing systems.
- **Material Recycling**: Recycling polymers, especially thermoset plastics, is challenging due to mixing, different types, and additives/coatings that are hard to remove.

Future Directions:

- **Energy-Saving Methods**: Future research should focus on developing more energy-saving methods and improving resource utilization in additive manufacturing.
- **Unified Evaluation Model**: There is a need for a unified evaluation model integrating economic, environmental, and sustainability assessments to address current assessment gaps.
- **Sustainable Design Practices**: Incorporating environmental considerations at the design stage can significantly reduce AM's environmental impacts.
- **Standardization**: Standardizing materials and processes will help overcome barriers to adoption and enhance environmental performance.
- **Material Eco-Friendliness**: Investing in research to improve the eco-friendliness of materials and enhance recycling processes is crucial.
- **Workforce Training**: Developing training programs to equip the workforce with digital skills for operating advanced AM technologies is essential.
- **Collaboration and Innovation**: Encouraging collaboration among stakeholders and fostering innovation in business models and production methods will enhance sustainability.

- **Research Focus**: Future research will focus on addressing model limitations, including a more complex supply chain network and uncertainty incorporation.
- **Plastic Recycling**: Finding new, cheap, and effective ways to recycle plastics, improving sorting technologies, and developing methods to remove additives/coatings will advance recycling efforts.

By addressing these challenges and focusing on future directions, additive manufacturing can become more sustainable and environmentally friendly, contributing to a circular economy and reducing environmental impact.

1.8 Conclusion

In conclusion, the investigation into sustainable additive manufacturing (AM) and the ecological ramifications of conventional manufacturing methodologies highlights the paramount necessity for environmentally sustainable production techniques within the industrial sector. Sustainable AM presents considerable benefits in comparison to traditional methods by minimizing material wastage, decreasing energy usage, and enhancing resource efficiency. By adhering to principles such as diminishing environmental repercussions, emphasizing energy efficiency, augmenting material efficacy, and securing economic feasibility, sustainable manufacturing methodologies can alleviate ecological degradation while fostering innovation and satisfying market requirements for sustainable products. Progressions in green additive manufacturing technologies, including the incorporation of life cycle assessment (LCA), the creation of sustainable materials, energy-efficient AM procedures, and waste minimization strategies, exemplify the industry's dedication to sustainability. Nonetheless, challenges remain, encompassing environmental impact reduction, sustainability assessment, and technological constraints; however, prospective pathways emphasize energy-saving approaches, cohesive evaluation frameworks, sustainable design methodologies, and collaborative efforts for innovation.

By tackling these challenges and capitalizing on opportunities for progress, additive manufacturing can assume a crucial position in the shift towards a more sustainable and environmentally aware manufacturing paradigm. This shift is not only vital for diminishing environmental impacts but also for fostering economic development, encouraging innovation, and ensuring a more sustainable future for forthcoming generations.

References

1. Benarji, K., Kumar, Y.R., Paul, C.P., Jinoop, A.N., Bindra, K.S.: Parametric investigation and characterization on SS316 built by laser-assisted directed energy deposition. Proc. Inst. Mech. Eng. Part L J. Mater. Des. Appl. **234**, 452–466 (2020). https://doi.org/10.1177/1464420719894718
2. Benarji, K., Ravi Kumar, Y., Jinoop, A.N., Paul, C.P., Bindra, K.S.: Effect of heat-treatment on the microstructure, mechanical properties and corrosion behaviour of SS 316 structures built by laser directed energy deposition based additive manufacturing. Met. Mater. Int. **27**, 488–499 (2021). https://doi.org/10.1007/s12540-020-00838-y
3. Hegab, H., Khanna, N., Monib, N., Salem, A.: Design for sustainable additive manufacturing: a review. Sustain. Mater. Technol. **35** (2023). https://doi.org/10.1016/j.susmat.2023.e00576
4. Pusateri, V., Hauschild, M.Z., Kara, S., Goulas, C., Olsen, S.I.: Quantitative sustainability assessment of metal additive manufacturing: a systematic review. CIRP J. Manuf. Sci. Technol. **49**, 95–110 (2024). https://doi.org/10.1016/j.cirpj.2023.12.005
5. Alami, A.H., Ghani Olabi, A., Alashkar, A., Alasad, S., Aljaghoub, H., Rezk, H., Abdelkareem, M.A.: Additive manufacturing in the aerospace and automotive industries: recent trends and role in achieving sustainable development goals. Ain Shams Eng. J. 14 (2023). https://doi.org/10.1016/j.asej.2023.102516
6. Manco, P., Caterino, M., Rinaldi, M., Fera, M.: Additive manufacturing in green supply chains: a parametric model for life cycle assessment and cost. Sustain. Prod. Consum. **36**, 463–478 (2023). https://doi.org/10.1016/j.spc.2023.01.015
7. Javaid, M., Haleem, A., Singh, R.P., Suman, R., Rab, S.: Role of additive manufacturing applications towards environmental sustainability. Adv. Ind. Eng. Polym. Res. **4**, 312–322 (2021). https://doi.org/10.1016/j.aiepr.2021.07.005
8. Salvi, H., Vesuwala, H., Raval, P., Badheka, V., Khanna, N.: Sustainability analysis of additive + subtractive manufacturing processes for Inconel 625. Sustain. Mater. Technol. **35**(2023). https://doi.org/10.1016/j.susmat.2023.e00580
9. Danish, M., Gupta, M.K., Irfan, S.A., Ghazali, S.M., Rathore, M.F., Krolczyk, G.M., Alsaady, A.: Machine learning models for prediction and classification of tool wear in sustainable milling of additively manufactured 316 stainless steel. Results Eng. **22**(2024). https://doi.org/10.1016/j.rineng.2024.102015
10. Jayawardane, H., Davies, I.J., Gamage, J.R., John, M., Biswas, W.K.: Sustainability perspectives—a review of additive and subtractive manufacturing. Sustain. Manuf. Serv. Econ. **2** (2023). https://doi.org/10.1016/j.smse.2023.100015
11. Gonçalves, A., Ferreira, B., Leite, M., Ribeiro, I.: Environmental and economic sustainability impacts of metal additive manufacturing: a study in the industrial machinery and aeronautical sectors. Sustain. Prod. Consum. **42**, 292–308 (2023). https://doi.org/10.1016/j.spc.2023.10.004
12. Baniasadi, H., Chatzikosmidou, D., Seppälä, J.: Innovative integration of pyrolyzed biomass into polyamide 11: sustainable advancements through in situ polymerization for enhanced mechanical, thermal, and additive manufacturing properties. Addit. Manuf. **78** (2023). https://doi.org/10.1016/j.addma.2023.103869
13. Gao, M., Liu, C., Li, L., Li, Q., Wang, Q., Liu, Z.: Emergy-based method for the sustainability assessment and improvement of additive manufacturing systems. Energy **290** (2024). https://doi.org/10.1016/j.energy.2023.130115
14. Singh, S., Mohanty, R.P., Mangla, S.K., Agrawal, V.: Critical success factors of additive manufacturing for higher sustainable competitive advantage in supply chains. J. Clean. Prod. **425** (2023). https://doi.org/10.1016/j.jclepro.2023.138908
15. Ahmed, A.A., Nazzal, M.A., Darras, B.M., Eltaggaz, A., Deiab, I.M.: Comparative sustainability assessment of powder bed fusion and solid-state additive manufacturing processes: the case of direct metal laser sintering versus additive friction stir deposition. Sustain. Mater. Technol. **39** (2024). https://doi.org/10.1016/j.susmat.2024.e00858

16. Haleem, A., Javaid, M., Rab, S., Singh, R.P., Suman, R., Kumar, L.: Significant potential and materials used in additive manufacturing technologies towards sustainability. Sustain. Oper. Comput. **4**, 172–182 (2023). https://doi.org/10.1016/j.susoc.2023.11.004
17. Domingues, J., Marques, T., Mateus, A., Carreira, P., Malça, C.: An additive manufacturing solution to produce big green parts from tires and recycled plastics. Procedia Manuf. **12**, 242–248 (2017). https://doi.org/10.1016/j.promfg.2017.08.028
18. Dutta, B., Babu, S., Jared, B.: Metal additive manufacturing, science, technology and applications of metals in additive manufacturing, pp. 1–10 (2019). https://doi.org/10.1016/b978-0-12-816634-5.00001-7
19. Prakash, K.S., Nancharaih, T., Rao, V.V.S.: Additive manufacturing techniques in manufacturing—an overview. Mater. Today Proc. **5**, 3873–3882 (2018)
20. Zhang, Y., Wu, L., Guo, X., Kane, S., Deng, Y., Jung, Y.G., Lee, J.H., Zhang, J.: Additive manufacturing of metallic materials: a review. J. Mater. Eng. Perform. **27**, 1–13 (2018). https://doi.org/10.1007/s11665-017-2747-y
21. Sergio, M., De Lima, F., Sankaré, S.: Microstructure and mechanical behavior of laser additive manufactured AISI 316 stainless steel stringers. J. Mater. Des. **55**, 526–532 (2014). https://doi.org/10.1016/j.matdes.2013.10.016
22. Das, A., Shukla, M.: Surface morphology, bioactivity, and antibacterial studies of pulsed laser deposited hydroxyapatite coatings on stainless Steel 254 for orthopedic implant applications. Proc. Inst. Mech. Eng. Part L J. Mater. Des. Appl. **233**, 120–127 (2019). https://doi.org/10.1177/1464420716663029
23. Jinoop, A.N., Paul, C.P.: Laser-assisted directed energy deposition of nickel super alloys: a review. J. Mater. Des. Appl. **0**, 1–25 (2019). https://doi.org/10.1177/1464420719852658
24. Jinoop, A.N., Paul, C.P., Mishra, S.K., Bindra, K.S.: Laser Additive Manufacturing using directed energy deposition of Inconel-718 wall structures with tailored characteristics. Vacuum **166**, 270–278 (2019). https://doi.org/10.1016/j.vacuum.2019.05.027
25. Gan, Z., Yu, G., He, X., Li, S.: Numerical simulation of thermal behavior and multicomponent mass transfer in direct laser deposition of co-base alloy on steel. Int. J. Heat Mass Transf. **104**, 28–38 (2017). https://doi.org/10.1016/j.ijheatmasstransfer.2016.08.049
26. Traxel, K.D., Bandyopadhyay, A.: First demonstration of additive manufacturing of cutting tools using directed energy deposition system: stellite™-based cutting tools. Addit. Manuf. **25**, 460–468 (2019). https://doi.org/10.1016/j.addma.2018.11.019
27. Rudolph, J.P., Emmelmann, C.: Self-learning calculation for selective laser melting. Procedia CIRP **67**, 185–190 (2018). https://doi.org/10.1016/j.procir.2017.12.197
28. Panda, S. K., Rath, K. C., Mishra, S. : The new age of manufacturing from concept to creation: innovative world of 3D printing. In: Evolutionary Manufacturing, Design and Operational Practices for Resource and Environmental Sustainability, pp. 223–244 (2024). https://doi.org/10.1002/9781394198221.ch20
29. Everton, S.K., Hirsch, M., Stavroulakis, P.I., Leach, R.K., Clare, A.T.: Review of in-situ process monitoring and in-situ metrology for metal additive manufacturing. Mater. Des. **95**, 431–445 (2016). https://doi.org/10.1016/j.matdes.2016.01.099
30. Wei, C., Li, L., Zhang, X., Chueh, Y.H.: 3D printing of multiple metallic materials via modified selective laser melting. CIRP Ann. **67**, 245–248 (2018). https://doi.org/10.1016/j.cirp.2018.04.096
31. Dai, D., Gu, D.: Influence of thermodynamics within molten pool on migration and distribution state of reinforcement during selective laser melting of AlN/AlSi10Mg composites. Int. J. Mach. Tools Manuf **100**, 14–24 (2016). https://doi.org/10.1016/j.ijmachtools.2015.10.004
32. Wang, P., Gammer, C., Brenne, F., Niendorf, T., Eckert, J., Scudino, S.: A heat treatable TiB2/Al-3.5Cu-1.5Mg-1Si composite fabricated by selective laser melting: microstructure, heat treatment and mechanical properties. Compos. Part B Eng. **147**, 162–168 (2018). https://doi.org/10.1016/j.compositesb.2018
33. Al Rashid, A., Koç, M.: Additive manufacturing for sustainability and circular economy: needs, challenges, and opportunities for 3D printing of recycled polymeric waste. Mater. Today Sustain. **24** (2023). https://doi.org/10.1016/j.mtsust.2023.100529

Dr. K. Benarji is an accomplished Assistant Professor in the Mechanical Engineering Department at Shri Vishnu Engineering College for Women's, Andhra Pradesh. I received my doctoral degree from the Department of Mechanical Engineering, NIT Warangal, India, in July 2021, The thesis title is **"Processing and Characterization of Laser directed energy deposition Based Additive Manufactured SS 316 and its Composites"**. M.Tech in Production Engineering from NIT Rourkela. His research primarily focuses on laser additive manufacturing, metal cutting, and sustainable manufacturing processes, with notable expertise in laser surface treatment and material characterization. Dr. Benarji has published several articles in high-impact international journals which consists of 5 SCI journals, 2 Scopus, 1 Book chapter and 3 international conferences and also received awards for his research contributions, including a Best Paper award at an international conference.

Dr. Jinoop Arackal Narayanan is a Fellow of the Institute of Materials, Minerals and Mining (FIMMM), Chartered Engineer (CEng) and a Lecturer (Assistant Professor) in Mechanical Engineering at the School of Computing, Engineering & Digital Technologies, Teesside University, UK. Prior to this role, he held positions at the Welding and Additive Manufacturing Centre, at Cranfield University, and the Laser Additive Manufacturing Laboratory at Raja Ramanna Centre for Advanced Technology (RRCAT), Indore, India. His research interests span metal additive manufacturing, surface engineering, process optimisation, advanced materials processing, thermo-mechanical treatments, and both in-situ and ex-situ material characterisations. He has authored over 90 publications, including peer-reviewed journal articles, book chapters, and conference papers/ presentations at both national and international levels. Notably, he authored the first comprehensive textbook on Additive Manufacturing in India, published by McGraw-Hill Higher Education for undergraduate and postgraduate scholars. In addition to his research, Dr. Narayanan serves as an editor and reviewer for several esteemed international journals. He is actively involved in collaborative projects on metal additive manufacturing with international universities, research labs, and industries. He is currently working on research projects funded by Innovate UK and the UK Department for Science, Innovation and Technology.

Shashikant Das obtained his B.Tech from NIT Calicut in 2011 and his M.Tech from IIT Kharagpur in 2015. Further, he obtained his PhD from the Department of Mechanical and Industrial Engineering, IIT Roorkee, in 2022. His area of research is thermal comfort, where he focused on the impact of a hot climate on the human body in terms of analytical and experimental investigation. In his analytical work, he has gone through steady-state, transient state, and exergy analyses of the human body. In experimental work, he has gone through the effect of a hot environment on occupants' performance, sensation, and physiology. He has three years of academic experience in teaching subjects like heat transfer, thermodynamics, air conditioning, I-C engines, power plants, and fluid mechanics at the UG level.

Dr. Sujit Mishra is an Assistant professor and Head of the Mechanical Engineering Department, SoET, at Centurion University of Technology and Management, Paralakhemundi campus. His expertise in various subject areas, such as Machine Design, Fluid mechanics, Computational Fluid Dynamics, Computer-Aided Drafting, Computer-Aided Manufacturing, and Computer-Aided Engineering at both UG and PG levels. He has published more than 13 International journals, 3 book chapters, and one book, 4 Patents. He is also a part of an active Life Member of the Indian Society of Technical Education (ISTE), IMS.

Chapter 2
Advances Sustainability on Material Science and Production

Subhra Gantayat, Soumya Dash, and Md. Kamruzzaman

Abstract Sustainable development, spanning social, ecological, and economic realms, necessitates preserving the integrity of resource exploitation, guiding investments, and directing technological progress. Over recent decades, integrating environmental considerations into material and product design has emerged as a critical imperative. Materials have historically fuelled technological advancements, from silicon powering the information age to high-strength aluminum alloys propelling commercial aviation. In the perspective of sustainable development, materials will persist in their pivotal role, enabling progress across various sectors. The transition of solar power into electricity directly hinges on the existence, efficiency, and manufacturability of photovoltaic materials. Sustainable progress hinges on the comprehensive engagement of the materials community; without it, the realization of sustainable development goals remains unattainable. This paper discovers the intersection of sustainable growthas well as materials science, highlighting key areas where materials science is poised to play a crucial part. With the sustainable development agenda assuming increasing importance across public and industrial domains, it stands as a fundamental objective driving societal progress.

Keywords Sustainable development · Sustainable developments goal (SDG) · Sustainability · Social · Economical · Environmental · Sustainable materials

S. Gantayat (✉)
Department of Physics, KISS Deemed to be University, Bhubaneswar, Odisha, India
e-mail: subhra.gantayat@kiss.ac.in

S. Dash
CV Raman Polytechnic, Bhubaneswar, Odisha, India
e-mail: soumyadash1991@gmail.com

Md. Kamruzzaman
Dhaka University of Engineering and Technology (DUET), Gazipur, Bangladesh, India
e-mail: kamruzzaman@duet.ac.bd

© The Author(s), under exclusive license to Springer Nature Switzerland AG 2025
K. C. Rath et al. (eds.), *Engineering, Technology and Management*, Information Systems Engineering and Management 54, https://doi.org/10.1007/978-3-031-93654-8_2

Preface
Material science is an interdisciplinary field of researching and discovering materials. Materials play a significant role in sustainable development by helping to reduce environmental impact and resource consumption which includes sustainable building materials, green design, energy efficiency waste management. In other way, materials contribute to sustainable development include: producing clean energy, providing access to clean water, sequestering carbon and ensuring a dependable supply of raw and recycled materials.

This chapter has been written in a very simple and lucid manner describing about material science in sustainable development, utilization of wastes for sustainable development, sustainable materials in daily life different suggestive proposals for sustainable development.

2.1 Introduction

The world's current economies heavily rely on materials and energy. Nevertheless, defining sustainability remains a topic of considerable controversy. Numerous debates have ensued regarding the precise meaning and the criteria that should define sustainability. As suggested earlier, sustainability can be conceptualized as a procedure of alteration where source manipulation; direction of investment, development of technology, and institutional variation harmonize to improve both present and forthcoming capabilities to encounter human requirements as well as objectives [1]. In 1987, sustainable progress was initially well-defined as "development that encounters the desires of the existing deprived of negotiating the capability of forthcoming peers to encounter their individual needs" [2]. This definition encapsulates a trajectory of development that shifts us from our current state towards a more sustainable one, utilizing today's technology as the baseline and focusing on technological advancements as the means of development. Sustainable growth entails the utilization of locally available materials that are energy-efficient and long-lasting, fostering environments conducive to healthy and comfortable living throughout the entire lifecycle. This lifecycle encompasses materials production, design, construction, operation, and maintenance processes. The extensive objective of sustainability is to achieve effective source consumption, incorporating energy conservation as well as the utilization of environmentally responsive materials [3]. The study of sustainability should comprehend the valuation of environmental, social, and economic impacts across the entire lifecycle of a creation. Sustainable development covers a broad spectrum of concepts, including ecological stewardship, materials organization, green engineering, renewable & unsoiled energy tools, as well as water and air management. It represents a process whereby policies aimed at recurrent developments in the economy, environment, as well as society-the "triple bottom line" are informed by scientific investigation. Given the inherent interconnectedness of the 3 constituents of the triple bottom line, sustainable development matters are categorized by complex interrelations. At a conceptual level, the contributions to

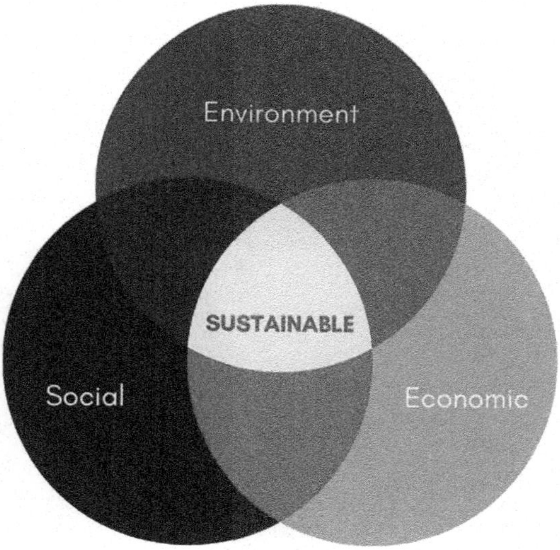

Fig. 2.1 A schematic diagram: sustainable development overlapping economical, social and environmental aspects

sustainability can be categorized into three dimensions: Economic, Social, and Environmental. A schematic diagram illustrating the overlap of these three crucial aspects for sustainable development is depicted in Fig. 2.1.

In order to generate a further effective utilization of resources, economic sustainability involves setting amounts for goods as well as services by utilizing actual cost pricing and full cost accounting procedures. Social sustainability emphasizes enhancing the value of life of the human beings and the environment in which they are living, encompassing aspects such as culture, well-being, edification, and intergenerational equity. By encouraging the usage of renewable sources and minimizing the usage of the water, energy, materials, and land at every step of the project, environmental sustainability encompasses the idea that sustainable development should save the environment rather than contaminate it [4].

This study aims to highlight the crucial role of materials in sustainable development, encouraging technologists, policy planners, and industry leaders to support this understanding in defining and planning the future roles of advanced materials within the context of sustainability.

2.2 Material Science in Sustainable Development

Materials scientists and designers have numerous opportunities to integrate sustainable development into their works. Figure 2.2 seems to illustrate the interconnectedness between sustainable development and materials science which highlights various key aspects described as follows:

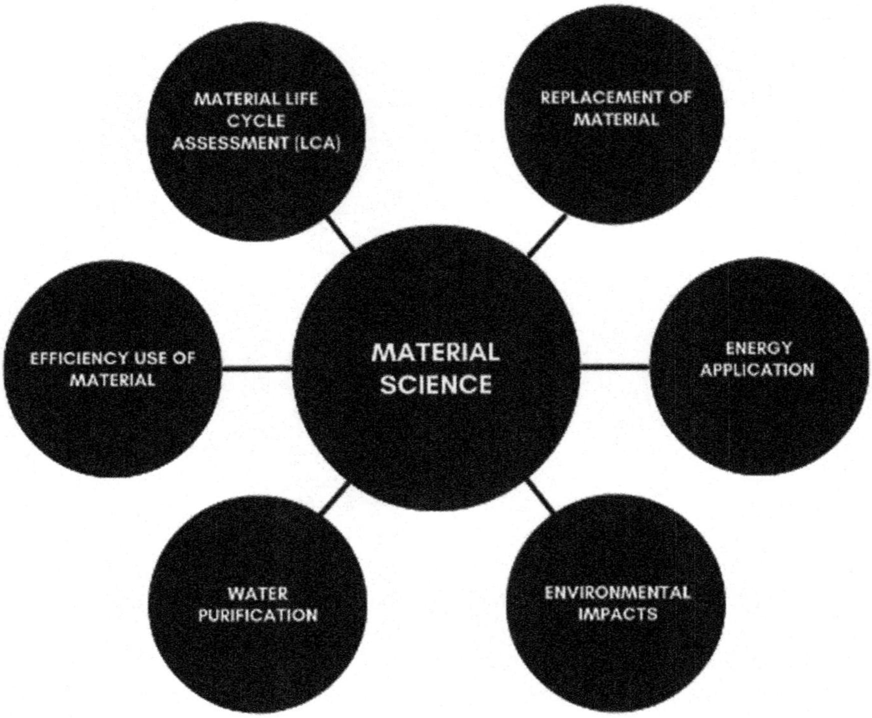

Fig. 2.2 The interconnectedness between sustainable development and materials science

- Efficient Use of Materials: This involves strategies like conservation, substitution, reuse, repurposing, and recycling to minimize resource consumption and waste generation.
- Materials Life-Cycle Assessment (LCA): LCA appraises the environmental consequences of materials throughout their whole life time, from extraction to removal, helping to identify areas for improvement.
- Replacement of Materials: Considering factors such as scarcity, resource availability, and materials economics, researchers look for alternative materials to replace those that are unsustainable or environmentally harmful.
- Energy Applications: Materials science plays a crucial role in supporting alternate energy technologies, such as solar panels and wind turbines, as well as in improving the efficiency of energy production and consumption processes.
- Mitigation of Undesirable Environmental Impacts: Materials science contributes to mitigating issues like corrosion, pollution, and toxic waste generation associated with technology and economic activities, thereby reducing their negative environmental footprint.
- Water Purification: Development of materials for water purification processes is essential for ensuring access to fresh and safe drinking water, a critical factorin sustainable development and public health.

These connections emphasize how advancements in materials science can directly contribute to achieving sustainability goals by addressing resource efficiency, environmental impacts, and societal needs.

2.3 Utilization of Wastes for Sustainable Development

Pollution can be reduced, greenhouse gas emissions can be decreased, and valuable resources can be recovered when garbage is used effectively. India can enhance the general standard of living, generate employment, and support sustainable development by implementing waste-to-wealth initiatives.

2.3.1 Domestic Plastic Waste/Vehicle Waste

Addressing the mounting plastic waste in our ecosystem requires a strategic approach, focusing on effective recycling and reuse methods. With substantial quantities of plastic waste generated daily in the medical sector-comprising syringes, medicine covers, tubing, gloves, glucose bottles, as well as bags there's a pressing necessity to incorporate recycling initiatives. Notably, these biomedical plastics find a new purpose in road construction projects, exemplifying the potential for innovative solutions to tackle environmental challenges [5]. Combining plastic waste polyethylene terephthalate (PET) with crumb rubber waste offers a dual benefit, enhancing the strength of both rubber and concrete materials [6].

The rapid increase in waste generation, driven by factors such as growing population, urbanization, and industrialization, poses significant socio-economic and environmental challenges. To address this issue, one element of integrated waste management recycling has gained prominence. This approach involves repurposing solid waste for numerous determinations, as well as construction of building, as a feasible strategy for managing the solid waste. The advancement of technologies and practices for deployment of municipal waste that is solid in nature as building materials offers several benefits. It not only produces employment chances and provides a means of livelihood but also decreases the price of building construction and the demand for raw materials. Additionally, repurposing waste for construction enhances public health and environmental well-being, thereby contributing to socio-economic and environmental development [7].

Awoyera et al. conducted research aimed at enhancing the flexural ability of concrete tiles by integrating fibers from waste polyethylene terephthalate (PET) bottles. Contrasts of verified properties against usual concrete tiles revealed that adding 10 and 15% of these fibers improved the properties of the floor tiles [8]. Mustafa et al. studied the usage of leftover plastic from vehicles as an additional for sand in concrete construction. Through impact testing, they found that replacing sand with 5, 10, and 15% waste plastics amplified the energy preoccupation volume of concrete [9]. Zhou et al. demonstrated that solid waste i.e. phosphor gypsum, processed through recurrent pressing hydration, could be transformed into hardened tiles with satisfactory mechanical strength [10]. Esmaeili and Aslani investigated the replacement of cement in concrete with Copper Mine Tailings (CMT), resulting in significantly enhanced mechanical properties [11]. Surtiyeni et al. developed a composite by utilizing domestic waste comprising compounds that is organic through hot pressing, and tested its constraints. Composite exhibited excellent fire resilient properties with strength as compared to materials that are used for wood building [12].

2.3.2 Water Waste

The textile industry, while employing a vast number of individuals worldwide, is also recognized as one of the most polluting sectors. Consequently, researchers are actively seeking solutions and innovations to develop new textile materials that promote sustainability within the present landscape of fashion and textile industry [13]. Ensuring a consistent resource of safe consumption water is becoming a progressively stimulating task worldwide. Moreover, the treatment of both water and wastewater, together with industrial wastewater, is serious and vital for safe protecting the atmosphere and humanoid well-being. The disposition of ecologically sustainable carbon nano-materials, as well as carbon nanotubes, graphene, graphene oxide-relied nano-materials, and carbon and graphene quantum dots, has emerged as a priority. These materials are applied as adsorbents for wastewater management and decontamination, particularly for industrial and pharmaceutical-laden wastes [14].

Cong et al. have reported on CNTs and carbon allotropes by a cylinder-shaped nanostructure, which can effectively eliminate pharmaceutical-derived contaminants [15]. Mishra et al. have demonstrated the elimination of dyes from wastewater generated by the textile industry using CNTs [16]. Upadhyayula et al. have documented that CNT-relied composites show notable capabilities for remediating wastewater contaminated with various organic, inorganic, and biological pollutants [17]. Gupta et al. have shown that CNTs and CNT-relied composites have the capability for adsorbing dyes from aqueous structures, including MWCNTs and SWCNTs [18]. Perreault et al. have investigated the application of graphene nanomaterials for treating wastewater due to their significant physicochemical characteristics [19]. Nasrollahzadeh et al. discovered that various carbon quantum dots (CQDs) and graphene quantum dots (GQDs) relied photocatalysts are effective in removing extremely toxic pollutants, including dyes and heavyweight metals [20].

The extreme task to sustainable development may indeed lie in the need to change human behavior. While technology plays a crucial role, it alone cannot provide the solution. Governments, policymakers, and technologists must work together to progress national and worldwide strategies that promote sustainability. Furthermore, the contrast between the locked loop system of natural ecosystem, wherever waste is minimal, and the exposed loop countryside of humanoid civilization highlights

Table 2.1 Utilization of various waste material

Sl. no	Type of waste	Utilization of waste material	References
1	Biomedical plastic waste (Glucose bottle, syringes, gloves, medicine covers etc	Road construction	Lamba et al. [5]
2	Waste plastic with crumb rubber waste	Concrete/road construction	Lamba et al. [5]
3	Fibres of waste PET bottles with concrete	Improve the flexural ability concrete tiles	Awoyera et al. [8]
4	Concrete with copper mine tailings (CMT)	Enhanced mechanical property	Esmaeili et al. [11]
5	Vehicle waste plastic with concrete	Increase energy absorption capacity	Mustafa et al. [9]
6	Domestic waste	Excellent fire resilient property	Surtiyeni et al. [12]
7	Municipality solid waste	Used as building material	Soni et al. [7]
8	Water waste	Helps in removal of dyes from waste water	Nasrollahz et al. [14]
		Elimination of organic, inorganic and biological water pollutants	Nasrollahz et al. [14]

the essential for systemic change. As the worldwide populace continues to grow, it raises important questions about not just how many persons planet can support, but just how many can be supported sustainably while maintaining an ample worth of existing [21]. Different waste materials used for sustainable development has been depicted in Table 2.1.

2.4 Sustainable Materials in Daily Life

Materials investigators currently utilize an extensive variety of elements from the periodic table to drive advancements in advance technology. Though, intelligences of metal shortages, such as rare earths, are becoming increasingly common and have garnered attention in the general use. While industry traditionally focuses on cost, performance, and carbon footprint when selecting materials for new product designs, the importance of chemical toxicity in sustainable development cannot be overlooked. Therefore, there is a growing call for the incorporation of toxicity metrics into materials informatics databanks to better inform decision-making processes. The modern construction sector faces a multitude of encounters, principally stemming from the rapid growth in urban populace and the depletion of normal resources necessary for manufacture material production. Moreover, heightened awareness of weather alteration is prompting companies to reassess their strategies and prioritize the growth of extra sustainable building materials.

2.4.1 Agro Waste Materials

The agro wastes such as rice husk ash, bamboo leaves and ash of sugarcane bagasse have promising result in the field of sustainable materials which is used in construction unit. Many researchers have suggested that the clay bricks can be prepared by using the rice husk ash (RHA) and sugarcane bagasse ash (SBA) which leads to the development of light weight bricks. These bricks are efficiently used in the construction purposes as they reduce buildings' structural load. In addition to this, the combination of SBA and RHA with the bricks enhance the weather resistance and improve efflorescence resistanceleading to durability of the structures.

Additive agro-waste products like sawdust when blended with different concentration of fly ash and red mud for preparing insulation bricks provides encouraging results. The unique properties of these insulating bricks like compressive strength, porosity and water absorption rates, find them efficient for different applications in construction purposes. Besides insulation bricks, thermal insulation materials are also produced for buildings by using agro waste. Natural fibres like flax, wood, hemp, straw, and coconut are utilized for making insulation material which are eco-friendly. Lightweight composites are manufactured from agro wastes (wheat and barley straw fibre) for building insulation. Building reinforcement materials are produced by applying agro waste as cited by researchers. Agro waste vegetable fibres are also used for cement based reinforcement material which significantly increases the durability of the cementitious materials.

Agro-waste also show their potentiality in the production of bio-based plastics. Agro-waste residues like corn stalks, sunflower stalks and bagasse fibers are employed for manufacturing thermoplastics. This process gives a substitute of traditional wood fibers for the advancement of more sustainable and ecological plastic materials. As compared to petroleum–based plastic, bio based plastics provides better mechanical strength, durability and reduce environmental impact. Researchers revealed the manufacture of particle board from agro-waste which replaces the conventional wood byproducts like shavings and sawdust. The overall quality and performance of the particleboards will be enhanced and dependency on traditional wood source will be reduced by this inventive use of agro-waste [22].

2.4.2 Acoustic Materials

In modern time, advancement in technology in material science, production process, chemistry, and nanoscience have created remarkable development in engineered materials. Particularly, development of sustainable acoustic materials plays a vital role for improvement in acoustical comfort.

Their adoption is speedily expanding across various industries, including architecture, automotive, aerospace, and construction, as they offer effective solutions to address noise pollution and improve overall environmental quality. Indeed, acoustical quality and comfort are increasingly recognized as vital components of sustainable built environments. Neglecting acoustical considerations in both indoor and outdoor spaces can lead to various issues for users, with hearing loss and non-hearing-related syndromes. To address these concerns, sound absorbing materials are usually employed to enhance acoustical comfort in built surroundings. These materials absorb sound energy, thereby reducing reverberation and sound levels within enclosed spaces. By integrating such materials into design strategies, architects and engineers can generate environments that endorse both sustainability and user well-being. For enhanced sustainability, numerous acoustic materials have been established by utilizing recycled materials such as expanded polystyrene, agricultural residue, used paper and card board, rubber from tyres and carpets and textile waste. Many of acoustic materials are too recyclable. Sustainable polymer made from corn or sugar cane, soybean or other vegetables oils have been widely used to manufacture sound absorbing materials. Furthermore, use of use of nano fibres made from recyclable polyamide6(PA6) and water soluble polyvinylalohol (PVA) also produce good sound absorbing fibrous materials [23].

2.4.3 Tribology of Materials

Tribology plays a significant role in sustainability, encompassing studies on friction, wear, and lubrication. The acceptance of eco-friendly lubricants and additives, which are biodegradable and have little toxicity, helps mitigate health and pollution issues. Reusing worn parts after re-manufacturing, along with utilizing bio lubricants derived from biomass and other waste materials, contributes to the circular economy. Reducing friction and wear not only minimizes CO_2 emissions but also extends the lifespan of products by enhancing wear resistance and dropping the energy essential for manufacturing novel products. This results in greater efficiency, material savings, reduced energy losses, decreased pollution, and ultimately a more sustainable lifestyle, aligning with the objectives of sustainable development goals. Tribological studies show a vital role in minimizing the particles which are released from brakes, vehicle tires (including microplastics and minor polymer particles), and train wheels. Research has indicated that over 20% of the energy consumed in transport, industrial, power generation, and domestic usage is attributed to tribological contacts. Therefore, tribology is essential for sustainable growth, as its results directly impact numerous goals of the sustainable development [24].

2.4.4 Solar PV Cells

Amongst several sustainable and renewable energy skills, solar photovoltaic (PV) cells stand out as the furthermost communal and mature green energy schemes [25], addressing the world's increasing energy needs. Solar PV systems find application in various important sectors, including "Large-Scale Solar PV," "Residential Solar PV," "Green Hydrogen," "Water Desalination," and "Transportation." Large-scale solar PV installations are deployed in power plants with capacities exceeding 1 MW, often reaching the Gig-scale. Residential applications include rooftop solar PV schemes, PV-powered heating, ventilation, and air conditioning units, and PV-powered water pumps. Solar PV power desalination systems, accessible worldwide on a commercial scale, show increasing potential over period. Furthermore, substantial development has been prepared in utilizing solar PV for the production of green hydrogen, constructing green buildings, and powering electric vehicles in the transport segment.

2.4.5 Bamboo

Global influence of deforestation and natural resource depletion has spurred efforts to find results through research and development. In this pursuit, exploiting bamboo as a versatile feedstock for numerous applications has emerged as a promising approach. Bamboo is indeed a versatile and valuable resource for the pulp and paper industry. Its long fibers make it an excellent raw material for producing high-quality paper. Bamboo fibers offer a range of benefits that make them an excellent choice for clothing and medical supplies. Their moisture-wicking, antibacterial, and eco-friendly properties make them a sustainable option for the clothing industry. Textiles manufactured from bamboo-fibre are used in blankets, t-shirts, underwear bedding, socks, towels, and many other textile products.

Bamboo is indeed emerging as a sustainable alternate to conventional building materials like synthetic polymers, metal alloys, steel, and cement. These traditional materials, while essential for modern infrastructure, contribute significantly to environmental degradation due to their energy-intensive production processes and pollution. In contrast, bamboo offers a range of advantages which includes, higher yield, light weight, economic nature, fast growth, having similar properties to steel and concrete, speedy construction and a natural sink for carbon dioxide that make it an attractive and eco-friendly option for construction. Bamboo culms are an outstanding building material for houses, roof trusses, shelters, wall cladding, columns, boards, pillars, tools, scaffolding, flooring, ceilings, walls, windows, furniture, doors and fences, particularly in rural and disaster-prone areas. Their lightweight, flexible, and resilient nature, combined with their sustainability and affordability, make bamboo a critical resource for building safer, more durable, and environmentally friendly structures.

Bamboo fibers are increasingly being recognized for their potential in reinforcing various polymers, leading to the advancement of a vast range of bamboo fiber-reinforced composites like phenolic resin-based bamboo fibre reinforced composites, polyester-based bamboo fibre reinforced composites, polyvinyl chloride-based bamboo fibre reinforced composites, polypropylene-based bamboo fibre reinforced composites, polystyrene-based bamboo fibre reinforced composites and epoxy-based bamboo fibre reinforced composites. These composites combine the natural strength and flexibility of bamboo fibers with the durability and versatility of thermoplastic and thermoset polymers, resulting in materials that are both strong and eco-friendly.

Bamboo has indeed emerged as a valuable resource for enhancing food security, benefiting both humans and livestock. The young shoots of bamboo are highly nutritious, being low in fat yet rich in essential nutrients like potassium, carbohydrates, dietary fibers, vitamins, and bioactive compounds. These shoots can be consumed in different ways, including fermented, boiled, raw, canned, marinated, and even in liquid or frozen forms. The high cluster of fiber and bamboo proteins makes it an excellent food source for livestock. Bamboo leaves, in particular, serve as a nutritious feed for animals such as cattle, sheep, goats, and chickens. One of the key benefits of bamboo as a feedstock is its ability to absorb and accumulate calcium from the soil.

The increasing costs and diminishing fossil fuels reserves, like petroleum, coupled with growing concerns about environmental impact and, sustainability have driven significant research into alternative biofuels. Bamboo has appeared as an encouraging candidate in this field. Fuel characteristics of Bamboo's make suitable for generating biofuel. It has a high calorific value and volatile content, which contribute to its efficiency as a fuel source. Various products of bio-energy can be obtained from bamboo fiber including pyrolysis products charcoal, firewoodbriquettes biofuel, gasification products pellets and biomass. These products not only offer a renewable energy source but also help to reduce the environmental footprint associated with traditional fossil fuels.

Vast use of bamboo charcoal as energy resource, for cooking and heating are found especially in tropical and subtropical regions. In addition to this, bamboo charcoal act as an outstanding gas adsorbent helps to trap poisonous gas and contaminants. Porous structure of bamboo charcoal allows it to efficiently filter out contaminants from air and water by improving their quality. Every part of the bamboo plant such as rhizome, roots, leaves, culm, and seeds are widely used in pharmaceutical units for medical purposes.

The presence of phytosterols and lignans in bamboo shoots gives them cancer resistant properties, making them useful in controlling the early stages of cancer infection. Additionally, bamboo shoots are beneficial for loss of weight, cholesterol balance, and immunity. The leaves and stems of the species Bambusa bambos are traditionally applied to treat a range of conditions, including leucoderma (a skin disorder), inflammatory ailments, bronchitis, gonorrhea, and fever. Bamboo extracts also exhibit a broad array of pharmaceutical properties, including antidiabetic, antioxidant, anticancer, anti-inflammatory, antimicrobial, and antihypertensive effects. These properties mark bamboo a valued resource in traditional treatments for conditions such as hypertension, excessive sweating, and even paralysis [26].

2.4.6 CNT and CNT Based Composites

The enthusiasm within the research community regarding the widespread potential application of carbon nanostructured materials is certainly justified. These materials possess exceptional properties that create them highly promising for many uses, which includes sustainable energy conversion, wastewater treatment, bio-medical devices, non-bio-medical areas. Carbon nanotubes (CNTs), cylindrical nanostructures congregation of carbon atoms, offer a promising solution for eliminating pharmaceutical-derived contaminants and their metabolites from the environment. CNT-based composites have acquired remarkable observation in the remediation of wastewater due to their exceptional ability to separate a broad range of inorganic, organic, and biological pollutants. For instance, CNT sheets can be used to eliminatepoisonous heavy metals like cadmium, lead, arsenic, and cobalt, which are often challenging to eliminate using conventional methods. Additionally, novel CNT-based photo catalysts and nanocatalysts have emerged as highly efficient catalysts in wastewater treatment. These materials can act as ideal catalyst supports, enhancing the degradation of pollutants through advanced oxidation processes.

CNTs are considered superior nanosorbents compared to traditional sorbents like zeolite, clay, activated carbon, and diatomite due to their exceptional mechanical, structural and chemical properties. In addition to CNT based nanosorbents, other carbon allotropes like graphene quantum dots (GQDs), carbon nanodots (CNDs), carbon quantum dots (CQDs), graphene, and graphene oxide (GO), are gaining significant attention for their versatile applications, particularly in environmental remediation. Graphene nanocomposites can be applied as Nano sorbents effectively to remove a harmful and poisonous mix of contaminants, including organic compounds, heavy metals, dyes, and pharmaceuticals. Furthermore, they are also highly effective in removing radionuclides (e.g., uranium, cesium) and heavy metals (e.g., lead, mercury, cadmium) from wastewater.

Carbon nanodots (CNDs), carbon quantum dots (CQDs), graphene quantum dots (GQDs), and their various composites and hybrids, including N-doped CDs/GQDs and carbon nitride dots, offer an extensive application in non-biomedical areas, like optics, catalysis, and environmental engineering. GQDs and CQDs significantly boost the catalytic and photocatalytic performance of conventional semiconductors, making them powerful tools for environmental remediation. Besides, CQDs (Carbon Quantum Dots) and GQDs (Graphene Quantum Dots) indeed hold significant promise in biomedical fields like gene delivery, medicinedistribution, bioimaging as well as biosensing.

It has been shown that Graphene oxide (GO) nanosheets have indeed demonstrated potential as effective selective barriers for water permeability and the combination of superhydrophobicity (water-repelling) and superoleophilicity (oil-attracting) makes GO nanosheets highly useful in various water treatment and separation processes. Graphene also can be engineered to form nanoscale pores, creating separation membranes for separating small molecules, ions, and even gases from water [14].

The growing ultimatum for energy has spurred investigators to focus on developing safe energy resources for humankind. One significant benefit of discovering biological and chemical methods to produce green and effectual electrical energy is the decrease of dependency on soon-to-be-depleted fossil fuels. Biofuels represent one of the developing segments getting significant attention in this regard. Various carbon-based nanomaterials, such as graphene, carbon nanoparticles, nanotubes, nanorod etc. can be utilized to build and develop anextensive range of biofuels, thereby offering environmentally friendly energy resources. Carbon-relied nanomaterials, such as single-walled carbon nanotubes (SWCNT) and multi-walled carbon nanotubes (MWCNT), have gained attention for their superior catalytic activity in biodiesel conversion. Their large surface area and outstanding conductivity make them efficient catalysts, improving the reaction kinetics and overall conversion rates. Similarly, graphene and its derivatives have developed as encouraging agents for improving biogas production. Graphene's exceptional properties, including its high surface area, outstanding thermal and electrical conductivity, and chemical stability, make it an ideal material for facilitating microbial attachment and improving the biodegradation procedures.

Multi-walled carbon nanotubes (MWCNT) are indeed a promising category of electrically conductive material with the potential to enhance biogas and methane production rates. In addition to their role in biogas production, MWCNTs have also been examined as nano-additives in diesel fuel. When compared to ordinary diesel, diesel mixed with CNTs (Carbon Nanotubes) shows a lower Brake-Specific Fuel Consumption (BSFC). This means that fewer fuel is essential to produce the equivalent amount of energy, improving fuel efficiency. Moreover, as the concentration of CNTs in the diesel increases, the BSFC decreases even further, indicating better combustion efficiency and potentially reduced emissions. Carbon nanotubes (CNTs) are highly effective in improving the performance of diesel fuel when used as additives. One of the significant advantages of CNT-blended fuel is the reduction in ignition time within the combustion chamber. The faster ignition helps to prevent the buildup of waste gases, which can lead to incomplete combustion and increased emissions during the premixed combustion phase. Moreover, CNT-blended diesel has been observed to generate higher in-cylinder pressure compared to regular diesel. This increase in pressure is attributed to the superior igniting properties of CNTs, which facilitate a more rapid and complete combustion process. Another notable benefit of CNT-blended fuel is the reduction in carbon monoxide (CO) emissions. This improvement can be linked to the secondary atomizing effect of the CNT droplets in the fuel.

To optimize biofuel production, advanced skills with ML, AI, robotics, smart metering, modeling and simulation software, numerical and mathematical tools, statistical procedures, and optimization tools can be deployed. These skills aim to improve manufacture parameters, minimize time of production, diminish material wastage, and improved biofuels manufacture processes. Undoubtedly, biofuels will linger to show an acute character in the renewable energy segment and serve as energetic sustainable energy resolutions [27].

2.4.7 Buildings Facade Materials

Buildings stand as significant energy consumers and environmental polluters. Numerous studies have been conducted to measure and analyze energy consumption in buildings. With building sectors accounting for over 40% of entire concluding energy consumption and utilizing 30% of resources that are natural, they also contribute to 70% of electricity usage and emit more than 30% of greenhouse gases in urbanized nations. As per studies, more than 60% of CO_2 emissions can be attributed to the operation of buildings. This indicates that buildings not only require vast amounts of energy but also contribute heavily to environmental degradation. Finding effective solutions to mitigate the adverse ecological influences of buildings is crucial. The façade, an integral component of a building's architecture, holds substantial impending for plummeting energy consumption and alleviating its adverse environmental effects.

"Photovoltaic materials," "Thermochromic materials," and "Photostrictive materials" emerge as promising replacements for integrating into building facades. Yun et al. conducted a study on a photovoltaic facade in UK, determining its efficacy as a pre heating device during a natural ventilation and winter structure in the season of summers. Some researchers conducted a review of various studies focused on the use of photovoltaic (PV) and thermoelectric façades for energy saving in buildings. Their research highlighted that a combined photovoltaic-thermoelectric façade could result in energy savings of up to 22%.

Researchers conducted simulations of various smart window technologies, including electrochromic, photochromic and thermochromic windows to assess their effectiveness in reducing energy consumption in buildings. Among these, electrochromic windows which are specifically controlled by operative temperature has been proved to be the most effective. They also explored the application of Thermochromic Mortar (TM) facade systems, revealing that buildings incorporating this technology experience a 3% reduction in energy consumption [28].

2.4.8 Smart Textiles Materials

To minimize environmental impact caused by textile industries, research has developed new sustainable materials in clothing for every day consumers and specialized professionals. In this regard, new smart fabrics have been developed to furnish the high-performance characteristics. To fulfil the demands of modern consumers and professionals, the new developed fabrics provides unique properties like wash resistance, flexibility, durability, comfortable feel and lightweight components. The sensors, conductive fibers, and other advanced materials are embedded with these textiles to monitor different parameters which are related to consumers need. Due to this capability, smart textiles get highly useful in different areas such as sports, health care and even everyday clothing.

Incorporation of generators, nanogenerators, and triboelectric nanogenerators (TENGs) with smart wearable devices provides interesting applications in different fields. These devices convert the mechanical energy (body movements) into electrical energy and offer sustainable power source for wearable items. This capability is mostly useful in smart textiles, for internal security, environmental protection, mobile communication, biomedical sensing and, health monitoring where constant supply of energy is very much essential.

Textile industry also adds the development of self-cleaning textiles for substantial improvement. Coating of this fabrics with nanocrystalline titanium dioxide (TiO_2), helps to breakdown organic materials when exposed to sun light. These textiles having self-cleaning properties provides lots of advantages like reduce water consumption, energy efficiency and reduced use of chemicals (detergents).

Researcher's findings also highlighted the superior performance of Gore-Tex fabric matched to another natural fabric such as cotton fabricand synthetic fabrics like polyester, attributed to its breathable and waterproof properties. Graphene, known for its outstanding electrical conductivity, flexibility, and strength, is increasingly being used as an electrode material in smart textiles. When integrated into garments, graphene offers several advantages such as improved garment performance and skin compliance and health safety. This can also be utilized in the construction of smart textiles which are eco-friendly, resilient, and washable. Additionally, experimentations on self-cleaning textiles was conducted by cotton, which was improved with TiO_2 by the $TiOSO_4$ and $Ti(O\text{-}i\text{-}Pr)_4$ impregnation procedure. Their conclusions exhibited talented fallouts for producing individual protective clothing, like coats, suits, and overalls [13]. Table 2.2 depicts utilization of sustainable materials in daily life.

2.5 Suggestive Proposals for Sustainable Development

- Digital technologies show an essential part in the growth of manufacturing firms by facilitating the deployment of product-service systems. These systems, which address the sustainability concept from economic, environmental, and social perspectives, contribute to enhancing companies' competitiveness. By transferring ownership and operational accountability to suppliers, firms can not only improve the quality and effectiveness of their solutions but also strengthen their client relationships [29].
- Industrial waste holds significant potential as a source of minor raw materials. Many waste categories comprise valued metals at concentrations higher than those found in principal ores, sometimes by factors of ten or even a hundred. Metals like cobalt, nickel, etc. rare earth elements, among others, are crucial for transitioning to a completely sustainable industry. Consequently, extracting and purifying these metals from industrial waste often proves to be more cost-effective than processing primary ores, creating recycling an appealing and suitable investment [30].

Table 2.2 Utility of sustainable materials

Sl. No	Sustainable material	Utility in daily life	References
1	Agro waste	(i) Sustainable construction material (ii) Thermal insulation materials and reinforcement materials for buildings (iii) Insulation bricks (iv) Bio-based plastics	Maraveas [22]
2	Acoustic materials (i) Sustainable polymer (ii) Nano-fibres (iii) Expanded polystyrene, agricultural residue, used paper and card board, rubber	Helps to improve acoustical comfort in architecture, automotive, aerospace and construction industry	Arenas [23]
3	Tribology of materials and lubricants	Water quality gets improved by minimizing pollution, removing dumping and reducing hazardous chemicals and materials released	Enrico Ciulli [24]
	Industrial tribology	Water-related ecosystems which comprises rivers, mountains, forests, wetlands, aquifers and lakes is protected as well as restored by increasing recycling and safe reuse globally	
	Bio-tribology	The no. of demises and illnesses which occur due to hazardous chemicals and air, water as well as soilpollution and contamination is reduced	
4	Bamboo	(i) For paper and pulp industry, it is utilized as a raw material (ii) Used for building and construction materials (iii) Used in medical and pharmacological industry (iv) Used for biofuel and bio-energy production (v) Used for textile and apparel industry (vi) Food and feed stock production	Chaudhary et al. [26]

(continued)

Table 2.2 (continued)

Sl. No	Sustainable material	Utility in daily life	References
5	CNTs and CNT based composite	Used in waste water treatment for elimination of aqueous pollutants and pharmaceutical derived contaminants	Nasrollahzadeh et al. [14]
	CNT sheets	Remove toxic heavy metals like cadmium, lead, arsenic and cobalt	Nasrollahzadeh et al. [14]
	Carbon quantum dots, carbon nano dots and Graphene quantum dot	Non bio medical areas and bio medical fields	Nasrollahzadeh et al. [14]
	Graphene oxide nano-sheets	Effective selective barriers for separation process	Nasrollahzadeh et al. [14]
6	Biofuel from SWCNT and MWCNT	(i) Biodiesel: Transportation, reduce the risk of environmental contamination, and reduce the increase of atmospheric carbon CO_2 levels (ii) Biogas: Electricity generation, heating and cooking fuel (iii) CNT blended fuel/diesel: reduce carbon monoxide emission and faster ignition	Khan et al [27]
7	Building façade's smart materials (i) Photovoltaic façade material (ii) Thremochromic mortar(TM) material façade and electrochromic smart window (iii) Photostrictive façade material	(i) Preheating device in winters and natural ventilation in summer (ii) Reduce buildings' energy consumption (iii) Alleviate negative environmental effect of building's architecture	Balali and Valipour [28]
8	Smart textile materials (i) Gore-Tex fabrics (ii) Graphene (iii) Self-cleaning textiles using modified cotton with TiO_2	(i) Superior performance due to its breathable and waterproof properties as compared to natural fabrics (cotton and polyster) (ii) Used in the design of smart textile are flexible washable and environmentally friendly, improved garment performance, skin compliance and health safety (iii) Personal protective clothing like suits, coats and overalls	Provin et al. [13]

- In contemporary times, plastics are becoming increasingly prevalent in various aspects of everyday life, including wrapping in food and beverage firms, cosmetics, pharmaceuticals, and additional manufacture segments requiring effective and safe delivery of product to the communal. However, the post-utilization disposal of produced plastic waste poses significant adverse influences on the environment if not accomplished properly. The detrimental effects of plastic waste extend to animal life, land, water, air quality, and public health. To address these challenges, several scientifically based plastic waste management strategies have been established, including recycling, incineration, bioremediation, and landfill disposal. These approaches purpose to promote a clean and sustainable environment by effectively managing plastic waste disposal [31].
- Achieving sustainability in solid waste management necessitates strategic planning to divert waste away from landfills. This endeavor calls for international assistance, particularly for minor revenue countries. Works can aid municipalities in these regions by providing methodologies to assess their resources, determine optimal facility locations for waste management, optimize transportation logistics, and foster integration with the private segment [32].
- Sustainable mobility and planning of transportation are integral components of sustainable development efforts aimed at mitigating environmental impacts. To address this, research endeavors should delve into innovative mobility solutions, including shared mobility facilities, the electrification of transport, and the encouragement of energetic methods of transportation. By embracing these approaches, we can work towards creating an additional sustainable, equitable, as well as accessible forthcoming for all.
- AI, can help manufacturing companies reduce the complexity of supply and distribution chains. By utilizing the vast data processing capabilities available through AI, along with data generated by supply chain actors and collected via the digital systems of smart factories, companies can design processes and products that are more environmentally and socioeconomically sustainable.
- Management of water plays a vital role in sustainable growth, influencing humanoid wellbeing, ecosystems, and economic endeavors. Consequently, research efforts should concentrate on pioneering administration of water approaches, including water recycle, green infrastructure, wastewater-to-energy initiatives, and storm water managing. These endeavors aim to tackle pressing issues such as scarcity of water, quality of water degradation, and flooding, thereby fostering sustainable water resource management practices.

2.6 Conclusion and Scope of Future Work

Technological innovation has garnered significant attention from scholars, practitioners, and policymakers as a crucial tool for achieving sustainable development. The importance of energy based materials to sustainable development is emphasized in various articles in this issue. Discussions revolve around the character of materials in achieving greater sustainability, with several suggestions proposed to make the construction industry more environmentally friendly. These suggestions include reducing the energy consumption of buildings over their lifetime, minimizing waste in construction processes, utilizing waste materials for construction purposes, decreasing emissions in the manufacture of building materials, and appropriating carbon by incorporating carbon encompassing materials into construction projects. Moreover, building with more durable materials is also highlighted as a sustainable approach. Outstandingly, these procedures are not only environmentally advantageous but also economically advantageous. For instance, reducing embodied energies in construction processes can lead to cost savings. Technological advancements, particularly in materials science, underpin these sustainability efforts. Innovative novel materials assist architects and engineers to construct structures that offer greater value, providing enhanced usability, livability, aesthetics, and sustainability. Overall, technology plays a vital character in achieving a higher level of sustainability by enabling economic reductions, reusing materials, and implementing recycling practices. Through technological innovation, it becomes feasible to realize a more sustainable constructed environment that meets the requirements of current and forthcoming compeers.

Scope of the future works are as the follows:

- In Indian framework, there is a requirement of new experimentation to explore the implementation of agro-waste as a substitute for sustainable production materials. This initiative could lead to state-of-the-art resolutions that discourse both environmental concerns and the demand for creation materials.
- Technological innovation in automation, particularly with advanced microprocessor-relied control schemes, enhances the performance of various technologies like solar panels, integrated heating, ventilation, and air conditioning (HVAC) schemes, and electrical systems.
- Consideration should be given to non-conventional materials for sustainable construction, particularly those utilizing bio-construction systems reinforced with cellulose fibers. An example of such a system could involve the use of a composite material consisting of metakaolin, lime, hemp hurds, and waste paper pulp instead of conventional concrete. This approach promotes sustainability by reducing reliance on traditional building materials while incorporating renewable resources.
- A digital computer-based sustainable design tool for building material applications in specific building projects involves an integrated design approach. This approach includes Life Cycle Analysis (LCA) based on computer-aided design (CAD) models. However, current technological limitations prevent the tool from providing fully accurate or optimal solutions. Advances in algorithms and the incorporation of more realistic data into the engine can enhance the analysis precision of these tools, potentially revolutionizing the current landscape of building design technology.

References

1. Stavins, R.N., Pfaff, A.: Readings in the field of natural resource and environmental economics (1999). SSRN 168969
2. Boar, A., Bastida, R., Marimon, F.: A systematic literature review: relationships between the sharing economy, sustainability and sustainable development goals. Sustainability **12**, 6744 (2020)
3. Green, M.L., Espinal, L., Traversa, E., Amis, E.J.: Materials for sustainable development. MRS Bull. **37**, 303–309 (2012)
4. Koltun, P.: Materials and sustainable development. Prog. Nat. Sci. Mater. Int. **20**, 16–29 (2010)
5. Lamba, P., Kaur, D.P., Raj, S., Sorout, J.: Recycling/reuse of plastic waste as construction material for sustainable development: a review. Environ. Sci. Pollut. Res.Pollut. Res. **29**, 86156–86179 (2022)
6. Leng, Z., Padhan, R.K., Sreeram, A.: Production of a sustainable paving material through chemical recycling of waste PET into crumb rubber modified asphalt. J. Clean. Prod. **180**, 682–688 (2018)
7. Soni, A., Das, P.K., Hashmi, A.W., Yusuf, M., Kamyab, H., Chelliapan, S.: Challenges and opportunities of utilizing municipal solid waste as alternative building materials for sustainable development goals: a review. Sustain. Chem. Pharm. **27**, 100706 (2022)
8. Awoyera, P.O., Olalusi, O.B., Iweriebo, N.: Physical, strength, and microscale properties of plastic fiber-reinforced concrete containing fine ceramics particles. Materialia. **15**, 100970 (2021)
9. Mustafa, M.A.-T., Hanafi, I., Mahmoud, R., Tayeh, B.A.: Effect of partial replacement of sand by plastic waste on impact resistance of concrete: experiment and simulation. In: Structures, pp. 519–526 (2019)
10. Zhou, J., Sheng, Z., Li, T., Shu, Z., Chen, Y., Wang, Y.: Preparation of hardened tiles from waste phosphogypsum by a new intermittent pressing hydration. Ceram. Int. **42**, 7237–7245 (2016)
11. Esmaeili, J., Aslani, H.: Use of copper mine tailing in concrete: strength characteristics and durability performance. J. Mater. cycles waste Manag. **21**, 729–741 (2019)
12. Surtiyeni, N., Rahmadani, R., Kurniasih, N., Khairurrijal, Abdullah, M.: A fire-retardant composite made from domestic waste and PVA. Adv. Mater. Sci. Eng. **2016**, 7516278 (2016)
13. Provin, A.P., de Aguiar Dutra, A.R., Machado, M.M., Cubas, A.L.V.: New materials for clothing: rethinking possibilities through a sustainability approach-a review. J. Clean. Prod. **282**, 124444 (2021)
14. Nasrollahzadeh, M., Sajjadi, M., Iravani, S., Varma, R.S.: Carbon-based sustainable nanomaterials for water treatment: state-of-art and future perspectives. Chemosphere **263**, 128005 (2021)
15. Cong, Q., Yuan, X., Qu, J.: A review on the removal of antibiotics by carbon nanotubes. Water Sci. Technol. **68**, 1679–1687 (2013)
16. Mishra, A.K., Arockiadoss, T., Ramaprabhu, S.: Study of removal of azo dye by functionalized multi walled carbon nanotubes. Chem. Eng. J. **162**, 1026–1034 (2010)
17. Upadhyayula, V.K.K., Deng, S., Mitchell, M.C., Smith, G.B.: Application of carbon nanotube technology for removal of contaminants in drinking water: a review. Sci. Total. Environ. **408**, 1–13 (2009)
18. Gupta, V.K., Kumar, R., Nayak, A., Saleh, T.A., Barakat, M.A.: Adsorptive removal of dyes from aqueous solution onto carbon nanotubes: a review. Adv. Colloid Interface Sci. **193**, 24–34 (2013)
19. Perreault, F., De Faria, A.F., Elimelech, M.: Environmental applications of graphene-based nanomaterials. Chem. Soc. Rev. **44**, 5861–5896 (2015)
20. Nasrollahzadeh, M., Baran, T., Baran, N.Y., Sajjadi, M., Tahsili, M.R., Shokouhimehr, M.: Pd nanocatalyst stabilized on amine-modified zeolite: antibacterial and catalytic activities for environmental pollution remediation in aqueous medium. Sep. Purif. Technol.Purif. Technol. **239**, 116542 (2020)

21. Cash, D.W., Clark, W.C., Alcock, F., Dickson, N.M., Eckley, N., Guston, D.H., Jäger, J., Mitchell, R.B.: Knowledge systems for sustainable development. Proc. Natl. Acad. Sci. **100**, 8086–8091 (2003)
22. Maraveas, C.: Production of sustainable construction materials using agro-wastes. Materials (Basel) **13**, 262 (2020)
23. Arenas, J.P., Sakagami, K.: Sustainable acoustic materials (2020)
24. Ciulli, E.: Vastness of tribology research fields and their contribution to sustainable development. Lubricants. **12**, 33 (2024)
25. David, A., Joel, O.O.: Design and construction of a plastic shredder machine for recycling and management of plastic wastes. Int. J. Sci. Eng. Res. **9**, 1379–1385 (2018)
26. Chaudhary, U., Malik, S., Rana, V., Joshi, G.: Bamboo in the pulp, paper and allied industries. Adv. Bamboo Sci. 100069 (2024)
27. Khan, N.R., Rashid, A.B.: Carbon-based nanomaterials: a paradigm shift in biofuel synthesis and processing for a sustainable energy future. Energy Convers. Manag. X. 100590 (2024)
28. Balali, A., Valipour, A.: Identification and selection of building façade's smart materials according to sustainable development goals. Sustain. Mater. Technol. **26**, e00213 (2020)
29. Lamperti, S., Cavallo, A., Sassanelli, C.: Digital servitization and business model innovation in SMEs: a model to escape from market disruption. IEEE Trans. Eng. Manag.Manag. **71**, 4619–4633 (2023)
30. Ippolito, N.M., Amato, A., Innocenzi, V., Ferella, F., Zueva, S., Beolchini, F., Vegliò, F.: Integrating life cycle assessment and life cycle costing of fluorescent spent lamps recycling by hydrometallurgical processes aimed at the rare earths recovery. J. Environ. Chem. Eng. **10**, 107064 (2022)
31. Jaiswal, S.K., Mukti, S.K., Rath, K.C.: E-waste control and its recycling to build sustainable society in the global context. In: Handbook of Research on Applications of AI, Digital Twin, and Internet of Things for Sustainable Development, pp. 200–222. IGI Global (2023)
32. Swain, S., Muduli, K., Kommula, V.P., Sahoo, K.K.: Innovations in internet of medical things, artificial intelligence, and readiness of the healthcare sector towards health 4.0 adoption. Int. J. Soc. Ecol. Sustain. Dev. **13**, 1–14 (2022)

Dr. Subhra Gantayat is presently working as Assistant Professor in the Department of Physics in KISS Deemed to be University. She has completed her Ph.D. in 2016 in the area of Nano composite materials from KIIT Deemed to be University Bhubaneswar, Odisha. To her credit She has published 13 research papers in various national and international journals of repute. Her area of research focuses on polymer nanocomposite, sustainability and environmental impact, material science and sustainable production. Dr. Gantayat's research interest encompasses the structure-synthesis—characterization relationships exhibited by functional materials. She is keen to design and develop eco-friendly sustainable materials in the form of nanoparticle, nanostructures, fibers and nanocomposites using ease and cost effective processes. She has about 30 years of teaching experience in physics, Engineering physics, Physics of semiconductor Devices Electromagnetic field theory, Mathematical Physics.

Soumya Dash is currently pursuing a Ph.D. in Mechanical Engineering at the School of Mechanical Engineering, KIIT University, Bhubaneswar, India and also serves as an Assistant Professor in the Department of Mechanical Engineering at C V Raman Polytechnic, Bhubaneswar. Her research focuses on improving productivity, sustainability and quality in manufacturing systems, with specialized interests in polymer composites, advanced manufacturing processes and operations management. Soumya is dedicated to optimizing manufacturing workflows, implementing quality assurance techniques, and explore sustainable practices in industrial practices. Her work is aimed at developing eco-friendly manufacturing strategies that meet contemporary industrial needs. She has published papers in national and international conferences and journals. Through her academic pursuits and research, she is committed to promoting sustainable practices and fostering innovation, with the goal of contributing to industrial excellence and advancing in the field of mechanical engineering.

Prof. Md. Kamruzzaman is presently working as Professor in Department of Mechanical Engineering, Dhaka University of Engineering and Technology, Gazipur, Bangladesh. He received his Ph.D. degree in Production Engineering from Bangladesh University of Engineering and Technology in the year of 2010. Good no of research scholars have been awarded Ph.D. degree under his supervision. He has published many research papers in different International journals of repute. He has presented no of papers in different International conferences in Bangladesh and abroad. He has reviewed no of Ph.D. thesis of different University and number of research papers of International Journals. His research interests are in non-traditional machining, hard machining on cryogenic cooling, surface characterization of machined surface, optimization of machinability parameters and Machine tools, Manufacturing, automobile, robotic and automation. He has been teaching manufacturing related courses since last twenty-six years.

Chapter 3
Circular Economy Governance and Regulatory Framework

Lakshmi Prasad Panda, Kali Charan Rath, N. V. Jagannadha Rao, Chitrasen Samantra, and Kamalakanta Muduli

Abstract The transition to a circular economy (CE) necessitates robust governance and regulatory frameworks to ensure sustainable resource management, waste reduction, and economic resilience. This study delves into the critical elements of CE governance by examining policy instruments such as national strategies, legislative mandates, economic incentives, and multi-stakeholder partnerships. The key to this governance is the harmonization of efforts across governmental levels and sectors, facilitated by advanced regulatory tools such as extended producer responsibility (EPR), eco-design directives, and green public procurement. This study emphasises the significance of monitoring and reporting systems in tracking progress and ensuring accountability. Through an analysis of global case studies from the European Union, China, Japan, and other leading regions, this study identifies best practices and regulatory innovations while addressing persistent challenges. The findings emphasize the need for adaptive governance structures that can integrate technological advancements and respond to dynamic economic and environmental contexts. This comprehensive evaluation offers strategic recommendations for policymakers,

L. P. Panda (✉)
Department of Humanities, Government College of Engineering Kalahandi, Bhawanipatna, Odisha, India
e-mail: pandalakshmiprasad@gmail.com

K. Charan Rath
Department of Mechanical Engineering, Gandhi Institute of Engineering and Technology University, Odisha, Gunupur, India

N. V. J. Rao
School of Management Studies, Gandhi Institute of Engineering and Technology University, Odisha, Gunupur, India

C. Samantra
Department of Mechanical Engineering, Parala Maharaja Engineering College, Berhampur, Ganjam, Odisha, India

K. Muduli
Mechanical Engineering Department, Papua New Guinea University of Technology, Lae, Morobe Province, Papua New Guinea
e-mail: kamalakanta.muduli@pnguot.ac.pg

© The Author(s), under exclusive license to Springer Nature Switzerland AG 2025
K. C. Rath et al. (eds.), *Engineering, Technology and Management*, Information Systems Engineering and Management 54, https://doi.org/10.1007/978-3-031-93654-8_3

industry stakeholders, and researchers to create effective regulatory environments that promote CE agenda and foster sustainable and resilient economic systems.

Keywords Circular economy · Governance · Regulatory framework · Resource efficiency · Waste management · Extended producer responsibility · Eco-design · Policy innovation

Preface

This chapter explores the essential governance and regulatory frameworks required to drive the transition to a circular economy (CE). It examines key policy instruments such as national strategies, legislative mandates, and economic incentives, as well as the role of multi-stakeholder partnerships in ensuring effective resource management and waste reduction. The chapter highlights the importance of harmonizing efforts across sectors and governmental levels, supported by advanced regulatory tools like extended producer responsibility (EPR), eco-design directives, and green public procurement. By analyzing global case studies and identifying best practices, this chapter offers critical insights into adaptive governance and provides strategic recommendations for fostering sustainable, resilient economic systems.

3.1 Introduction

The transition to a circular economy (CE) marks a pivotal shift from the traditional linear economic model, which focuses on extracting resources, producing goods, and disposing waste [1, 2]. Circular economy is a transformative approach to economic development that prioritizes sustainability by designing products for longevity, optimizing resource efficiency, and minimizing waste. In contrast, CE is a transformative approach to economic development that prioritizes sustainability by designing products for longevity, optimizing resource efficiency, and minimizing waste. Central to CE are principles such as eco-design, which ensures that products are durable and recyclable, and lean manufacturing, which reduces waste and enhances productivity [3]. It emphasizes extending the product life through the repair, remanufacturing, and recycling of materials to maintain closed-loop cycles [4]. CE also promotes circular supply chains and reverses logistics to recapture the value of the used products. Business models have shifted towards services rather than ownership, fostering the efficient use of resources. Collaboration across sectors, supported by robust policies, economic incentives, and consumer education, has driven this shift. Innovation and technology play crucial roles in tracking and optimizing resource use, whereas environmental impact assessments and lifecycle analyses guide sustainable practices. The economic viability of CE is demonstrated through cost savings, new revenue streams, and reduced risks, making it a comprehensive strategy for sustainable development [5, 6].

3.1.1 *Importance of Governance and Regulatory Frameworks in CE*

Governance and regulatory frameworks are crucial for a successful circular economy (CE). They provide structure and incentives for sustainable practices, guiding stakeholders towards resource efficiency and waste reduction. Effective frameworks establish clear policies, enforce compliance through regulations, and facilitate collaboration among stakeholders, ultimately accelerating the shift to a CE [7, 8].

3.1.2 *Objectives*

(a) Assess the impact of existing regulatory frameworks, including extended producer responsibility (EPR) and eco-design directives, on promoting circular economic practices and enhancing resource efficiency.
(b) Evaluate the effectiveness of governance structures in coordinating multi-sectoral efforts and fostering collaboration among stakeholders to achieve sustainable development goals through circular economic initiatives.

3.2 Overview of Circular Economy Principles and Their Benefits

Circular economy principles aim to reduce waste and increase resource efficiency by designing durable, repairable, and recyclable products. By promoting reuse, recycling, and sharing, CE practices minimize raw material extraction, environmental impact, and greenhouse gas emissions. This approach fosters economic resilience through innovation, job creation, and new markets for recycled materials [9, 10].

Circular economy principles advocate a fundamental change in how economic activities, are approached emphasizing sustainable resource management and waste reduction across the entire lifecycle of products and services. A key aspect of these principles is designing products for longevity and guaranteeing their reusability, repaired, or recycled at the end of their lives. The hypothetical data-set illustrates a comparison between linear and circular economy approaches in terms of resource efficiency and waste reduction.

Some of the Indian companies that are embracing circular economy principles:

(a) **Mahindra & Mahindra—Farm Equipment Sector**: Mahindra & Mahindra, a prominent Indian automotive manufacturer, has implemented circular economy practices in the farm equipment sector. The company focuses on designing tractors and agricultural machinery with durable components that are easy to repair and maintain. Mahindra also promotes the reuse of components through remanufacturing processes, extending the lifecycle of agricultural equipment and reducing waste generation. By adopting circular economy principles, Mahindra and Mahindra aimed to enhance resource efficiency, reduce environmental impacts, and support sustainable agricultural practices in India [11, 12].

(b) **ITC Limited—Paperboards and Packaging Business**: ITC Limited, a diversified conglomerate in India, has integrated circular economy principles into its paperboard and packaging businesses. The company focused on sourcing wood pulp from sustainably managed forests and promoting the use of recycled fibers in paperboard products. ITC Limited also implements efficient packaging designs that minimize material use and optimize recycling. By adopting closed-loop recycling systems, ITC Limited reduces waste generation and conserves natural resources, thereby contributing to environmental sustainability in the paper and packaging industry.

(c) **Tata Steel—Steel Manufacturing**: Tata Steel, one of India's largest steel manufacturers, has embraced circular economy strategies to optimize resource use and reduce the environmental impact of steel manufacturing. The company focuses on recycling and reusing steel scrap and by-products from the manufacturing processes. Tata Steel also explored innovative technologies, such as converting steel slag into construction materials, to minimize waste and maximize resource recovery. By promoting circular economy principles in steel production, Tata Steel aims to achieve operational efficiency, reduce its carbon footprint, and support sustainable development in the industrial sector.

(d) **Infosys—Information Technology Services**: Infosys, a leading global provider of information technology services in India, has implemented circular economic initiatives to promote sustainability across operations. The company focuses on energy efficiency measures, waste reduction strategies, and sustainable procurement practices in its information technology (IT) infrastructure and services. Infosys also invest in renewable energy projects and green building initiatives to minimize environmental impacts and carbon emissions. By incorporating the ideas of the circular economy into its business practices, Infosys aims to achieve resource efficiency, reduce its ecological footprint, and lead to sustainable technological solutions.

(e) **Godrej & Boyce—Consumer Products and Industrial Engineering**: Godrej and Boyce, a diversified conglomerate with an interest in consumer products and industrial engineering, has adopted circular economy principles to enhance sustainability in its manufacturing processes and product offerings. The company has focused on designing energy-efficient appliances, promoting material recycling, and implementing eco-friendly packaging solutions. Godrej and Boyce have also engaged in community initiatives for waste management and environmental conservation. By embracing circular economy practices, Godrej and Boyce aimed to reduce the environmental impact, improve resource efficiency, and contribute to sustainable development in India and globally.

(f) **Renault's—Automotive Industry**: Renault focuses on three main pillars: remanufacturing, recycling, and energy recovery. Renault's remanufacturing operations involve refurbishing the components and parts of end-of-life vehicles to extend their lifecycles and reduce waste. They also prioritize recycling materials such as metals and plastics from scrapped vehicles. By integrating circular economy principles into their manufacturing and supply chain processes, Renault aimed to minimize resource consumption, lower greenhouse gas emissions, and enhance sustainability across operations.

The adoption of circular economic principles varies significantly across Indian industries, each facing unique challenges and exhibiting different success stories. The automotive and electronics industries focus on material reuse and e-waste management, whereas the textile industry pioneers sustainable fashion and recycling. The construction industry is pioneering in the use of recycled materials and sustainable practices. Continued innovation, regulatory support, and consumer awareness are being critical for overcoming challenges and advancing circular economy principles across these sectors (Table 3.1). The advantages of the circular economy over the linear economic approach are place in (Table 3.2).

3.3 Circular Economy (CE) Governance

Effective governance is crucial for advancing circular economy initiatives. Governments, industries, and stakeholders must collaborate to create policies, standards, and partnerships that promote sustainable practices and reduce waste. Examining circular economy governance involves reviewing existing research on policy development, implementation, and impact assessment. This includes national and international policies, legislative measures, and regulatory mechanisms that promote sustainable production, consumption, waste minimization, and resource efficiency.

Facilitating Circular Business Models: Financial support can assist businesses in transitioning towards circular business models, such as leasing, sharing, and product-as-a-service models. This includes subsidies for implementing circular supply chains or incorporating recycled materials into the production processes.

Addressing Market Failures: Economic incentives can correct market failures that hinder the adoption of circular economy practices, such as externalities associated with waste disposal and environmental degradation. Subsidies can internalize these costs and incentivize sustainable practices.

Promoting Job Creation and Economic Growth: Investments in circular economy initiatives can spur job creation in sectors such as recycling, remanufacturing, and sustainable manufacturing. This contributes to economic growth while also addressing environmental challenges.

Overall, economic incentives and subsidies play a crucial role in overcoming barriers to circular economy adoption, stimulating innovation, creating market demand, and

Table 3.1 Comparative analysis of Indian industries adopting circular economy principles

Industry	Key initiatives	Challenges	Success stories	Future outlook
Automotive	Recycling and reusing materials	High initial costs for setting up recycling facilities	Mahindra and Mahindra's zero-waste initiatives	Increased use of recycled materials in manufacturing
	Eco-design for longevity and recyclability	Managing end-of-life vehicles effectively	Tata Motors' use of recycled materials in vehicle production	Development of efficient recycling technologies
	Extended Producer Responsibility (EPR) schemes	Consumer awareness about recycled products		
Textiles	Sustainable fashion through up cycling and recycling	Collection and sorting of textile waste	Arvind Mills' zero liquid discharge system	Growth in sustainable fashion market
	Using organic and recycled fibers	High costs of sustainable materials	Raymond's use of recycled fibers in garments	Increased consumer demand for sustainable products
	Circular supply chains	Maintaining product quality with recycled materials		
Electronics	E-waste recycling and refurbishment	Informal e-waste recycling sectors	Infosys' e-waste management program	– Stricter regulations and better enforcement
	Product-as-a-service models	Technological obsolescence	Dell's global recycling programs	– Innovation in product design to enhance longevity
	Urban mining for precious metals	Collection and recycling infrastructure		
Construction	Use of recycled and sustainable materials	Regulatory compliance	Godrej Construction's use of fly ash and recycled concrete	Expansion of green building initiatives

(continued)

Table 3.1 (continued)

Industry	Key initiatives	Challenges	Success stories	Future outlook
	Demolition waste recycling	Cost and quality concerns of recycled materials	UltraTech Cement's sustainable practices	Adoption of circular construction practices
	Designing buildings for disassembly and reuse	Awareness and adoption among builders and contractors		

Table 3.2 Advantages of circular economy approach over liner economy approach

Aspect	Linear economy approach	Circular economy approach
Product design	90% designed for single-use; materials difficult to separate or recycle	20% designed for durability and disassembly; materials selected for recyclability
Resource use (Tonnes)	100,000 tonnes extracted annually; 20% recycled	70,000 tonnes extracted annually; 70% recycled
Waste generation (Tonnes)	30,000 tonnes of waste annually; 10% recycled	10,000 tonnes of waste annually; 90% recycled
Environmental impact (CO_2 emissions)	200,000 tonnes CO_2 emissions annually	120,000 tonnes CO_2 emissions annually
Economic benefits (Cost savings)	5 million spent annually on waste disposal and resource procurement	3 million saved annually from material recovery and reuse

fostering a sustainable and resilient economy. The effective design and implementation of these incentives require collaboration between governments, businesses, and stakeholders to maximize their impact in promoting sustainable development goals and achieving long-term environmental benefits.

3.3.1 Regulatory Measures

Regulations are necessary for the governance of the circular economy in order to guarantee accountability, encourage sustainable behaviors, and ease stakeholder engagement. They support a more sustainable economic model by reducing waste, establishing standards, and providing incentives for resource efficiency.

3.3.1.1 Extended Producer Responsibility (EPR)

Extended Producer Responsibility (EPR) is an environmental policy that holds manufacturers accountable for their products' entire lifecycle, emphasizing proper disposal and recycling. This approach encourages the creation of durable, recyclable, and eco-friendly products through various mechanisms like take-back programs, producer-funded recycling infrastructure, and regulatory compliance. EPR aims to promote environmental protection, resource efficiency, and innovation while fostering consumer engagement through education and accessible return systems. Notable examples include the EU's WEEE Directive for electronics and Germany's Packaging Act. Despite challenges such as implementation costs, EPR plays a crucial role in advancing a circular economy by promoting sustainable production and consumption practices.

Simple Hypothetical Example

Let us say that A company sold 10,000,00 units of smartphones in 2019, maintain the objective of a 70% recycling rate. Suppose the company collected 6,00,000 units of smartphones with a EPR fee per unit ₹2 if the target is met, ₹5 if not met.

Target Number of Units to be Recycled: 1,000,000 × 0.70 = 7,00,000 units.

Shortfall in Recycling: 7,00,000–6,00,000 = 1,00,000 units.

The company collected 6,00,000 units of smartphones with a EPR Fee per unit ₹2, if the target is met, since the producer did not meet the recycling target total ***EPR Fee*** *is 1,000,000 units × ₹5 = ₹50,00,000.*

If Extended Producer Responsibility (EPR) targets are not met, producers face various consequences including financial penalties such as increased fees and fines. Regulatory actions can result in legal proceedings operational restrictions, and reputational damage, thereby affecting brand image and market position. Environmentally, failure leads to increased waste and lower recycling rates, exacerbating pollution and resource depletion. Operational costs increase owing to remediation efforts, and companies miss out on the benefits of recycled materials. Additionally, consumer backlash, including boycotts and demands for accountability, can pressure companies to improve their practices. Governments may respond to stricter regulations and increased monitoring, thereby further enforcing compliance. Overall, non-compliance with EPR targets not only impacts producers financially and operationally but also hinders environmental sustainability and circular economic goals. The benefits and incentives of Extended Producer Responsibility (EPR) are multifaceted. Environmentally, EPR significantly reduces e-waste through higher recycling rates, contributing to a cleaner and more sustainable environment. For producers, financial incentives such as savings from meeting or exceeding recycling targets encourage the design of more durable and recyclable products, fostering innovation and sustainable manufacturing practices. Additionally, EPR enhances consumer awareness and participation in recycling programs, as individuals become more informed about take-back schemes and the importance of recycling, leading to a more engaged and environmentally-conscious consumer base.

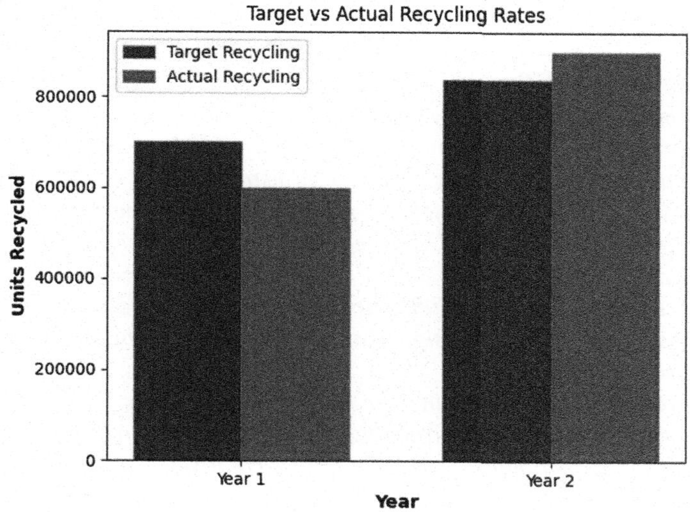

Fig. 3.1 Target versus actual recycling rates

Figure 3.1 shows how the target and recycling of materials changed over the two periods. The bar chart illustrates the target versus actual recycling rates for two years under an Extended Producer Responsibility (EPR) scheme in a hypothetical country.

- **Year 1**
  - Target Recycling: 700,000 units.
  - Actual Recycling: 600,000 units.
- **Year 2**
  - Target Recycling: 840,000 units.
  - Actual Recycling: 900,000 units.

3.3.1.2 Eco-Design Directives

Eco-design directives are regulatory frameworks aimed at improving the environmental performance of goods over the course of their lifetime, from design to disposal. These directives typically mandate manufacturers to consider environmental impacts at the design stage, promoting the development of products that are more energy-efficient, durable, recyclable, and environmentally friendly [13].

Key elements of eco-design directives

(a) ***Energy Efficiency***: *Products must meet specific energy consumption standards to reduce their environmental impact during use. Example: Energy labels for household appliances such as refrigerators and washing machines.*

(b) ***Resource Efficiency***: *Encourages the use of materials that are sustainable, recyclable, and have a lower environmental footprint. Example: Designing products with fewer materials and avoiding hazardous substances.*
(c) ***Durability and Reparability***: *Products should be designed to last longer and be easier to repair, reducing the need for frequent replacements. Example: Standardized parts and easy disassembly instructions.*
(d) ***Recyclability***: *Products must be designed to facilitate recycling at the end of their life cycle. Example: Using materials that can be easily separated and recycled.*
(e) ***Reduction of Hazardous Substances***: *Limiting the use of harmful chemicals and materials in product design. Example: Restricting substances such as lead, mercury, and cadmium.*
(f) ***Lifecycle Analysis***: *Assessing the environmental impact of a product from production to disposal, ensuring that eco-design considerations are integrated at every stage. Example: Conducting lifecycle assessments (LCA) to identify and mitigate environmental impacts.*

(i) **European Union's eco-design directive**

A framework was established by the EU Eco-Design Directive (2009/125/EC) to impose mandatory ecological requirements on energy-related items and energy-using devices. This regulation seeks to advance the EU's sustainability objectives and enhance the environmental performance of its products. Key aspects include:

- **Energy Labelling**: Products must display an energy label that indicates their efficiency. **Standby Power Reduction**: Limits the amount of energy that products can consume in standby mode.
- **Resource Efficiency**: Resource efficiency seeks to extend product lifespans and reduce waste and also encourage the use of materials that are easier to recycle and have a lower environmental impact. This is essential for sustainable business practices aimed at cutting waste and pollution, conserving resources, ensuring stable supply chains, and minimizing disruption risks. In eco-industrial parks, green supply chain management within a CE framework enhances environmental performance and resource efficiency. SMEs also see resource efficiency as a means to gain competitive advantages and enter new markets.

(ii) **Japan's Top Runner Program**

Eco design is an essential for sustainable development, incorporates environmental considerations into product design through eco-friendly materials, energy-efficient processes, and waste minimization. It boosts environmental performance and lowers carbon footprints across various industries, including fashion. In footwear, it balances functionality and comfort with sustainable materials and production methods.

Japan's Top Runner Program sets energy efficiency standards for various products, using the best-performing model on the market as a benchmark. Manufacturers must ensure that their products meet or exceed these benchmarks, thereby promoting continuous in product design.

Benefits of eco-design directives

(a) **Environmental Protection**: Reduces the ecological footprint of products, conserve resources, and minimizes waste.
(b) **Economic Savings**: Consumers benefit from reduced energy bills and longer-lasting products.
(c) **Innovation**: Encourages manufacturers to innovate and improve product design, leading to more competitive markets.
(d) **Regulatory Compliance**: Helps countries and regions meet their environmental targets and commitments under international agreements.

To illustrate the impact of the eco-design directive using hypothetical data, letu's consider the implementation of an eco-design directive targeting household refrigerators in a hypothetical country. The directive mandates that all new refrigerators must satisfy certain energy-efficiency standards and should be designed for durability and recyclability [14, 15].

The Eco-design Directive emphasizes three key elements energy efficiency, durability, and recyclability. First, new refrigerators must achieve an energy efficiency rating of A++ or higher, to ensure a significant reduction in energy consumption. Second, to enhance longevity and reduce waste, refrigerators must have a minimum lifespan of 15 Lastly, to promote environmental sustainability, at least 80% of the materials used in manufacturing these refrigerators must be recyclable, encouraging the reuse of resources and minimizing landfill contributions. These measures collectively aim to reduce the environmental impact while fostering sustainable consumer habits. Suppose a company sold 10,00,000 units of refrigerators with an average energy consumption per unit of 500 kWh/year and a lifespan of 10 years with 50% recyclable materials in year 1. After the implementation of eco-design directives, in the 5th year, if the company sold 12,00,000 units with an average energy consumption per unit of 300 kWh/year and an average lifespan of 15 years with 80% recyclable materials the impact assessment is as follows.

Impact Assessment

(a) **Energy Consumption Reduction**

Before Implementation of Eco-design Directive

Total Energy Consumption: 10,00,000 units × 500 kWh/year = 500,000,000 kWh/year.

After Implementation of Eco-design Directive

Total Energy Consumption: 12, 00,000 units × 300 kWh/year = 360,000,000 kWh/year1.

(b) **Energy Savings**

500,000,000 kWh/year − 360,000,000 kWh/year = 140,000,000 kWh/ year.

(c) **Environmental Benefits (Reduction in CO_2 Emissions)**:

Assuming 0.5 kg CO_2/kWh:

140,000,000 kWh/year × 0.5 kg CO_2/kWh = 70,000,000 kg CO_2/year.

Equivalent to removing approximately 15,000 cars from the road annually (assuming an average car emits 4.6 metric tons of CO_2 per year).

(d) **Economic Benefits for Consumers (Energy Cost Savings)**

Assuming electricity cost of **₹0.10 per kWh** for 140,000,000 kWh/year × ₹0.10 per kWh = ₹14,000,000 saved annually of ₹140,000,000.

(e) **Increased Durability (Waste Reduction)**

Before: 10,00,000 units ÷ 10 years = 1,00,000 units/year end-of-life units.

After: 1,200,000 units ÷ 15 years = 80,000 units/year end-of-life units.

Reduction in Waste: 100,000 − 80,000 = 20,000 units/year.

(f) **Improved Recyclability (Recyclable Material Increase)**

Before: 10,00,000 × 0.50 = 500,000 units recyclable.

After: 1,200,000 × 0.80 = 960,000 units recyclable.

Increase in Recyclable Materials: 960,000 − 500,000 = 460,000 units.

This hypothetical example illustrates how eco-design directives can lead to significant environmental and economic benefits by reducing energy consumption, lowering CO_2 emissions, saving costs for consumers, reducing waste, and increasing product recyclability.

The Fig. 3.2 illustrates the significant impact of the eco-design directive on refrigerators in a hypothetical country and show improvements across key categories. After implementation, the energy consumption decreased from 500 million kWh/year to 360 million kWh/year, leading to a reduction in CO_2 emissions from 250 to 180 million kg/year. This enhanced energy efficiency translates to annual consumer savings of ₹14 million. Additionally, the recyclability of refrigerators improved, with the number of recycled units increased from 500,000 to 960,000. These results demonstrate the effectiveness of the defect in promoting energy efficiency, reducing environmental impact, and increasing economic benefits for consumers, while enhancing product recyclability.

By adopting and expanding these eco-design initiatives, the automotive companies of Mahindra and Mahindra can significantly enhance their sustainability efforts, reduce environmental impacts, and lead the automotive industry towards a more circular and sustainable future. Mahindra and Mahindra incorporated eco-design principles into their operations, where' a hypothetical table with data showcasing various eco-design initiatives, their current status, and target outcomes for future improvement (Table 3.3).

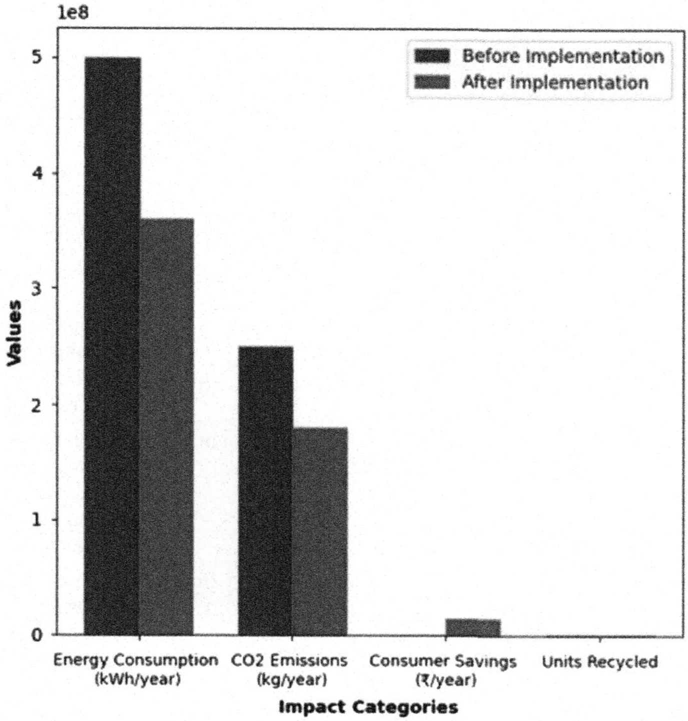

Fig. 3.2 Impact of eco-design directive on refrigerators in a hypothetical company

Eco-design directives play a crucial role in the promotion of sustainable product design and consumption. By integrating environmental considerations into the design process, these directives help to reduce the negative impacts of products on the environment, support resource efficiency, and drive innovation in sustainable manufacturing practices.

3.3.1.3 Waste Management Regulations

Waste management involves the collection, processing, recycling, and disposal of waste to reduce environmental impact and protect public health. It includes waste reduction, proper handling of hazardous materials, and promoting sustainable practices. Policies that mandate recycling targets, landfill bans on certain materials, and stringent waste management practices. Comprehensive waste management regulations aim to reduce landfill use, increase recycling rates, and promote sustainable waste management. These regulations include mandatory recycling programs, landfill diversion targets, and strict penalties for non-compliance. Comprehensive waste management regulations aim to reduce landfill use, increase recycling rates, and

Table 3.3 Eco-design initiatives at Mahindra & Mahindra

Initiative	Current status (2024)	Target outcome (2030)	Description
Use of recycled materials	30% of materials are recycled	60% of materials are recycled	Increasing the percentage of recycled materials in vehicle production
Modular design	20% of products have modular designs	50% of products have modular designs	Designing products with interchangeable parts to extend product life
Lightweight materials	15% reduction in vehicle weight	30% reduction in vehicle weight	Using lightweight, durable materials to improve fuel efficiency and reduce emissions
Energy-efficient manufacturing	25% reduction in energy use	50% reduction in energy use	Implementing energy-efficient processes and technologies in manufacturing
Renewable energy use	20% of energy from renewables	50% of energy from renewables	Increasing the use of renewable energy sources in manufacturing facilities
Water usage reduction	10% reduction in water use	40% reduction in water use	Implementing water-saving technologies and recycling water in production processes
End-of-life management	40% of components recyclable or reusable	80% of components recyclable or reusable	Designing products so that a high percentage of components can be easily recycled or reused at the end of their life
Sustainable packaging	25% reduction in packaging waste	60% reduction in packaging waste	Using sustainable materials and designs to reduce packaging waste

(continued)

Table 3.3 (continued)

Initiative	Current status (2024)	Target outcome (2030)	Description
Eco-friendly materials	35% of materials are eco-friendly	70% of materials are eco-friendly	Increasing the use of biodegradable, non-toxic, and sustainable materials
Digital twin technology	Limited use in design phase	Full integration in design and production phases	Using digital twin technology to optimize product design and manufacturing processes, reducing waste and improving efficiency

promote sustainable waste management. These regulations include mandatory recycling programs, landfill diversion targets, and strict penalties for non-compliance [13, 14].

The key components involved households and businesses separating recyclable materials from general waste, with the goal of achieving a 60% recycling rate by Year 5. Landfill diversion targets reduce the amount of waste sent to landfills by 30% within the same timeframe. Producers are required to design products that are easier to recycle, with Extended Producer Responsibility (EPR) fees imposed based on product recyclability. Penalties for non-compliance include fines and increased fees for excessive landfill waste. Before implementation, the baseline year data showed the total waste generated at 1,000,000 tons/year, a recycling rate of 25%, 750,000 tons/year sent to landfills, and an EPR compliance rate of 50%. After five years, the total waste generation increased slightly to 1,100,000 tons/year, the recycling rate improved to 65%, landfill waste decreased to 525,000 tons/year, and EPR compliance reached 95%.

(a) **European Union (EU) waste framework directive**

The EU Waste Framework Directive is a landmark legislation that prioritizes environmental protection and public health by minimizing waste's harmful effects. It establishes a waste hierarchy that emphasizes prevention, reuse, recycling, and recovery over disposal methods like landfilling. Key provisions include Extended Producer Responsibility, recycling and recovery targets, and promoting circular design principles. These measures have driven up recycling rates, reduced landfill volumes, and spurred innovation in eco-design and waste management technologies, playing a vital role in advancing sustainable practices and a circular economy in the EU.

Japan's 3Rs Initiative (Reduce, Reuse, Recycle)

Japan's 3Rs Initiative is a comprehensive policy framework aimed at promoting sustainable waste management and resource efficiency. Key elements include:

Legal Framework: The Basic Act for Establishing a Recycling-Based Society has been passed by Japan to support its 3Rs plan.

Extended Producer Responsibility (EPR): Japan's EPR systems cover a wide range of products, including packaging, automobiles, and electronics. Producers are responsible for recycling costs and meeting recycling targets, which drive them to design products for easier disassembly and recycling [15].

Public-Private Partnerships: This initiative involves collaboration between the government, industry, and academia to develop and implement innovative recycling technologies and practices.

Technological Innovation: Japan has invested in advanced recycling technologies, such as material recovery from electronic waste and plastic recycling, to recover valuable resources from waste streams.

Effectiveness: Japan's 3Rs initiative has led to high recycling rates, with some municipalities achieving rates that exceed 80%. The initiative has also promoted sustainable consumption patterns and minimized the environmental impacts associated with waste generation and disposal [16, 17].

Japan's 3Rs principle demonstrates how effective regulatory practices, such as comprehensive legal frameworks, extended producer responsibility, setting targets, promoting eco-design, and fostering public-private partnerships, can drive significant progress towards a circular economy. By learning from these examples, other regions and countries can adapt and implement similar strategies to achieve sustainability goals and effectively reduce their environmental footprints.

3.3.1.4 Economic Instruments

Economic instruments are mechanisms used to change behavior and promote environmental sustainability, such as tradeable permits, taxes, and subsidies. They encourage companies and consumers to embrace waste reduction and resource efficiency techniques in a circular economy by establishing financial incentives.

(a) **Tax Incentives and Subsidies**: Tax incentives and subsidies are powerful financial tools that by governments use to encourage businesses to adopt circular economic practices. These incentives reduce the financial burden on companies implementing sustainable practices, promoting the use of recycled materials, and investing in recycling infrastructure. Financial incentives for businesses that adopt circular practices, such as tax breaks for using recycled materials, or subsidies for setting up recycling plants.
The key components of tax incentives and subsidies for circular practices include tax breaks for using recycled materials and subsidies for recycling plants. Businesses receive tax deductions or credits for incorporating recycled or sustainable materials in their production processes, encouraging manufacturers to reduce the demand for virgin resources and minimize waste. Additionally, financial support, such as grants and low-interest loans, is provided for setting up and operating recycling facilities, which lowers the initial investment barrier and increases the recycling capacity and efficiency. Suppose that in a country, before

the implementation of taxes and incentives, let's assume that 200 businesses used recycled materials, each paying an average of ₹10 million/year in taxes. There are 50 recycling plants, with an annual investment of approximately ₹100 million. After five years, the number of businesses using recycled materials rose to 500, with average taxes reduced to ₹ 8 million/year, resulting in savings of ₹2 million/year per business. The number of recycling plants has doubled to 100, with annual investments increasing to ₹ 200 million. These data illustrates the significant positive impact of tax incentives and subsidies on promoting sustainable practices and enhancing the recycling infrastructure.

(b) **Pollution and Waste Taxes**: Taxes on waste generation and pollution discourage disposal and promote recycling. Pollution and waste taxes have been instrumental in driving significant environmental improvement over the last five years. Initially, the country had substantial emissions and waste generation, with low recycling rates. With the introduction of pollution and waste taxes, aimed at CO_2 emissions and waste disposal respectively, the nation saw a marked reduction in environmental impact. In a hypothetical country, the data shows that CO_2 emissions have dropped from 1,000,000 tons annually to 800,000 tons, a reduction of 200,000 tons per year. Waste generation has also decreased from 500,000 tons to 400,000 tons annually, with the recycling rate increasing from 30% to 50%. These outcomes reflect the effectiveness of financial incentives in prompting businesses and individuals to adopt cleaner practices and manage resources responsibly. The tax revenues generated, totalling ₹80 million annually from pollution and waste taxes combined, underscore the financial sustainability of these measures while supporting on-going environmental initiatives. Figure 3.3 the bar chart shows the impact of pollution and waste taxes over five years, comparing the baseline year to year 5.

CO_2 Emissions: Reduced from 1,000,000 to 800,000 tons/year.

Waste Generated: Reduced from 600,000 to 400,000 tons/year.

Recycling Rate: Increased from 40% to 50%.

Tax Revenue: Increased from $500 million to $600 million.

Overall, taxes led to lower emissions and waste, higher recycling rates, and increased tax revenues.

(c) **Deposit-Refund Schemes**: Deposit-refund schemes are effective systems designed to promote recycling and reduce waste, particularly for beverage containers. Consumers pay a small deposit when purchasing a product, which is refunded when they return the empty container to designated collection points. This financial incentive encourages recycling, significantly boosting recycling rates while reducing litter in public spaces. Successful implementations, such as Germany's P-fund system, showcase how these programs can achieve tangible environmental benefits by minimizing waste and conserving resources.

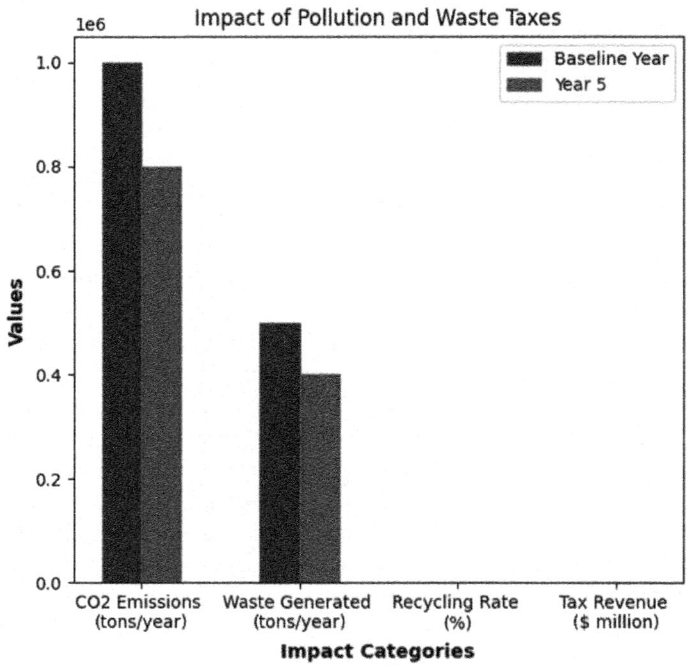

Fig. 3.3 The impact of pollution and waste taxes

3.3.1.5 Public Procurement

Green Public Procurement (GPP): Governmental commitment to purchase goods and services that meet certain environmental criteria, encouraging markets for sustainable products. Green Public Procurement (GPP) is a strategic approach in which public authorities incorporate has environmental considerations into their procurement processes to promote sustainability. GPP has lower environmental impacts throughout its life cycles [21]. Because of this approach, the public sector is a driver of market demand for environmentally responsible products and practices. It aims to reduce the carbon footprint, conserve resources and foster motivation in eco-friendly technologies. Effective GPP implementation requires clear environmental standards, collaboration with suppliers to meet those standards, and monitoring to ensure compliance. Ultimately, GPP contributes to building a more sustainable economy by leveraging public institutions purchasing power to support environmentally sound practices and solutions [18, 19].

Circular Procurement: Policies that prioritize products designed for longevity, reparability, and recyclability in public sector purchasing. Circular procurement is a strategic approach in which purchasing decisions are guided by principles aimed at promoting a circular economy [20]. Circular procurement focuses on prioritizing products and services designed to conserve resources, minimize waste, and reduce environmental impact throughout their lifecycle. It involves adopting criteria that encourage suppliers to provide durable, repairable, reusable, and recyclable goods. By integrating circular principles into procurement strategies, organizations contribute to sustainable development goals while driving innovation in product design and business models that enhance resource efficiency and waste reduction [21]. Circular procurement relies on stakeholder collaboration, including suppliers and policymakers, to align practices with circular economy principles. This proactive approach fosters resilient, sustainable economies by closing material loops and promoting resource efficiency.

3.3.1.6 Research and Innovation Support

Securing funding for Research and Development (R&D) is crucial for driving innovation and technological advancement across industries. Governments play a vital role in allocating budgets to support R&D through grants and contracts, addressing societal needs, spurring economic growth, and maintaining global competitiveness. Private companies also invest in R&D to innovate, enhance efficiency, and gain market leadership. Collaboration between industry and academia further enriches R&D efforts. Securing funding requires rigorous proposal development, demonstrating potential impact, and aligning with strategic priorities. Successful funding strategies involve navigating complex landscapes, fostering partnerships, and ensuring a clear pathway from research to practical application or commercialization, maximizing the transformative impact of R&D investments.

Establishment of innovation hubs and partnerships between government, academia, and industry to foster collaboration on circular economy projects, and create an environment where creativity, entrepreneurship, and technological advancement thrive [22–24]. These hubs serve as focal points where diverse stakeholders-such as startups, established companies, academic institutions, and government bodies-come together to share resources, expertise, and ideas. By facilitating cross-disciplinary interactions and providing access to specialized facilities and funding opportunities, innovation hubs stimulate the development of ground breaking solutions and products. Partnerships within these hubs are integral, enabling participants to pool their strengths, mitigate risks, and accelerate their innovation cycles. They also bridge the gaps between research and market applications, promoting the commercialization of new technologies and fostering economic growth. Successful innovation hubs and partnerships not only nurture a culture of continuous learning and experimentation but also play a crucial role in addressing complex societal challenges through collaborative problem-solving and sustainable development initiatives [25].

3.3.1.7 Education and Awareness

Promoting the ideas of a circular economy requires education and awareness because it enables people and organizations to make wise choices regarding waste management and resource usage. Fostering a culture of sustainability encourages group efforts and propels the shift to more conscientious production and consumption methods.

Public Awareness Campaigns: *Public awareness campaigns are targeted initiatives to educate the public about the benefits of a circular economy and encourage sustainable consumption patterns and engage the public on specific issues, behaviours, or societal challenges, aiming to drive behavioural change and foster informed decision-making. They rely on clear objectives, tailored messaging, and multichannel approaches to effectively reach diverse audiences.*

Educational Programs: *Educational programs are organized efforts aimed at imparting knowledge and skills through structured curricula and engaging in activities, led by qualified instructors. These programs empower participants to learn, develop skills, and make informed decisions, contributing to their personal growth and professional advancement.*

3.3.1.8 Legislative Frameworks

Laws and regulations supporting circular economy practices have been enacted at various levels of the government to promote sustainable resource use, waste reduction, and environmental conservation. These include policies on extended producer responsibility (EPR), eco-design standards, waste management regulations emphasizing recycling and reuse, and incentives for green procurement and innovation. At the national level, such laws aim to integrate circular economic principles into economic strategies, whereas regional and local governments focus on implementing tailored measures to encourage circular practices within their communities and industries. Policy innovation entails adopting novel or significantly improved processes or products by public organizations to enhance efficiency and productivity, especially during crises like pandemics. It is crucial for tackling global economic and social challenges; its complexity often hampers implementation.

Circular Economy Roadmaps and Action Plans: *Circular economy roadmaps and action plans are strategic frameworks that outline specific steps and objectives for transitioning towards sustainable economic models* [26]. *They focused on minimizing waste, enhancing the effectiveness of resources, and encouraging sustainable production and consumption patterns through policies, stakeholder collaboration, and targeted investments in research and infrastructure. These roadmaps aim to create resilient economies that reduce the environmental impact while fostering innovation and creating new economic opportunities.*

National and Regional Legislation: *National and regional legislation comprises laws and regulations enacted by governments to govern diverse facets such as the economy, environment, and social welfare. These frameworks establish standards, guidelines, and enforcement mechanisms to ensure compliance, protect public interest, and promote sustainable development and societal well-being within specific geographical areas.*

3.3.1.9 International Collaboration

Participation in international agreements refers to the commitment and involvement of countries in treaties, conventions, or agreements that address global issues and promote cooperation regarding shared challenges. These agreements cover a wide range of topics, including environmental protection, human rights, trade, security, and sustainable development. By joining these agreements, countries agree to abide by specific obligations, collaborate with other nations, and work towards achieving common goals. Participation is crucial for fostering diplomatic relations, addressing trans boundary issues, and advancing global governance efforts aimed at effectively and collectively tackling complex global challenges.

Collaboration with other countries to share best practices, technologies, and strategies for implementing circular economic policies. Cross-border cooperation involves collaborating between neighbouring regions and countries to address shared challenges, promote regional development, and achieve mutual benefits through joint projects, policy alignment, and capacity building. This fosters integration, stability, and sustainable development by leveraging the collective strengths and resources across borders.

3.4 Challenges and Opportunities

Identifying regulatory barriers to the adoption of circular economy (CE) practices involves recognizing challenges that hinder the transition towards sustainable resource management and waste reduction. Several regulatory barriers impede the adoption of CE:

(a) **Fragmented Regulatory Landscape**: Complex and fragmented regulatory frameworks across different jurisdictions or sectors can create inconsistencies and uncertainties in businesses seeking to implement CE. Varying standards, definitions, and compliance requirements complicate compliance and increase administrative burden.

(b) **Lack of Harmonization**: Absence of harmonized standards and regulations at regional or global levels can hinder market integration and limit economies of scale for circular products and services. Divergent regulatory requirements may discourage cross-border trade of recycled or remanufactured goods.

(c) **Costly Compliance Requirements**: Regulatory compliance costs, such as fees for environmental permits, certifications, and audits, can be prohibitive for small and medium-sized enterprises (SMEs) or startups adopting CE practices. High initial investments in eco-design and recycling infrastructure may deter businesses from transitioning to business models.
(d) **Inadequate Extended Producer Responsibility (EPR) Frameworks**: Weak or ineffective EPR schemes may fail to incentivize Manufacturers should create items that are recyclable, repairable, and long-lasting. Insufficient financial incentives or unclear responsibilities for end-of-life product management can undermine efforts to close material loops and promote resource recovery [8].
(e) **Limited Public Procurement Support**: Insufficient integration of circular criteria into public procurement policies limits the demand for circular products and services. The lack of clear guidelines or incentives for government agencies to prioritize environmentally preferable purchasing options may impede the market uptake of sustainable innovation.
(f) **Regulatory Inertia and Resistance to Change**: Resistance to updating or adapt existing regulations to accommodate CE principles and innovations can slow down progress. Regulatory inertia may stem from vested interests, lack of awareness of CE benefits, and perceived risks associated with new technologies or business models.
(g) **Monitoring and Enforcement Challenges**: Inadequate monitoring mechanisms and enforcement capacities can undermine the effectiveness of CE regulations. Weak enforcement of recycling quotas, illegal dumping, or non-compliance with eco-design requirements may weaken the incentives for businesses to invest in circular practices.
(h) **Complexity in Cross-Sector Collaboration**: Promoting circularity often requires collaboration across multiple sectors, including industry, waste management, energy, and transport. Regulatory barriers that hinder intersectoral cooperation, information sharing, and joint investment in infrastructure can impede holistic solutions to resource efficiency and waste reduction.

Identifying these regulatory barriers is essential for policymakers, businesses, and stakeholders to develop targeted strategies and policy interventions to effectively address these challenges. Overcoming these barriers requires a coordinated approach to streamline regulations, enhance incentives, strengthen enforcement, and foster innovation in support of a transition towards a circular economy.

3.5 Conclusion

Based on extensive research into circular economy governance and regulatory frameworks, the key findings underscore the pivotal role of policies in advancing resource efficiency and waste reduction. Regulatory tools such as extended producer responsibility (EPR) and eco-design directives have proven effective in incentivizing industries to adopt sustainable practices, emphasizing product longevity and recyclability. These policies not only drive environmental benefits by minimizing waste generation and promoting recycling but also stimulate economic growth through innovation in green technologies and job creation in the recycling and remanufacturing sectors. However, challenges such as regulatory complexity, high compliance costs, and varying enforcement capacities persist, necessitating streamlined approaches and enhanced collaboration across sectors and regions. Moving forward, opportunities lie in further integrating circular economy principles into national and international policy agendas, bolstering incentives for circular innovation, and strengthening monitoring mechanisms to ensure sustainable outcomes across global supply chains.

3.5.1 Future Scope of Research

Research in advancing circular economy governance and regulatory frameworks includes prioritizing regulatory coherence across sectors and jurisdictions to streamline policies that promote circularity. Incentivizing businesses through financial incentives and grants to adopt circular design principles and invest in sustainable technologies is crucial, alongside fostering cross-sector collaboration among government, industry, academia, and civil society to accelerate innovation and capacity-building. Robust monitoring frameworks should track progress of recycling rates and resource efficiency metrics to inform evidence-based policymaking. Informing customers about the advantages of sustainable consumption and investing in a circular infrastructure is also essential. Addressing global supply chain challenges through international cooperation and expanding research on socio-economic impacts and policy effectiveness will further support the transition to a circular economy, enhancing resilience and sustainability on a global scale.

References

1. Andrews, D., Newton, E.J., Adibi, N., Chenadec, J., Bienge, K.: A circular economy for the data centre industry: Using design methods to address the challenge of whole system sustainability in a unique industrial sector. Sustainability **13**(11), 6319 (2021)
2. Arora, R., Mutz, D., Mohanraj, P.: Innovating for the circular economy: Driving sustainable transformation. CRC Press, (2023)
3. Bechtel, N., Bojko, R., Völkel, R.: Be in the loop: circular economy & strategic sustainable development, (2013)

4. Barnett, K., Hordacre, A.L., Molloy, S., Spoehr, J., Vas Dev, S.: Connecting ideas: Collaborative innovation for a complex world, (2010)
5. Corbin, R.A., Thomas, R.: Community colleges as incubators of innovation: Unleashing entrepreneurial opportunities for communities and students. Taylor &Francis (2023)
6. Chung, S.-W., Murakami-Suzuki, R.: A comparative study of e-waste recycling systems in Japan, South Korea and Taiwan from the EPR perspective: Implications for developing countries. Kojima. Chiba **21**, 11–23 (2008)
7. Dahmani, N., Belhadi, A., Benhida, K., Elfezazi, S., Touriki, F.E., Azougagh, Y.: Integrating lean design and eco-design to improve product design: From literature review to an operational framework. Energy & Environ. **33**(1), 189–219 (2022)
8. de Jesus, A., Lammi, M., Domenech, T., Vanhuyse, F., Mendonça, S.: Eco-innovation diversity in a circular economy: Towards circular innovation studies. Sustainability **13**(19), 10974 (2021)
9. Dong Y, Zhang F, Fu L (2019) Ecological design: The role of extended producer responsibility system. J. Coast. Res., **93**(SI), 354–361 (2019)
10. Dolva, C.: Green Public Procurement (GPP): How widespread is Green Public Procurement in Norway, and what factors are seen as drivers and barriers to a greener procurement practice? (2008)
11. Ei, A., Abdelhafeez, A.: Rethinking waste management: A holistic sustainability framework for a circular economy. J. Sustain. Dev. Green Technol., **2**(2), 26–33 (2023). https://doi.org/10.54216/jsdgt.020203
12. Galanton, N.:The circular supply chain: A paradigm shift towards sustainable resource. Vector European, (1) (April), 70–77 (2024). https://doi.org/10.52507/2345-1106.2024-1.15
13. Grobler, L., Schenck, R.: The notion of environmental citizenship in household waste management in the circular economy. In: Proceedings of the WasteCon 2021 conference of the institute of waste management of Southern Africa, pp. 9–11. (2021)
14. Gremyr, I., Hildenbrand, J., Sarasini, S., Raharjo, H.: A framework for developing and assessing eco-innovations. Eco-innovation and the development of business models: Lessons from experience and new frontiers in theory and practice, pp. 55–79 (2014)
15. Gupt, Y., Sahay, S.: Review of extended producer responsibility: A case study approach. Waste Manag. & Res. **33(7)**, 595–611 (2015)
16. Hapuwatte, B.M., Jawahir, I.S.: Closed-loop sustainable product design for circular economy. J. Ind. Ecol. **25**(6), 1430–1446 (2021)
17. Mahardika, S., Pratamo, N.B.A.: The problem of plastic waste, a comparison of the systems in Indonesia and the Netherlands: The Urgency of State intervention and legal instruments. Environ. Pollut. J. **4**(1), 933–948 (2024)
18. Marinina, O., Kirsanova, N., Nevskaya, M.: Circular economy models in industry: Developing a conceptual framework. Energies **15**(24), 9376 (2022)
19. Modak, P.: Circular economy practices in India. An Introd. Circ. Econ., 555–575 (2021)
20. Matuszak-Flejszman, A., Paliwoda, B.: Effectiveness and benefits of the eco-management and audit scheme: Evidence from polish organisations. Energies **15**(2), 434 (2022)
21. Ogunmakinde, O.E., Egbelakin, T., Sher, W.: Contributions of the circular economy to the UN sustainable development goals through sustainable construction. Resour. Conserv. Recycl. **178**, 106023 (2022)
22. Preston, F.: A global redesign? Shaping the circular economy. (2012)
23. Ranta, V., Aarikka-Stenroos, L., Mäkinen, S.J.: Creating value in the circular economy: A structured multiple-case analysis of business models. J. Clean. Prod. **201**, 988–1000 (2018)
24. Rawat, M., Raju, K.D.: Emerging role of green public procurement policy in achieving sustainable development: A case study of India. In: Sustainability in public procurement, corporate law and higher education, pp. 88–107. Routledge (2024)
25. Sillanpaa, M., Ncibi, C.: The circular economy: Case studies about the transition from the linear economy. Academic Press (2019)
26. Zemlyak, S., Gusarova, O., Khromenkova, G.: Entrepreneurial initiatives, education and culture: hubs for enterprise innovations and economic development. Sustainability **15**(5), 4016 (2023)

Dr. Lakshmi Prasad Panda is a dedicated academician and researcher, presently serving as Assistant Professor in the Department of Humanities at Govt. College of Engineering Kalahandi, Bhawanipatna, Odisha. With a strong commitment to academic excellence, Dr. Panda has consistently delivered high-quality research outputs, publishing numerous papers in reputable journals and conferences. His research expertise has been further demonstrated through successful PhD guidance, nurturing the next generation of scholars. Dr. Panda's relentless efforts and passion for knowledge dissemination have earned him recognition within the academic community, solidifying his position as a respected educator and researcher in his field.

Dr. Kali Charan Rath is a distinguished academician and researcher, currently serving as Associate Professor in the Department of Mechanical Engineering at Gandhi Institute of Engineering and Technology University, Gunupur, Odisha. With a steadfast dedication to academic excellence, Dr. Rath has established himself as a prolific researcher, publishing extensively in esteemed national and international conferences, journals, and book chapters. His scholarly contributions have been further amplified through editorial roles in multiple books, demonstrating his expertise and leadership in the field. As a devoted PhD supervisor, Dr. Rath is guiding numerous scholars, fostering the growth of innovative research and talent. His tireless efforts, unwavering commitment, and passion for knowledge dissemination have earned him a reputation as a respected educator, researcher, and mentor in the academic community.

Dr. N. V. Jagannadha Rao is a renowned academician and scholar, presently serving as Professor at the School of Management Studies, Gandhi Institute of Engineering and Technology University, Gunupur, Odisha. With a wealth of experience in management education, Prof. Rao has been instrumental in shaping the academic landscape, fostering a culture of excellence and innovation. His academic expertise spans various facets of management, and he has published research papers, articles, and case studies in reputable journals and conferences. As an accomplished educator, Prof. Rao has guided numerous students, imparting knowledge, and inspiring the next generation of business leaders. His dedication, passion for teaching, and commitment to academic excellence have earned him a respected reputation in the academic community.

Dr. Chitrasen Samantra is a dynamic academician and researcher, currently serving as Assistant Professor in the Department of Mechanical Engineering at Parala Maharaja Engineering College, Berhampur, Odisha. With a strong academic background and research expertise, Dr. Samantra has made significant contributions to the field of mechanical engineering. He has published numerous research papers in reputable national and international journals and conferences, showcasing his scholarly pursuits. As an educator, Dr. Samantra is committed to imparting knowledge, guiding students, and fostering a culture of innovation and excellence. His dedication, passion for teaching, and research acumen have earned him recognition within the academic community, solidifying his position as a respected educator and researcher.

Dr. Kamalakanta Muduli is a distinguished academician and researcher, presently serving as Associate Professor in the Mechanical Engineering Department at Papua New Guinea University of Technology, Lae. With a strong international academic presence, Dr. Muduli has established himself as an expert in his field, boasting an impressive research portfolio. His scholarly contributions have been showcased through numerous publications in reputable international journals and conferences. Dr. Muduli's academic excellence is further underscored by his ORCID ID: 0000-0002-4245-9149, demonstrating his commitment to research transparency and innovation. As a dedicated educator, he continues to inspire and guide students, fostering a culture of academic excellence and global collaboration.

Chapter 4
Design and Implementation of an Ultrasonic Braking System Using PLC for Autonomous Vehicle Collision Avoidance

Sivasankara Raju Rallabandi, Naresh Babu Vemuru, Kaparapu Nagaraju, Harikiran Vuddagiri, Thimothy Pandi, and Seepana Praveen Kumar

Abstract This endeavor is centered on the advance of a vehicle rate controller that is combined by an automated braking system. The work encompasses a range of electronic components, including sensors, relays, control systems, programmable logic controllers (PLCs), signal transmitters, receivers, and ladder circuits. These components are crucial for implementing autonomous braking and speed control. WPL Soft is a software utility specifically designed for programming Programmable Logic Controllers (PLCs). It is utilized for doing circuit design and programming operations. Understanding the principles of ride control and the functioning of autonomous braking is essential, particularly when utilizing instruments like ultrasonic and infrared (IR) sensors. These sensors are essential for detecting obstacles and ensuring immediate activation of the braking system to prevent collisions. The proposal is lying on the concept of the Automatic overspeed control system (AOCS). The maneuver of the system relies on the exposure of stress in the clutch and braking mechanisms. Once the vehicle exceeds a predetermined limit, the microcontroller

S. R. Rallabandi (✉)
Department of Mechanical Engineering, Aditya Institute of Technology and Management, Tekkali, Andhra Pradesh, India
e-mail: sivaraju80@gmail.com

N. B. Vemuru
Department of Mechanical Engineering, Geetanjali Institute of Science and Technology, Gangavaram, Kovur, Andhra Pradesh, India

K. Nagaraju · H. Vuddagiri
Department of Mechanical Engineering, Avanthi Institute of Engineering and Technology, Makavarapalem, Andhra Pradesh, India

T. Pandi
Department of Mechanical Engineering, Lendi Institute of Engineering and Technology, Vizianagaram, Andhra Pradesh, India

S. Praveen Kumar
Ural Federal University, Russian Federation, Russia

© The Author(s), under exclusive license to Springer Nature Switzerland AG 2025
K. C. Rath et al. (eds.), *Engineering, Technology and Management*, Information Systems Engineering and Management 54, https://doi.org/10.1007/978-3-031-93654-8_4

takes control by engaging the clutch and brakes to reduce the speed to a safe and reliable level. This feature ensures that the vehicle maintains control and reduces the occurrence of accidents caused by excessive speed. The system is designed to enhance vehicle safety by automatically controlling speed and activating brakes when the vehicle exceeds a safe speed limit or encounters obstacles. This technology has the potential to significantly reduce the probability of accidents, providing a reliable and efficient solution for modern autos.

Keywords Ultrasonic braking system · Programmable logic controller · Autonomous vehicle · Sensor integration · WPL soft · Ladder logic programming · Vehicle safety

Preface

This chapter focuses on the development of a vehicle speed controller integrated with an automated braking system, utilizing a range of electronic components such as sensors, relays, control systems, and programmable logic controllers (PLCs). The chapter discusses the use of WPL Soft for designing and programming PLCs, alongside the principles of ride control and autonomous braking. Key to the system's functionality are ultrasonic and infrared sensors, which detect obstacles and trigger the braking mechanism to prevent collisions. Based on the Automatic Overspeed Control System (AOCS), the system automatically adjusts speed and engages braking when a vehicle exceeds safe limits, enhancing vehicle safety and reducing accident risks.

4.1 Introduction

Automation is significantly influencing the role of individuals in several systems, including the field of transportation. Vehicle speed control devices are more common. This device utilizes an ultrasonic sensor to detect obstacles or dangers in advance and alerts the driver about the possibility of a collision. As the driver keeps up with an automobile, their velocity navigation system (SCS) is going to reveal them on an LED screen how far away from obstacles their vehicle is, telling them to slow down [1–3].

Autonomous Emergency Braking (AEB) is a safety system in automobiles that automatically activates the brakes to avert or lessen the bang of a accident. The system employs a variety of sensors, including cameras, radars, and LIDAR, to detect potential collisions and function independently without requiring any input from the driver. Clever Momentary Assisted Control, as part of the AEB architecture, involves a sophisticated system that utilizes sensors to identify imminent collisions. Furthermore, it incorporates sophisticated algorithms to quickly ascertain the most advantageous timing and approach for engaging the brakes with utmost efficiency.

Momentary Assisted Control is a system that provides temporary assistance to the driver at critical situations, ensuring swift and precise braking intervention to prevent or minimize the impact of a potential collision. This technology seeks to augment the overall efficacy of AEB systems by integrating intelligence with momentary aid, hence enhancing their reliability in accident prevention and overall vehicle safety [4].

In the present day, makers of cars, as well as motorcycles, are also expecting technology and machinery that guarantee safety even at high speeds. The key to solving such problems lies in the utilization of a proper braking device and the implementation of controlled speed reduction techniques. The implementation of AEB in vehicles is the contemporary trend to enhance the safety of the car. There are other options as well, like ABS, which have benefits and are valuable. AEB integrates sensors and brake controllers to mitigate the potential harm caused by excessive momentum on individuals. Although computerized braking systems can effectively prevent crashes, their main objective is to minimize the damage and costs that a vehicle incurs prior to colliding with an object [5–8].

According to research on medical professionals make evidence-based decisions and create individualized treatment programs. Real-time monitoring and control systems provide prompt interventions, modifications, and enhanced patient safety [9, 10]. The integration of alternative therapies and palliative care guarantees a comprehensive approach to cancer therapy. Collaborative initiatives and data consolidation enhance progress in cancer research and population health investigations [11, 12].

Advocated sensor fusion to improve autonomous braking systemsoperation of each capacitive and ultrasonic sensor's dwellings in order to determine the difficulties in the context of a security programme [13]. In order to govern the device's operations and complete its register, they use C programming on a 32-bit arduino with an ARM chip (LPC2138). Put a plan into action to disclose and prevent accidents on the streets. Installing this technology into vehicles would make them smarter and more engaging, which should reduce the likelihood of accidents on the road. The ARM7 architecture is used to make this equipment is to ensure the safety of individuals [14].

Atechnique to enhance the security of automobiles. The solution provided by this machine can assist the operator by alerting them to potential obstacles and the presence of nearby vehicles that may collide with them [15]. Moreover, the implementation of advanced technology in autos results in slower device performance, which helps prevent accidents. This invention utilizes an ultrasonic sensor, a pressure motor, and an LCD. A clever approach (Hurdle Detection) for measuring distances in order to provide a more convenient driving environment, utilizing ultrasonic rays. By implementing stricter safety measures in vehicles, sensors can continue to work towards their objectives until they find a feasible solution [16]. Build and install an electro-hydraulic braking system with master cylinders and electric linear actuators; include a classical electronic braking force distribution system [17]; use incompressible braking fluid; incorporate a spring-loaded mechanism; and suggest an optimal control strategy for reducing vehicle speed based on Pontryagin's minimum principle. A study on a unified controller for autonomous cars that utilizes differential braking to improve their lateral maneuverability and rapid response in emergency situations [18]. The controller is specifically engineered to prioritize collision avoidance, vehicle stability maintenance, and precise adherence to a predetermined trajectory. It demonstrates its efficacy in averting instability and crashes.

The prime intentions of this research toward create ultrasonic braking technology that allows motors to halt when sensors identify an obstacle, so averting frontal crashes. This technology enables autos to detect and anticipate potential collisions with other vehicles or dangers, and respond by controlling the brake system to stop the vehicle. This duty is accomplished by employing ultrasonic sensors, which consist of an ultrasonic wave emitter and a high-speed wave receiver. The car is equipped with an ultrasonic wave emitter located in the front, which produces and sends out ultrasonic waves at a set distance in front of the vehicle. The reflected waves, or detection pulses, are used to measure the distance between the vehicle and any potential obstacles. An 8051 microcontroller processes these detection pulse statistics to regulate the DC motor, ensuring precise control over the vehicle's braking system. Additionally, a servomotor adjusts the vehicle's dampers, enhancing stability and preventing accidents by sensing and responding to boundary conditions. This advanced braking system significantly improves vehicle safety by autonomously managing speed and applying brakes when necessary, thereby reducing the risk of collisions and enhancing overall road safety.

4.1.1 Project Scope

Developing a thorough plan for the ultrasonic braking systems architecture, this will include state-of-the-art ultrasonic sensors, a PLC for control logic implementation, and an 8051 microcontroller. Quick and painless in corporation of the planned system into the vehicle's are current control and braking infrastructure. Maximize the vehicle's obstacle detection accuracy and coverage by strategically placing and deploying ultrasonic sensors. Accurate distance measuring and trustworthy obstacle recognition are guaranteed by the sensors' rigorous calibration and testing. Create complex control logic in WPL Soft that responds to sensor input by regulating the vehicle's speed and braking. A complex ladder logic system was implemented to process data in real-time and activate the brakes promptly. The system will be tested in a real- world setting with a wide range of speeds, obstacles, and environmental factors to ensure its effectiveness. Optimization of performance is achieved through iteratively refining the control logic and sensor parameters using empirical test findings. Comprehensive evaluation of the system is dependability and collision avoidance capabilities. Including fail-safe systems to increase safety and guaranteeing conformity with strict safety standards and regulatory requirements.

4.1.2 *Work Originality*

To create a strong and dependable braking system for driverless cars, this study mixes PLC technology with cutting-edge ultrasonic sensors and microcontroller systems in a novel way. In order to keep processes that are vital to safety running, PLCs' real-time processing capabilities are essential. The system greatly improves the vehicle's ability to avoid collisions by using cutting-edge ultrasonic sensors to identify obstacles in real-time with unprecedented accuracy. Utilizing WPL Soft for ladder logic programming, the research presents a control strategy that is both flexible and complex. It can adapt to various driving circumstances, guaranteeing the maximum level of vehicle safety. Demonstrating the system's operational performance and direct applicability to automobile safety advancements, the project goes beyond theoretical development and emphasizes practical implementation and thorough real-world testing. This work fills a critical gap in the development of modern automobile technology by improving the safety aspects of autonomous vehicles through an emphasis on the automatic prevention of head-on collisions.

4.2 Design and Implementation

In order to achieve the targeted deceleration, this study uses rule-based logic in a high-level control system to ensure that the tire's maximum capacity at the road friction limit is satisfied. By developing a set of spoken rules, Rule-Based Supervisory Control (RBSC) monitors how well the AB system is doing in reaching its goal. A rule base and an inference mechanism form the foundation of a rule-based system.

4.2.1 *Modeling and Control*

This section delves into the development and modelling of the control system for AEB. A system of nonlinear variable equations constitutes the kinetic model of the vehicle equipped with AEB. From here on out, call these two sets of laws the axial and lateral movements equations, respectively. One of primary goals is to master the autonomous straight-line brake manoeuvre. This motion can be studied by reducing the complexity of the laws of motion that characterise smooth chaotic transverse motion.

In the event of an emergency, this system is programmed to apply the brakes. Hierarchical structure characterizes the AEB control system. Specifically, the deceleration actions are planned and executed by the Rule Based Supervisory Control system, which is also responsible for providing high-level control. By carrying out instructions given to it at higher levels, the low-level control system efficiently handles slip dynamics. A flat, dry road makes up the landscape. From a control algorithm perspective, two separate low-level controllers have been designed due to the importance of wheel slip management in AEB applications. In order to make the emergency braking manoeuvre more stable, the two controls work together. Separated from the other controllers are a gain scheduled exponential cubic limiter and a robust shifting wheel slip controller. The controllers are examined in their respective domains.

The automated system, a goal state estimator, a controller, an interface, a human driver, and sensory apparatus comprise the feedback loop. The primary component of the loop that establishes the control law for controlling the driver is the controller. In order to facilitate communication between the autonomous system and the human driver, interfaces and sensors are included. In the feedback loop, the driver receives data from both the inside of the car and the outside regarding the autonomous system and traffic conditions. This information may contain useful details about driving conditions to help driver's decisions. However, it may also contain irrelevant information that could cause disturbances and divert the driver's attention from the road ahead [4].

The driver's state may change depending on what they think to be known. As a result, the system ought to continuously check the driver's status and identify any modifications. These differences may put the driver in an unsafe driving condition whether operating a vehicle or interacting with the assistance system. As a result, the interaction loop intended to regulate the driver state also incorporates a driver state controller. The driving conditions are also taken into consideration by the controller. A certain level of scenario criticality is necessary to accomplish this purpose.

4.2.2 Vehicle Speed and Spacing Control

Vehicle Speed Control and spacing direct refer to regulating a vehicle's rate and maintaining a safe reserve from other vehicles. This is achieved through coordinated throttle actuation, ensuring safety on highways and in traffic congestion [19].

Coordinated brake and throttle actuation enables precise speed adjustments and vehicle spacing maintenance. This integrated system enhances road safety, reduces collision risk, and improves traffic flow. It also promotes efficient driving, maximizes fuel efficiency, and minimizes wear and tear on vehicle components.

Considering the starting spacing (do), the host vehicle's rate (vhost), and a time constant (tsafe) that guarantees a safe braking manoeuvre for the driver allows us to calculate the relative distance between the two vehicles. With the help of Eq. (4.1), we can determine the ideal distance between vehicles.

$$\mathrm{d_{safe}} = \mathrm{d_o} + \mathrm{t_{safe}} \cdot \mathrm{v_{host}} \tag{4.1}$$

No matter what mode the vehicle is in, the ACC system has to give the right amount of acceleration to keep the desired speed or safe distance. The accelerator control module (ACC) sends signals to the throttle control module of an ICE [20].

The ACC block and throttle position sensor provide the acceleration value, which in turn controls the throttle valve.

4.2.3 PLC Programming

Experts in engineering, consulting, and electrical work can determine the PLC with little to no training or study [21]. When deciding whether to use a relay or solid-state electronics, there are a number of factors to consider [22, 23]. Transformations can be accomplished in a PLC, and signal lights on the PLC at key diagnostic points assist with regulation.

In addition to being stable and easy to maintain, it works well in today's environment. The PLC has several uses and is adaptable, which helps keep costs down [24]. Programmable Logic Controllers (PLCs) are crucial in industrial automation and control systems due to their robustness, flexibility, and efficiency. Here are the key reasons highlighting the importance of PLCs:

Reliability and Durability

Industrial-Grade Design: PLCs are PLCs are designed to endure challenging industrial conditions, such as high humidity, dust, and electrical noise, among other things.

Long Lifespan: They have a long operational life, making them a cost-effective solution for long-term automation needs.

Flexibility and Scalability

Modular Architecture: PLCs can be easily expanded by adding modules for additional I/O (input/output) points, communication interfaces, and special functions.

Versatile Programming: They offer support for a variety of programming languages, giving engineers the flexibility to select the most appropriate one for their specific application.

Real-Time Operation

Deterministic Control: PLCs operate in real-time, ensuring that control processes are executed within precise time constraints, which is critical for applications like automated braking systems.

Fast Processing: They can process input signals and execute control instructions rapidly, making them suitable for high-speed automation tasks.

Ease of Maintenance and Troubleshooting

Diagnostics and Monitoring: PLCs offer built-in diagnostic tools and real-time monitoring capabilities, allowing for quick identification and resolution of issues.

User-Friendly Interfaces: Programming software for PLCs, such as WPL Soft, provides intuitive graphical interfaces for designing and testing control logic.

Integration and Communication

Network Compatibility: PLCs supports are capable of supporting a wide range of communication protocols such as Ethernet/IP, Modbus, and Profibus. This allows for effortless integration with other industrial systems and devices.

Interoperability: To provide all-encompassing automation solutions, they can communicate with various control systems, sensors, actuators, and HMIs.

Programmability and Control

Customizable Control Logic: Engineers can program PLCs to execute complex control algorithms tailored to specific automation tasks, such as the automatic braking system in vehicles.

Adaptive Control: PLCs can adjust control parameters dynamically based on real-time facts from sensors and further input devices.

Safety and Compliance

Fail-Safe Operation: PLCs are designed to handle failures gracefully, ensuring the system reverts to a safe state to prevent accidents and equipment damage.

Standards Compliance: They adhere to industry standards and certifications, ensuring that automated systems meet safety and regulatory requirements.

Cost-Effectiveness

Reduced Downtime: Reliable operation and easy troubleshooting minimize downtime, enhancing overall productivity.

Efficiency Gains: Automation using PLCs can lead to significant efficiency improvements, reducing labor costs and increasing output quality.

Application in Autonomous Braking Systems

In the context of autonomous braking systems for vehicles:

- **Precise Control**: PLCs ensure accurate control of braking mechanisms based on sensor inputs, crucial for preventing collisions.
- **Real-Time Processing**: The ability to process sensor data and execute braking commands in real-time ensures timely response to obstacles.
- **Safety and Reliability**: The robust design of PLCs ensures the braking system operates reliably under various conditions, enhancing vehicle safety.
- **Manufacturing production lines:** PLCs are used to control the flow of materials, operate machines, and ensure that products are manufactured to quality specifications.
- **Building automation:** PLCs People-located controllers (PLCs) are utilized to control lighting systems, safety devices, ventilation, air conditioning, and heating (HVAC) systems.
- **Traffic control systems:** PLCs are used to control traffic lights and other devices that help to keep traffic flowing smoothly.
- **Water treatment plants:** PLCs are used to control the pumps, filters, and other equipment that is used to treat water.
- **Power plants:** PLCs are used to control the generators, turbines, and other equipment that is used to generate electricity.
- **Amusement rides:** PLCs are used to control the operation of amusement rides, such as roller coasters and Ferris wheels.

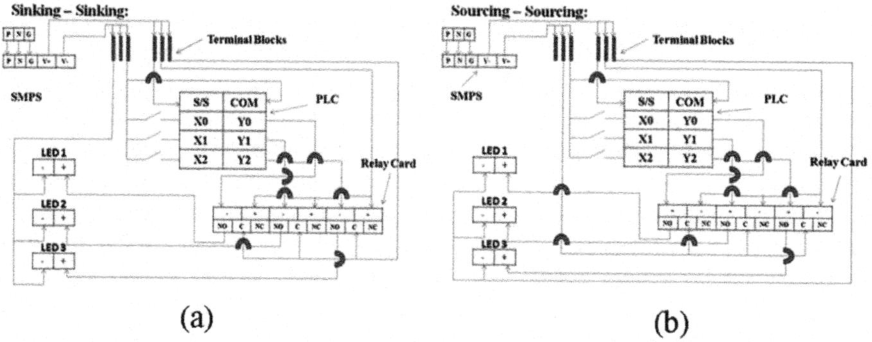

Fig. 4.1 PLC modules connected to programming interface (**a**) Sinking circuit diagram, (**b**) sourcing circuit diagram

Overall, PLCs are indispensable in modern automation due to their robustness, versatility, and ability to handle complex control tasks efficiently.

4.2.4 PLC Wiring

When discussing ideas related to electrical wiring, there are typically two main types of people who participate. They are sourcing in further sinking: Having a strong grasp of sinking and sourcing concepts is crucial when helping users choose the right input or output component for their PLC technology. When discussing input or output circuits, these words come up frequently [25, 26]. Sinking & Source is used exclusively with Digital DC circuits and shown in Fig. 4.1a and b respectively. If, the common pin is + polarity called as sourcing circuit and—polarity called a sinking circuit. PLC is generally having 2 different output types such as Relay type output and Transistor type output. In transistor type having two types such as NPN (Sinking) and PNP (Sourcing).

4.2.5 Ladder Logic and Programming PLCs

The phrase "programmable logic controller" refers to a device that is manufactured with several "input" terminals. These terminals enable the PLC to interpret logical states, specifically "high" and "low," that are received from a variety of sensors and switches. Furthermore, it is equipped with a multitude of output terminals that are capable of emitting "high" and "low" signals, which can be used to activate lights, solenoids, contactors, miniature motors, and other devices that are suited for on/off control. A programming language for programmable logic controllers (PLCs) was purposefully designed to have a striking resemblance to ladder logic diagrams. This was done with the idea of making the process of programming more straightforward. Thus, someone with knowledge in industrial electrical systems, like an industrial electrician or electrical engineer, who is well-versed in interpreting ladder logic diagrams, would have no trouble programming a Programmable Logic Controller (PLC) to carry out the same control operations. In industrial settings, computer systems known as programmable logic controllers, or PLCs, are utilized. Some programmable logic controllers (PLCs) offer the capability to send and receive low-level direct current (DC) voltage signals, which are typically utilized in logic gate circuits. It is possible that the signal connection and programming standards will vary slightly from one type of PLC to another; however, they are comparable enough to offer a broad overview of PLC programming in this context. The subsequent depiction presents a basic Programmable Logic Controller (PLC) as it might be observed from a frontal perspective. The PLC's internal circuitry can be powered by connecting it to two screw terminals labelled L1 and L2, which offer a connection to 120 V AC. Make use of the six screw connections situated on the left panel to attach the sensors. Each terminal represents a separate input "channel" and is labelled with an "X". The screw terminal located in the lower-left position serves as the "Common" connection.

As depicted in Fig. 4.2a, the programmable logic controller (PLC) contains an opto-isolator device that is a light-emitting diode. Through the use of this device, the circuitry of the computer is provided with an electrically isolated "high" logic signal. A phototransistor is used to read the light that is produced by an LED. One of the LEDs that can be seen on the front panel of the PLC will illuminate whenever an input is activated. Using a device called TRIAC, or electronic relay, the PLC processor generates return signals, as shown in Fig. 4.2b. In most cases, the "Source" terminal is linked to the L1 side of the 120 VAC power supply. Each time an input is received, a light emitting diode (LED) on the front panel of the PLC illuminates when the output is activated.

The X1 input of the PLC does not receive any electrical power when the pushbutton switch is not depressed. The transmission of electrical energy to the Y1 coil will be impeded by the presence of a usually release X1 contact in sequence with the Y1 coil in the provided programme. Consequently, the indicator lamp associated with the PLC output Y1 remains unlit as a result of the absence of electrical power. By pressing the pushbutton switch, power will be transmitted to the X1 input of the PLC.

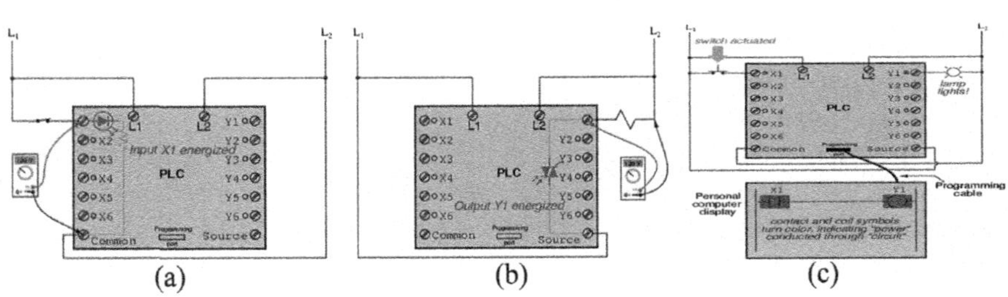

Fig. 4.2 **a** PLC input, **b** output device, **c** combined setup

In the event that the relay coil "X1" is activated, all X1 contacts that are depicted in the programme will be in the actuated state, as if relay links. The X1 input is activated, thereby enabling the transmission of power to the Y1 coil, the corresponding light will be turned on as well, which is known as the Y1 output.

The fact that the X1 contact, the Y1 inductor, the connected cables, and the "power" that are displayed on the personal computer are all simulated or virtual is an extremely important point to keep in mind. They do not have any physical existence, despite the fact that they are legitimate electrical components. These directives are contained within a computer program that is designed to seem very much like a genuinely authentic relay schematic diagram. As can be shown in Fig. 4.2c, it is of the utmost importance to acknowledge that the personal computer that is utilized for the purpose of displaying and modifying the programme of the PLC is not required for the PLC to operate continuously. In order to make it easier for you to comprehend the connection that exists between the conditions that exist in the actual world (such the status of the bulb and the closure of the switch) and the state of the program (which is represented by "power" during effective connections and essential coils), I have included the display of the personal computer in these pictures.

4.2.6 WPL Editor Window

The WPL (WinProLadder, vide in Fig. 4.3) Editor is a software tool specifically designed for the purpose of programming Programmable Logic Controllers (PLCs). It is frequently employed for the purpose of building and modeling control logic in industrial automation systems, such as vehicle speed controllers and autonomous braking systems. The WPL Editor is a robust tool that enables the design, testing, and implementation of control systems in different automation applications. It is particularly useful for developing sophisticated automotive safety functionalities, such as the autonomous braking system [27]. The main parts of the WPL Editor window are:

The menu bar has dropdown choices for working with files (New, Open, Save, Print), editing (Cut, Copy, and Paste), setting the view, and using simulation and debugging tools. The tool bar lets user quickly access commands that user use often, like starting or stopping the exercise, saving, opening a project, and compiling the ladder diagram. Project Explorer shows the current project's hierarchy, which includes all PLC programmes, subroutines, and files that are linked to them.

Ladder Diagram Workspace: This is where most of the ladder reasoning is made. It lets people move ladder parts like contacts, coils, timers, and counters around.

Instruction List Panel: This panel shows the instruction list (IL) version of the ladder diagram, which helps user understand the code that the graphical programming creates.

Status Bar: Shows details about the current state of the project, like where the pointer is and whether it is compiling or connecting to the PLC hardware. The

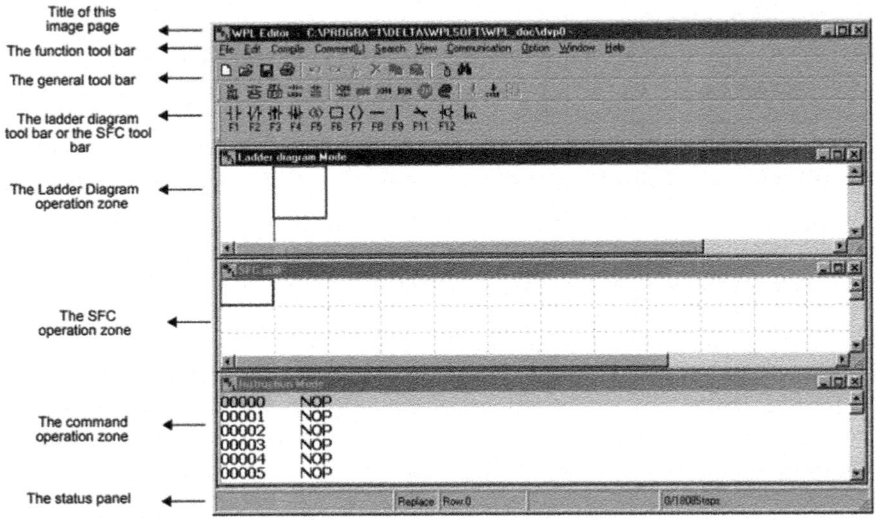

Fig. 4.3 WPL editor window

Element Toolbox has all the parts user need to make a ladder diagram, like input and output relays, timers, counters, comparison directions, and math operations.

Properties Window: This window shows the properties of the currently selected part in the ladder diagram and lets user set parameters and configure it. The output window shows notes from the compiler, such as errors and warnings, which help user figure out what's wrong with the programme.

Important Features and Tasks:

- **Making Ladder Diagrams**: To make a new ladder diagram, user'scan be drag parts from the toolbox onto the workspace and connect them in a way that makes sense.
- **Simulation and Debugging**: User can test and fix logic without the actual PLC hardware by simulating the ladder diagram in the WPL Editor. The user can monitor the progress of the procedure with the help of breakpoints.
- **Prior to uploading the program to a PLC**, it is necessary to build the ladder diagram in order to verify its logic and check for syntax errors.
- **Transferring the code to the PLC**: Once the program has been tested and found to function properly in simulation, it can be transferred to the PLC over a data link such as a serial, Ethernet, or USB connection.
- **Monitoring in Real Time**: The software is lets user see the state of the PLC's inputs, outputs, and internal registers while it's running.

4.2.7 *Simulation in WPL SOFT*

The first screen that appears when you open WPLSoft is the one that follows: the menu bar contains five buttons: the File button (F), the Communication button (C), the Option button (O), the Windows button (W), and the Help button (H). Following is an example of what will appear when you select "New" from the "File" menu. Edit (E), Compile (P), Comment (L), Search (S), and View (V) are some of the additional options that will be displayed in the function panel shown in Fig. 4.3. It is the user's responsibility to discuss these alternatives in the order in which they appear.

The "File" function is illustrated below, with the subsequent commands indicated, shown in Fig. 4.3. New: Establish a new document. First, go to the "File" menu and click on "New."

• Step 2: Click on the button in the toolbar. If want to use the third method, quickly press the "Ctrl" and "N" keys.

Open the documents that have become obsolete on the drive.

- From the "File" menu, select "Open" as the method to pursue.
- Method 2: To access the tool tip, click on the icon that is displayed there.
- Utilize the speedy key-in capability and enter the compound buttons by pressing the combination of the Ctrl and O keys. The contents of the file will be copied to the drive when you save it.
- Step one is to go to the "File" menu and select the "Save" option.Using the quick key-in feature, input the compound controls by pressing the combination of the control key and the letter S.
- "Save as" allows you to save the currently open file with an alternative name.
- Step one is to go to the "File" menu and choose the "Save as" option.

4.3 Proposed Working Model

The automatic format braking system and the vehicle's speed are anticipated to be managed by the proposed work. The process is comprised of three primary steps: - Identify the object (Hurdle) from the vehicle.

- Detect the object (Hurdle) from vehicle.
- Control the speed of vehicle.
- Automatic braking system.

The suggested method uses ultrasonic and infrared sensors to detect road impediments and alert the driver visually. Upon detection, the system displays the distance on an LED display and engages automatic braking if a potential accident is detected.

- **Hurdle Detection**

Ensuring the vehicle's speed and protecting a motorist and vehicle from numerous obstructions are the goals of this part of the suggested solution. To achieve this goal, ultrasonic sensors are mounted on the vehicle's front surfaces. The sensor consists of a single transmitter and a single receiver. The sensor's transmitter will continually broadcast the signals. The sensor's receiver will capture the reflected signals and transmit them to the microcontroller when they collide with an object and are reflected back. The signal is transmitted to the processing device via various ports in order to identify the object side. The capacity of an ultrasonic sensor is directly proportional to its range.

- **Car Speed Controlling System**

This system regulates the vehicle's pace using a microcontroller. Different instructions will be sent to the ports of the controller used to attempt to manage the vehicle's speed. To show the roadblock ahead of the car, the microcontroller receives instructions and, as encrypted, emits an 8-bit value. The vehicle's ultrasonic distance sensor and infrared sensors will generate the input signals. After processing data from the detection devices, the controller keeps the vehicle's speed constant according to the distance.

- **Ultrasonic Hurdle detection Circuit**

If you're in an empty room, the speed of sound is about 340 m per second. An ultrasonic sensor may detect the return pulse even if an audible echo would not be audible when a short 200 kHz pulse is sent through the air. The time it takes for an ultrasonic wave to travel in both directions allows us to calculate its distance. It is possible to locate the assortment since the ultrasonic sensor to the first barrier in its path by dividing the distance by two. The suggested solution also makes use of the ultrasonic piezzo transmitter and receiver because of how accessible, efficient, and inexpensive they are. At the outset of this planned study, the pulse will be sent. Even though the sensor range won't go beyond 50 cm, you can utilize the PIC output to directly drive an ultrasonic transmitter. With the help of a resonance circuit and a transistor, the ultrasonic transmitter can boost its detection range to 75 feet and generate a pulse at 200 kHz. About 20 V will be sent to the transmitter.

- **Activation of automatic braking system**

The LED display provides a direct indication of the distance between the automobile and the impediment whenever the obstacle is detected in front of the vehicle. As the car gets closer to the obstruction, the velocity of the vehicle reduces. The automatic braking system is engaged when the IR sensor detects a condition resembling an accident, which happens when the distance between two vehicles or between vehicles and an obstacle is extremely close.

4.4 Steps in Circuit Designing

The circuit designing process involves several steps, including requirement analysis, circuit simulation, and prototyping. These steps help ensure that the final circuit design meets the desired specifications and functions as intended.

To draw ladder diagram

- Method 1: Click on "F1–F12" function buttons
- Method 2: Click on symbols by using mouse

4.4.1 Circuit Diagram and Explanation

STEP-1: In this first step user need to start car as per logic, so the car should start when all doors are closed and the key should also kept starting the car, vide in Fig. 4.4a. So M50, M51, M52, M53 are the doors of the car and M1 is car as user designed in the first line, M is known as the memory allocation which act as dummy switch and M1 is load, when all switches are in ON position then the load(car) is applied. In the second line M1 (load) act as switch it act like key or acceleration in if user rotate the key then car be stop, so user used counter here.

STEP-2: As user observed in step 1 the car starts in step 1, in step 2 counters will act as switch and M1 act as acceleration then the light will be on and in second line M2 act as load (vide in Fig. 4.4b) if user stop key then it will reset the system so user latched RST to the 3rd line.

STEP-3: In this step the car speed should be increment when user press acceleration so X2 and Y2 will act as sensor and Y0 as an engine of the car, shown in Fig. 4.4c.

STEP-4: In this step (Fig. 4.4d) according to the logic when speed of the car reaches to speed 100kmph then the decrement of speed should be done automatically for every 50 s it should reducesupto 10kmph speed. So decrement had connected to this line and this should do according to timer so user added timer to the motor means M10.

STEP-5: In this step (Fig. 4.4e) when any obstacle come infront or behind the car then buzzer should be activated sudden and if car touch any object then air bags in the car should come out. Then automatically brake should activated and the must stop. In this M4 and M3 will works as sensor and Y3 will act as air bags and Y1 as buzzer. The model block diagram of PLC circuit and fabricated design board are shown in Fig. 4.4f and g, respectively.

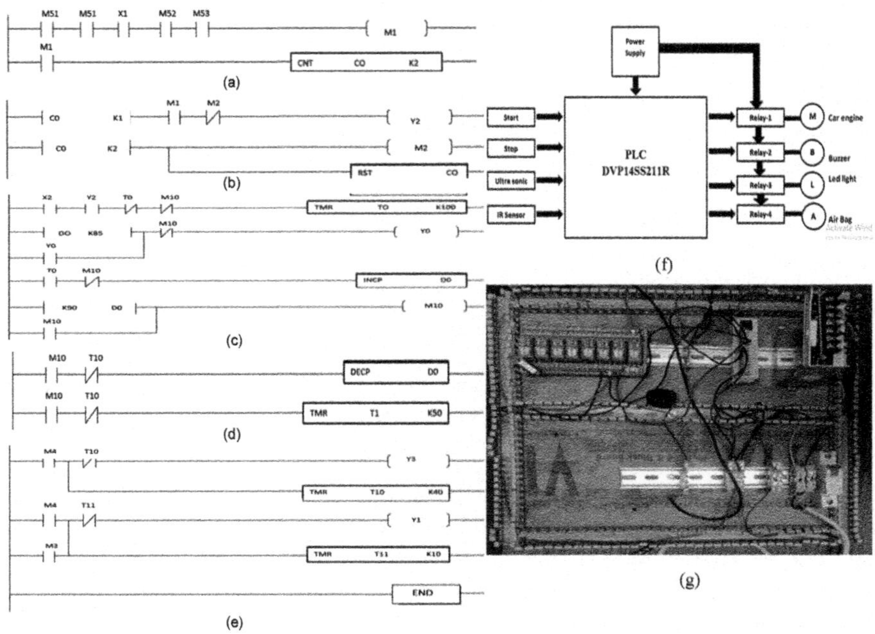

Fig. 4.4 PLC programming logic circuit of (**a**) starting and closing doors of car, (**b**) acceleration of PLC circuit, (**c**) incrementing acceleration sensor circuit, (**d**) decremented speed and timer sensor, (**e**) automatically brake and stop circuit, (**f**) block diagram of PLC circuit, (**g**) PCB design of activation of automatic braking system

4.5 Ladder Diagram Monitor

Before starting a "Ladder Diagram Monitor Start" operation under the "Communication" function just on function screen or clicking the icon, users must first open a PC ladder diagram window (as illustrated in Fig. 4.5a).

4.5.1 To Compile (P)

Make the transition from the Ladder Diagram code to the Command codes by following these steps.

Click on "Ladder Diagram -> Instruction" under the "Compile (P)" function in Step 1.

Step 2: Locate the "compile" icon in the toolbar and click on it.

Step 3: Press Ctrl + F10 to access the compound buttons, or use the fast key-in feature. Make the necessary changes to the command codes so they can be used with the ladder diagram programme. Using the "Compile (P)" function, go to "Instruction -> Ladder Diagram" as the first step.

Fig. 4.5 Detailed procedure of an operation: **a**) the ladder diagram, **b**) monitoring PLC window display

Procedure	The Ladder Symbols	Location of the Cursor	Input through the clicking on the Function Keys		Input through the Keyboard
1	—| |—	Row: 0, Line: 1	*Footnote 1	Device name: X Device number: 1	LD X1 ↵
2	—( )	Row: 0, Line: 2		Device name: Y Device number: 1	OUT Y1 ↵
3	—| |—	Row: 1, Line: 1		Device name: X Device number: 2	LD X2 ↵
4	|	Row: 1, Line: 2			F9
5	—( )	Row: 1, Line: 2		Device name: Y Device number: 2	OUT Y2 ↵
6	—| |—	Row: 2, Line: 1		Device name: X Device number: 1	LD X1 ↵
7	—| |—	Row: 3, Line: 1		Device name: M Device number: 0	LD M0 ↵
8	—[ ]	Row: 3, Line: 2	*Footnote 2	Application command MOV Operand 1: D Device value: 1 Operand 2: D Device value: 2	MOV D1 D2 ↵
9		Row: 4, Line: 0		Double click the mouse to input P0	P0 ↵
10	—|↑|—	Row: 4, Line: 1		Device name: M Device number: 1	LDP M1 ↵
11	|	Row: 4, Line: 2			F9
12	—[ ]	Row: 4, Line: 2		Counting command CNT Operand 1: C Device value: 0 Operand 2: K Device value: 100	CNT C0 K100 ↵
13	—|↓|—	Row: 5, Line: 1		Device name: M Device number: 1	LDF M1 ↵
14	—[ ]	Row: 6, Line: 1		Application command END	END ↵

(a)

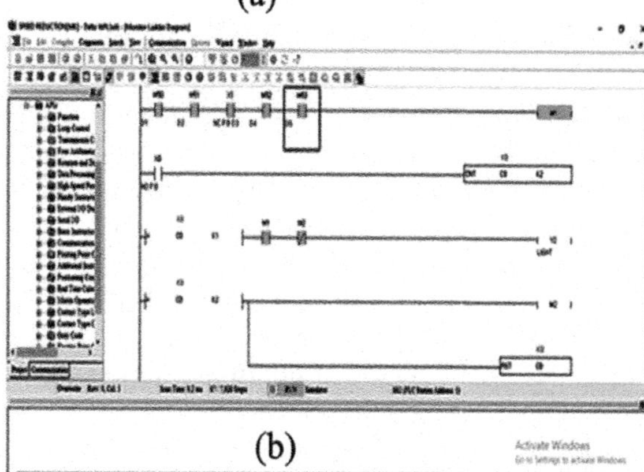

(b)

- Approach 2: Locate the symbol on the tool bar and click on it.
- Third Approach: Press Ctrl + F11 to quickly enter the compound buttons.Translate the SFC schematic into command codes (SFC -> Instruction). (In order to transform an SFC diagram into a ladder diagram, one must first transform it into command code, and then into a ladder diagram programme.)"In the "Compile (P)" menu, choose "SFC -> Instruction" as the first step. The second way is to use the icon on the toolbar, which is labelled.
- Step 3: Press Ctrl + F6 to access the compound buttons, or use the fast key-in feature. Make sure to check the repeated use of the "OUT, RST, SET, CNT, DCNT, and TMR" commands using the Double Loop Check.

In the "Compile (P)" menu, find "Double Loop Check" and click on it. After this, user need connect pc to the PLC to run our program to PLC by using Communication cable. Monitored contacts, output coils, or even application commands are all shown in green shown in Fig. 4.5b when monitoring begins. The green portion indicates that device in active condition when monitoring begins, if green color will not appear on the contact location, output coil location, or application command location if none of these components are currently in motion, however this is unlikely. There will also be a place where user can see what is in the register, like in Fig. 4.5. Finally run the program after connected to PLC carefully. These are the various steps used in the working of automated braking system depending on over speed after uploading program into PLC user can disconnect the cable it will run by itself. Before upload user need check in the simulation.

4.6 Conclusions and Future Scope

An ultrasonic braking system that utilizes PLC technology stands out as a groundbreaking innovation that enhances safety in the constantly changing world of autonomous vehicle technology. This research delves into the intricacies of combining ultrasonic sensors with microcontroller-driven PLCs, creating a strong framework capable of independently detecting impediments and applying the brakes. Vehicle collision avoidance systems are poised for a paradigm shift thanks to a symphony of safety protocols orchestrated by the PLC's real-time processing capabilities and the accuracy of ultrasonic sensors. Carefully crafted in WPL Soft, the ladder logic programming exemplifies the adaptability and complexity required for responsive braking and real-time obstacle recognition. This complex technological ballet not only makes autonomous vehicles safer overall, but it also guarantees that head-on collisions won't happen. Enhanced sensor fusion, the incorporation of artificial intelligence, and advanced driver assistance systems (ADAS) all hold great promise for the future. The potential for these innovations to usher in a whole safety ecosystem where cars can sense their surroundings, respond to potential dangers, and adjust their behavior accordingly is exciting. Investigating vehicle-to-everything (V2X) communication paves the way for novel possibilities in real-time data sharing, which in turn promotes an interdependent network of smart infrastructure and automobiles. This technology is kept at the forefront of automobile safety standards by complying with legal and regulatory requirements and by advances in human–machine interaction. To sum up, this study is a major milestone on the road to a future where driverless cars are incredibly dependable and safe. Combining PLCs with ultrasonic braking technology is a huge step forward in vehicle safety, marking the beginning of a new age in which accidents are far less likely to occur. There are limitless chances to improve, expand, and reimaging the limits of autonomous vehicle safety as we stand on the brink of this technological revolution. The road ahead is both thrilling and promising.

References

1. Al_Issa, H.A.: Sensors application using PIC16F877A microcontroller. Am. J. Remote Sens. **4**(3), 13 (2016). https://doi.org/10.11648/j.ajrs.20160403.11
2. Amin, N., Borschbach, M.: Quality of obstacle distance measurement using ultrasonic sensor and precision of two computer vision-based obstacle detection approaches. In: International Conference on Smart Sensors and Systems, IC-SSS 2015, pp. 1–6 IEEE (2017). https://doi.org/10.1109/SMARTSENS.2015.7873595.
3. Bhumkar, S.P., et al.: Accident avoidance and detection on highways. Int. J. Eng. Trends Technol. **3**(2), 247–252 (2012)
4. Chang, C.C. et al.: Distance measurement technology development at remotely teleoperated robotic manipulator system for underwater constructions. In: International Symposium on Underwater Technology, UT'04—Proceedings, pp. 333–338. IEEE (2004). https://doi.org/10.1109/ut.2004.1405598.
5. Correia, A.P., et al.: A control design approach for controlling an autonomous vehicle with FPGAs. J. Comput. **5**(3), 360–371 (2010). https://doi.org/10.4304/jcp.5.3.360-371
6. Gerdes, J.C., Hedrick, J.K.: Vehicle speed and spacing control via coordinated throttle and brake actuation. Control. Eng. Pract. **5**(11), 1607–1614 (1997). https://doi.org/10.1016/S0967-0661(97)10016-8
7. Gissinger, G.L., et al.: A mechatronic conception of a new intelligent braking system. Control. Eng. Pract. **11**, 163–170 (2003)
8. Gounis, K., Bassiliades, N.: Intelligent momentary assisted control for autonomous emergency braking. Simul. Model. Pract. Theory **115**, 102450 (2022). https://doi.org/10.1016/j.simpat.2021.102450
9. Hajiloo, R., et al.: Integrated steering and differential braking for emergency collision avoidance in autonomous vehicles. IEEE Trans. Intell. Transp. Syst. **22**(5), 3167–3178 (2021). https://doi.org/10.1109/TITS.2020.2984210
10. Hoomod, D.H.K., Al-Chalabi, S.M.M.: Achieving real-time tracking mobile wireless sensors using SE-KFA. J. Phys. Conf. Ser. **1003**(1), 012039 (2018). https://doi.org/10.1088/1742-6596/1003/1/012039
11. Khang, A. et al.: Robotic process applications and tools for manufacturing sector. In: Khang, A. et al. (eds.) Machine Vision and Industrial Robotics in Manufacturing, pp. 252–274. CRC Press, Boca Raton (2024). https://doi.org/10.1201/9781003438137
12. Patil, N. et al.: Brake system dynamics and control for an autonomous personal rapid transit vehicle. In: ASME International Mechanical Engineering Congress and Exposition, Proceedings (IMECE), pp. 1–9. American Society of Mechanical Engineers Digital Collection (2023). https://doi.org/10.1115/IMECE2023-113720.

13. Petri, A.M., Petreuş, D.M.: Adaptive cruise control in electric vehicles with field-oriented control. Appl. Sci. **12**(14), 7094 (2022). https://doi.org/10.3390/app12147094
14. Raju, R.S.S., Siva, VB.: Fabrication and tribological studies of Al-CSA composite using RSM. Int. J. Mater. Eng. Innov. **12**(2) 83–102 (2021). https://doi.org/10.1504/IJMATEI.2021.115596
15. Raju, S.R. et al.: Design of automated braking system of vehicle using PLC ladder and simulated with WPL software. In: AIP Conference Proceedings, p. 020044(6pages). American Institute of Physics Inc. (2023). https://doi.org/10.1063/5.0166076
16. Rallabandi, S. et al.: Sustainable waste management for solid waste dominance in metals, concrete brick mix of plasticizer with interlocking—a case study. In: Brojo Kishore, M. et al. (eds.) Application of Engineering, Technology, Management for Sustainable Manufacturing Sector, pp. 155–177. World Leadership Academy (2022)
17. Rallabandi, S.R., et al.: Development of a color-code sorting machine operating with a pneumatic and programmable logic control. Mater. Today Proc. (2023). https://doi.org/10.1016/j.matpr.2023.05.150
18. Rath, K.C. et al.: Artificial intelligence (AI)-enabled technology in medicine-advancing holistic healthcare monitoring and control systems. In: Computer Vision and AI-Integrated IoT Technologies in the Medical Ecosystem, pp. 87–108. CRC Press, Boca Raton (2024). https://doi.org/10.1201/9781003429609-6.
19. Rath, K.C. et al.: Enabling the future of manufacturing: integration of robotics and iot into smart factory infrastructure in industry 4.0. In: Machine Vision and Industrial Robotics in Manufacturing, pp. 101–128. CRC Press, Boca Raton (2024). https://doi.org/10.1201/9781003438137-6
20. Rath, K.C. et al.: The role of internet of things (IoT) technology in industry 4.0 economy. In: Khang, A. et al. (eds.) Advanced IoT Technologies and Applications in the Industry 4.0 Digital Economy, p. 28. CRC Press, Boca Raton (2024). https://doi.org/10.1201/9781003434269
21. Sairam, G.V., et al.: Intelligent mechatronic braking system. Int. J. Emerg. Technol. Adv. Eng. **3**(4), 3–8 (2013)
22. Raju, SR. et al.: Design, fabrication and development of mechanical arm for disable challenged persons. Dogo Rangsang Res. J. **13**(5(2)) 90–97 (2023)
23. Tahir, M.B., Abdullah, M.: Distance measuring (hurdle detection system) for safe environment in vehicles through ultrasonic rays. Glob. J. Res. Eng. Automot. Eng. **12**(1) 1–5 (2012)
24. Thakur, D., Thakare, A.P.: A review on Implementation of FPGA for automatic reverse braking system. Int. J. Sci. Res. ISSN (Online Index Copernicus Value Impact Factor **14**(1) 2319–7064 (2013)
25. Vuddagiri, H.K. et al.: Assessment of mechanical and tribological performance of hybrid Al/MoS2/Al2O3 composite by GFRA. Metall. Mater. Eng. **28**(1) 79–102 (2022). https://doi.org/10.30544/764
26. Webster, D.: A pulsed ultrasonic distance measurement system based upon phase digitizing. IEEE Trans. Instrum. Meas.Instrum. Meas. **43**(4), 578–582 (1994). https://doi.org/10.1109/19.310171
27. Yadav, R.R., et al.: Multilevel object sorting system using PLC controller. Int. Res. J. Eng. Technol. **6**(5), 1910–1915 (2019)

Dr. Sivasankara Raju holds a position as an associate professor at the Aditya Institute of Technology and Management in Tekkali, with a total experience of 16.4 years, which includes 12.3 years of teaching and 4.1 years of industrial experience. He published 32 Scopus-indexed documents, such as 15 articles (5 SCI, 7 Scopus, and 3 Web of Science), 13 conference papers, and 2 book chapters, along with 4 patents (1 granted, 1 filed, and 2 published) along with 4 Patents (1 Granted, 1 Filed and 2 Published). In his career, he received 7 prestigious awards, like a Certificate of Honour (2023), Best Innovation (2022), Best Paper (2021, 2020, and 2016), Young Researchers (2016), and Athletic Champion (2007). He acts as an editorial board member for six journals and as a reviewer for 12 journals. His research focuses on the evaluation of tribological behavior, the multicriteria decision-making (MCDM) approach, and the development of hybrid composites in metals and polymers.

Mr V. Naresh Babu is presently working as Assistant Professor and HMES (Polytechnic) at Geethanjali Institute of Science and Technology, Gangavaram with 11 years teaching experience and has published more than 5 International and National publications and attended International 3 conferences.

K Nagaraju is presently working as Assistant Professor at Avanthi Institute of Engineering and Technology, Makavarapalem with 12 years teaching experience and has published 3 International journals.

Dr Hari Kiran Vuddagiri is presently working as Associate Professor and Head of the Department at Avanthi Institute of Engineering and Technology, Makavarapalem with 15 years teaching experience and has published more than 20 International and National publications and attended International 6 conferences, 3 book chapters.

Dr. P. Thimothy is an accomplished academic and engineering professional with a profound expertise in Mechanical Engineering. He earned his Ph.D. from Andhra University College of Engineering and has amassed substantial teaching experience. Currently an Associate Professor at Lendi Institute of Engineering and Technology, he has previously served in various educational institutions. Dr. P. Thimothy is deeply committed to both teaching and research, with numerous publications in reputable journals showcasing his contributions to the field. His technical proficiency spans industry-standard software like AutoCAD, ANSYS, and CATIA, and he has undertaken multiple leadership roles on academic committees, underscoring his dedication to elevating the educational experience. Dr. P. Thimothy's commitment to fostering student growth and advancing innovative engineering practices reflects his passion for academic excellence and industry-relevant skills.

Seepana Praveenkumar has received his Bachelor's and Master's Degree from the Mechanical Engineering from 2011 to 2017. He obtained his Ph.D. from Ural Federal University, Russia in the area of energy systems and complexes on the year 2023. He is currently working in the department of Nuclear and Renewable Energy Sources. He published over 40 papers in the International, National and conferences in various SCI/SCIE/ESCI and Scopus indexed and also reviewer for various Elsevier and MDPI journals. His research interests are in the area of renewable energy preliminarily solar energy cooling technologies.

Chapter 5
Prediction of Truck Cabin Inspection Points Satisfying Tolerance Score

Nasim Ahmed, Vivek Verma, Sachin Kumar, and Kheelraj Pandey

Abstract This research focuses on improving the understanding and prediction of truck cabin inspection points that meet tolerance scores (PIST) using Coordinate Measuring Machines (CMM). The study integrates two main components: the development of an online digital platform for real-time aggregation and monitoring of PIST data from CMM inspections, and the creation of machine learning models for predicting PIST values. The platform provides continuous tracking of CMM data, enhancing the analysis and interpretation of inspection points. Machine learning models, built using Python-based regression techniques, are validated using performance metrics such as MAE, MSE, RMSE, and R^2, to predict PIST values with high accuracy. This approach streamlines the inspection process, reducing reliance on manual methods, while providing detailed visual insights into patterns affecting tolerance adherence. The ultimate goal is to optimize truck cabin quality, enhance inspection efficiency, and establish a robust, data-driven approach to PIST prediction and monitoring, introducing more precision and automation into the inspection process.

N. Ahmed · V. Verma (✉) · K. Pandey
Department of Mechanical Engineering, ASETL, Amity University, Lucknow, U.P, India
e-mail: vverma@lko.amity.edu

N. Ahmed
e-mail: nasim.ahmed@s.amity.edu

K. Pandey
e-mail: kpandey1@lko.amity.edu

S. Kumar
Department of Electronic and Communication Engineering, ASETL, Amity University, Lucknow, U.P, India
e-mail: skumar3@lko.amity.edu

© The Author(s), under exclusive license to Springer Nature Switzerland AG 2025
K. C. Rath et al. (eds.), *Engineering, Technology and Management*, Information Systems Engineering and Management 54, https://doi.org/10.1007/978-3-031-93654-8_5

Abbreviations

PIST	Percentage of Inspection Points Satisfying Tolerance
BIW	Body in White
CMM	Coordinate Measuring Machine
MAE	Mean Absolute Error
MSE	Mean Squared Error
RMSE	Root Mean Squared Error
R^2	R Squared Value

Preface
This chapter explores the enhancement of truck cabin inspection accuracy through the prediction of Tolerance Scores (PIST) using Coordinate Measuring Machines (CMM). The chapter introduces an innovative online platform for real-time data aggregation and employs machine learning models in Python to predict PIST with high precision. Validated through metrics such as MAE, MSE, RMSE, and R^2, this approach reduces manual inspection efforts and provides valuable insights into tolerance adherence patterns. The research establishes a scalable, data-driven framework that improves truck cabin quality and contributes to the automation of the automotive manufacturing inspection process.

5.1 Introduction

Truck cabins, forming the backbone of modern transportation, must meet stringent quality standards to withstand operational challenges such as long-haul journeys, varying weather conditions, and constant wear and tear. The rapid advancement of manufacturing technologies has significantly transformed the inspection processes in the automotive industry, particularly for heavy-duty vehicles like trucks. In the era of digital transformation and Industry 4.0, there is a growing need for more precise, automated, and data-driven solutions to replace conventional manual inspection methods. With increasing vehicle complexity, especially in truck cabin manufacturing, ensuring dimensional accuracy and quality has become a paramount concern. Accurate inspection not only affects the functionality of the truck but also contributes to safety, aerodynamic efficiency, and customer satisfaction.

The concept of Body-in-White (BIW) inspection, especially using Coordinate Measuring Machines (CMM), has evolved into a critical technique for evaluating the structural integrity of truck cabins. The Percentage of Inspection Points Satisfying Tolerance (PIST) score is a vital metric used in this context. It allows manufacturers to quantify the percentage of inspection points that fall within acceptable tolerance levels, providing a comprehensive evaluation of the truck cabin's dimensional accuracy. The goal is to ensure that as many inspection points as possible adhere to the predefined tolerance limits, resulting in superior build quality and structural stability.

As the automotive industry moves toward greater automation and smart manufacturing ecosystems, leveraging digital platforms for real-time monitoring of inspection data has become increasingly essential. Machine learning algorithms [1] offer the ability to predict PIST values with high precision, significantly reducing reliance on manual CMM assessments. This study aims to transform the truck cabin (BIW) inspection process by using digital innovation and machine learning methods. It seeks to develop a robust digital platform that serves as an integrated tool for monitoring and analyzing PIST data in real-time, harnessing machine learning algorithms to predict PIST values based on CMM inspection data accurately. This will reduce uncertainties inherent in manual assessments and streamline the inspection process with high precision.

In pursuit of these objectives, the project will utilize built-in machine learning libraries in Python to build regression-based models like Random Forest (RFR) and Gradient Boosting Regression (GBR) [1] to predict PIST values with high accuracy, optimizing the truck cabin inspection process, enabling higher quality standards, and enhancing quality control mechanisms. By seamlessly integrating digital analytics and machine learning tools into the platform, the project also strives to improve the ability to discern intricate patterns and trends within the data, further enabling proactive decision-making and ensuring consistent elevation in truck cabin quality.

5.1.1 BIW Manufacturing Process Layout

The Body in White (BIW) manufacturing process shown in Fig. 5.1a is a fundamental stage in automotive production, where the vehicle's structural framework is assembled before it moves on to painting and the addition of various other components. This process is critical because it provides the backbone of the vehicle, ensuring its structural integrity, safety, and performance. BIW manufacturing requires high precision in both engineering and assembly, and it involves several key stages, each focusing on different parts of the vehicle. Below is a detailed explanation of the BIW manufacturing process flow (Fig. 5.1b).

5.1.1.1 Front Wall Sub-assembly Process

The BIW process begins with the sub-assembly of the vehicle's front wall, which comprises essential components such as the front fenders, grille, and occasionally the headlights and their support structures. These components are generally assembled independently of the main body frame to maintain precision and ease of assembly in

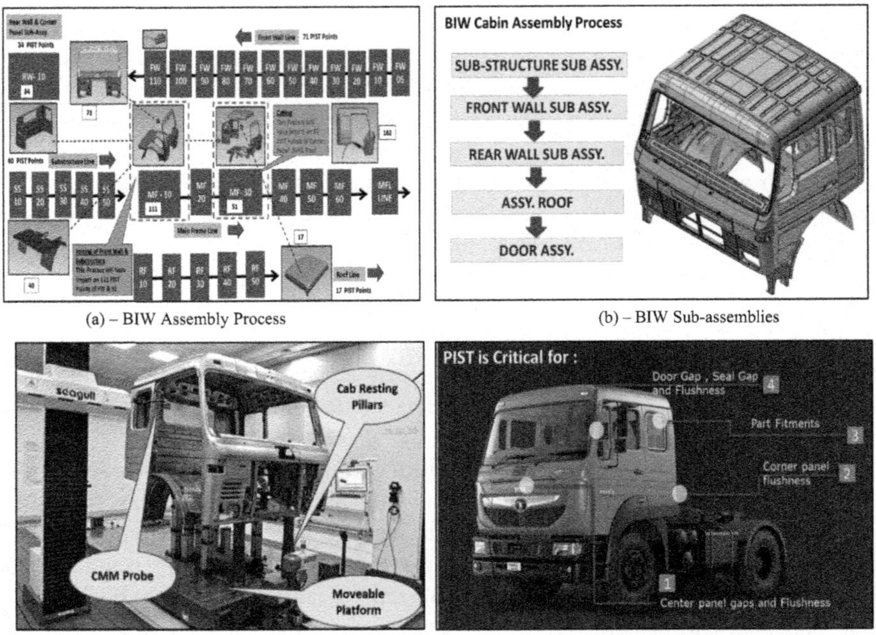

(a) – BIW Assembly Process (b) – BIW Sub-assemblies

(c) – CMM Lab PIST checking machine (d) – Benefits of PIST for Truck Cabin

Fig. 5.1 BIW assembly

subsequent stages. The front wall sub-assembly involves ensuring that each component is accurately positioned and aligned, as this section forms the front-facing structure of the vehicle. This precision is crucial for aesthetic and functional aspects, such as ensuring the correct fit of the engine hood and proper aerodynamic performance.

5.1.1.2 Substructure Sub-assembly Process

The substructure forms the lower part of the vehicle's body and is integral to the BIW process. It includes the floor panel, which provides the base for the cabin, and various other critical elements like the chassis mounting points and cross members. This part of the process is particularly significant because it sets the foundation for the vehicle's strength and stability. The substructure also serves as the base onto which suspension components and other mechanical systems will later be mounted, making it essential for supporting the weight of the vehicle and its passengers. Precision in assembling the substructure is vital to ensure the vehicle's durability and performance, especially under stress from road conditions or heavy loads.

5.1.1.3 Main Framer Station

The main framer station is where the core components of the vehicle's structure, including the front wall, substructure, roof, and side panels, come together. This is a pivotal point in the BIW process, as it is where the primary framework of the vehicle is built. At this stage, robotic and human-operated machinery work together to ensure that all components are aligned with utmost precision. Misalignments during this phase can result in compromised structural integrity or faulty vehicle dynamics. For instance, poor alignment can cause issues with door fitment, leading to wind noise or even safety risks. Therefore, achieving proper alignment at the main framer station is critical for ensuring the vehicle's overall structural cohesion and long-term durability.

5.1.1.4 Metal Finishing Line

After the vehicle's core structure is assembled, it moves to the metal finishing line for welding. Full MIG (Metal Inert Gas) welding is performed along the seams and joints of the structure. This welding process involves using an electric arc to fuse the metal components together, ensuring a strong and consistent bond. MIG welding is favored in automotive manufacturing for its ability to provide high-quality, durable welds. During this step, the entire vehicle body is meticulously welded to ensure that all parts are securely connected, enhancing the vehicle's rigidity and safety in case of collisions. The accuracy and quality of the welding at this stage are crucial for maintaining the structural integrity of the vehicle, preventing long-term issues like frame warping or stress fractures.

5.1.1.5 Assembly of Door and Engine Hood Enclosure on Cabin Body

Once the core structure has been welded, attention shifts to attaching critical external components such as the doors and the engine hood. This step involves carefully mounting the doors and the engine hood enclosure onto the cabin body. Special care is taken to ensure that components like hinges, handles, and latches are accurately installed to allow for smooth operation. Proper alignment is crucial in this stage to prevent issues like poor door sealing, which could affect aerodynamics, noise insulation, and waterproofing. The correct positioning of these parts also has a direct impact on the vehicle's overall aesthetics and customer satisfaction, as well-fitted doors and hoods enhance the perceived quality of the vehicle.

5.1.1.6 Inspection by Quality In-charge

Before the Body in White is considered complete, it undergoes a comprehensive inspection by the quality control team. This stage is crucial for ensuring that all parts of the vehicle body meet stringent quality standards. The inspection focuses on various parameters, including the accuracy of welding, the fitment of components, and the alignment of the vehicle's structure. Inspectors also check for any visual or functional defects that could compromise the vehicle's integrity or performance. Any issues discovered during this process are addressed immediately, ensuring that only high-quality bodies proceed to the next stage of vehicle assembly. The inspection process is thorough, as defects at this stage could have cascading effects on the vehicle's assembly and functionality in later stages.

5.1.1.7 Completion of the Process

After successfully passing all quality inspections, the vehicle's body is now considered a completed Body in White. This term refers to the bare structural body, without any paint, interior components, or powertrain elements. The BIW process concludes at this point, and the vehicle is now ready for the painting process and the addition of non-structural components. The high level of precision and quality control throughout the BIW process ensures that the body is structurally sound, ready to handle the rigors of the road, and capable of meeting safety standards. The completed BIW serves as the foundation for all subsequent processes, including painting, installation of the powertrain, interior trim, and the vehicle's final assembly.

The BIW manufacturing process is a meticulously planned and executed procedure that forms the backbone of vehicle production. It involves the careful assembly and welding of structural components, ensuring that the vehicle's frame meets high standards of safety, strength, and precision. By combining advanced engineering techniques with state-of-the-art technology, manufacturers are able to create robust vehicle bodies that are ready for the next stages of production. Each step in the BIW process, from the assembly of individual subcomponents to final quality inspection, plays a critical role in delivering vehicles that meet both aesthetic and functional requirements. Ultimately, the BIW process is foundational to ensuring that modern vehicles are not only visually appealing but also capable of delivering optimal performance and safety to consumers.

5.1.2 CMM Machine and its Key Components

A Coordinate Measuring Machine (CMM) as shown in Fig. 5.1c is a device that measures the physical dimensions of an object [2]. It is used in a variety of industries, including automotive, aerospace, and medical device manufacturing. CMMs are used to ensure that parts are manufactured to the correct specifications. These machines allow for the measurement of complex geometries with a high degree of accuracy, which is especially critical in industries where precision is paramount, such as truck cabin manufacturing. The importance of CMM technology lies in its ability to provide precise 3D measurements and coordinate data. This data is used to verify that manufactured parts meet their design specifications. Compared to manual methods of inspection, CMMs reduce human error, increase efficiency, and provide more consistent results [2]. They are crucial in quality control, where tolerance measurements can determine the overall safety and functionality of a vehicle. CMMs have evolved significantly over time, with modern machines being equipped with advanced software, higher resolution sensors, and automation features that enable real-time inspection processes. These machines can be integrated into a digital ecosystem, allowing the data to be analyzed and monitored through connected systems, thus contributing to the digitalization of the manufacturing process. There are a number of different components of a CMM machine, each of which plays a role in the measurement process [2]. These components include.

5.1.2.1 Base

The base of the CMM is the foundation of the machine. It is a heavy, stable platform that supports the other components of the machine. The stability of the base is essential as any vibrations or movements can affect the accuracy of the measurements.

5.1.2.2 Bridge

The bridge is the moving part of the CMM. It houses the measuring head and moves along the X-axis of the machine. Modern CMM bridges are designed to minimize deflections and ensure that even under load, the measurements remain accurate. Bridges can be made from various materials, including aluminum or ceramic, which offer different advantages in terms of weight and stiffness.

5.1.2.3 Column

The column is the vertical part of the CMM. It supports the bridge and moves along the Y-axis of the machine. Like the bridge, the column must be rigid and precise to ensure accurate movement along the axis.

5.1.2.4 Ram

The ram is the horizontal part of the CMM. It houses the measuring probe and moves along the Z-axis of the machine. The ram must be capable of delicate and precise movement to allow the probe to contact the object being measured at exact points.

5.1.2.5 Measuring Head

The measuring head is the part of the CMM that contains the measuring probe. It can be moved in all three axes to take measurements of the object. The measuring head often includes sensors that can record the precise position of the probe at any given moment.

5.1.2.6 Measuring Probe

The measuring probe is the part of the CMM that comes into contact with the object being measured. There are a variety of different types of measuring probes, each of which is designed for a specific type of measurement. Probes can be tactile, using physical contact to take measurements, or optical, using lasers or cameras to scan the surface without contact. Tactile probes are more commonly used in precision applications, while non-contact probes are ideal for measuring soft or delicate surfaces.

5.1.2.7 Controller

The controller is the computer that controls the CMM. It is responsible for moving the bridge, column, and ram, and for taking and recording measurements. Modern controllers are highly sophisticated and can interface with external systems for real-time data processing and feedback.

5.1.2.8 Software

The software is the program that runs on the controller. It allows the user to define the measurement routine, and to view and analyze the results. CMM software includes features like 3D visualization, automated reporting, and integration with other quality control systems. This software is essential for interpreting the large amounts of data that CMMs can generate, turning raw measurements into actionable insights.

The increasing automation of CMMs, coupled with software advancements, is revolutionizing the inspection process. The integration of CMMs with digital platforms allows for real-time monitoring, predictive analysis, and proactive decision-making in quality control. These technological advancements not only improve the efficiency of inspections but also significantly enhance the accuracy and reliability of the measurements. The CMM machine used in this study is manufactured by Accurate (Fig. 5.2), and its specifications are detailed in Table 5.1.

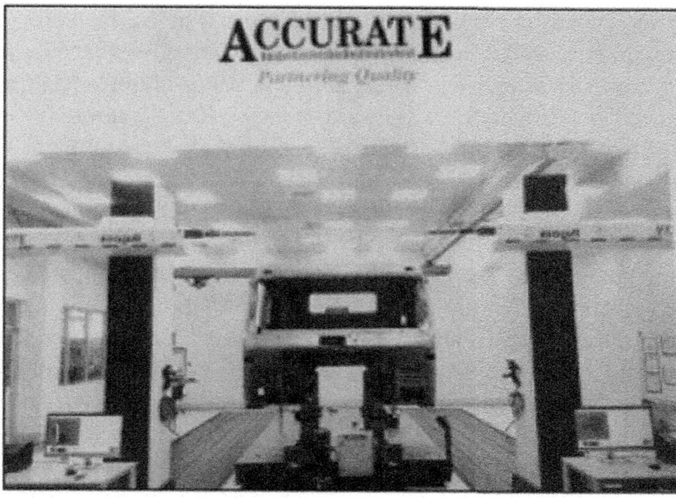

Fig. 5.2 CMM for truck cabin

Table 5.1 Accurate made CMM specifications

CMM parameters	Abbreviation	Value
Measuring range	X × Y × Z	7000 × 3000 × 2500
Machine length	L1	8500 mm
Machine width	W1	7300 mm
Machine height	H1	4260 mm
Ambient temperature	°C	18–26 °C
Humidity	[%]	40–70
Air supply	[Bar]	4.5 bar Pneumatic
Light illumination	Lux [lux]	400–600 lux

5.1.3 History and Evolution of CMM Inspection Methodology

The history of Coordinate Measuring Machine (CMM) inspection for truck cabins is deeply intertwined with the development of metrology and quality control in the automotive and manufacturing industries [3]. In the early days of automobile manufacturing, quality control relied on labor-intensive manual measurements and visual inspections. However, the introduction of CMMs in the mid-20th century marked a significant turning point, enabling faster and more accurate measurements with precise sensors and 3D capabilities [3]. As computer technology advanced, CMMs became integrated with digital systems, streamlining data analysis and reporting. Advancements in sensor technology further enhanced CMMs' capabilities, allowing them to capture intricate details [3]. Automation and robotics brought efficiency to CMM inspections, reducing human errors. In the era of Industry 4.0, CMMs are now seamlessly integrated into smart manufacturing ecosystems, facilitating real-time data exchange and adaptive manufacturing [3, 4]. This historical journey reflects the industry's pursuit of precision and quality assurance in truck cabin production, promising even more sophisticated inspection methods in the future.

As CMM technology continued to evolve, significant breakthroughs were made in the fields of multi-sensor integration and software optimization. Multi-sensor systems, which combine optical, laser, and tactile probes, allowed for the measurement of complex geometries that were previously challenging or impossible to inspect using traditional methods. These advancements expanded the applicability of CMMs in industries requiring high precision, such as aerospace and medical devices, and made CMMs indispensable tools in modern metrology. Furthermore, the development of virtual CMMs—software-based simulations of CMM processes—enabled the verification and validation of inspection routines before actual measurements were taken, thus reducing setup time and improving the efficiency of the entire inspection process. Virtual CMMs also allowed engineers to optimize probe paths and ensure collision-free inspections, thereby enhancing both accuracy and productivity.

With the advent of cloud computing and the Internet of Things (IoT), modern CMMs are becoming part of a connected ecosystem, enabling real-time monitoring and data analysis across multiple production facilities. This connectivity supports predictive maintenance, where CMMs can alert operators of potential issues before they result in downtime, thus ensuring continuous operation. Moreover, big data analytics applied to the vast amounts of data generated by CMMs allows manufacturers to identify trends and improve process controls, further elevating product quality. As CMM technology continues to advance, the focus is shifting towards full automation, where robotic arms and automated guided vehicles (AGVs) are increasingly being used to handle parts during inspection. This not only reduces labor costs but also enhances consistency by minimizing human intervention. Automated CMMs, integrated within flexible manufacturing systems, are capable of inspecting parts in-line, during production, rather than after the process, significantly reducing lead times and allowing for immediate corrective actions.

5.2 Body-In-White Pist Score

The BIW PIST score, or Body-in-White Percentage of Inspection Points Satisfying Tolerance score, is a quantitative metric used in quality control and assessment within the automotive manufacturing industry [4]. It is calculated as the percentage of inspection points on the BIW that are found to be within tolerance (usually ± 1.5 mm for truck cabins) during inspection using a Coordinate Measuring Machine (CMM) [4]. The BIW PIST score is more than just a quality metric; it is a key performance indicator in the optimization of the manufacturing process. A consistent and high PIST score directly correlates with the durability and overall performance of the truck cabin. A poor PIST score often signals issues with part alignment, structural integrity, or assembly precision, which can result in costly repairs, warranty claims,

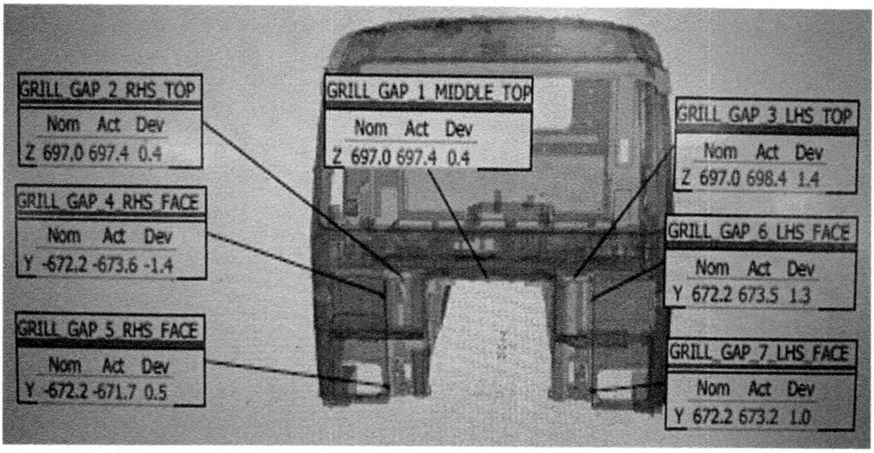

Fig. 5.3 CMM report

or even recalls. As such, continuous monitoring and improvement of the PIST score are integral to maintaining high production standards.

Moreover, in the context of large-scale manufacturing, especially in the automotive industry, the calculation of PIST involves determining the total number of inspection points (XTotal) on the BIW body and identifying how many of these inspection points (XOK) fall within the defined tolerance limits [4]. The PIST score is expressed as XOK/XTotalmultiplied by 100, providing a percentage value. A higher PIST score indicates a higher quality BIW assembly. PIST scores are used to track the quality of BIW assemblies over time and to identify areas where improvements can be made. There are four different types of PIST scores [4, 5].

5.2.1 A-rate PIST

This refers to the PIST of the BIW without hanging parts like doors and engine hoods as shown in Figs. 5.3 and 5.5(a).

5.2.2 B-rate PIST

This refers to the PIST of hanging parts like doors and engine hoods.

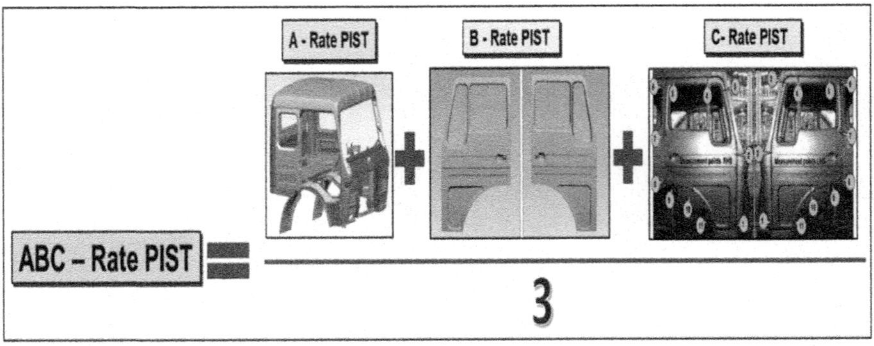

Fig. 5.4 Classification of BIW PIST scores

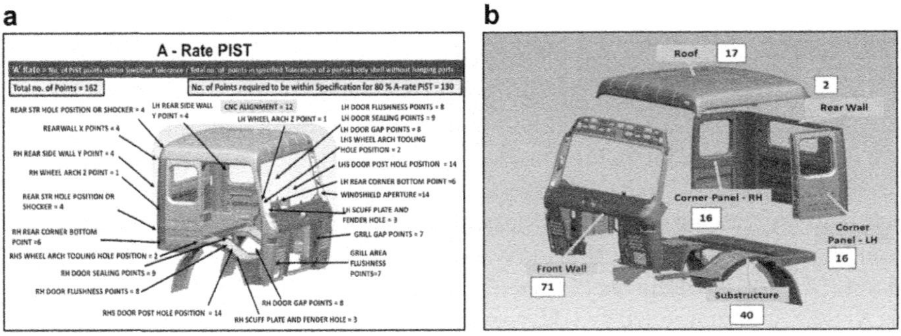

Fig. 5.5 **a** Location of PIST points on BIW **b** Distribution of PIST points on BIW

5.2.3 C-rate PIST

This refers to the door and engine hood gaps and flushness measured in a complete cabin made after assembly of A-rate BIW and B-rate hanging parts.

5.2.4 ABC Rate

ABC rate is average of all three A, B, and C rate PIST score as shown in Fig. 5.4.

The number of PIST points for A-rate, B-rate, and C-rate [4] in the Signa Truck BIW used in this study are provided in Table 5.2. Higher PIST score ensures proper door and seal gaps, promoting a more attractive appearance, better weatherproofing, and noise insulation [4] as shown in Fig. 5.4. PIST also contributes to the alignment and flushness of cabin components, positively impacting aerodynamics and structural integrity. A higher PIST score leads to improved overall vehicle finish, which can boost brand reputation and resale value [6].

Table 5.2 Number of PIST inspection points in Signa Truck

Type of PIST	No. of points	X	Y	Z
A-rate	162	35	84	43
B-rate	88	28	40	20
C-rate	34	12	12	10
Total	284	75	136	73

PIST scores provide valuable data for predictive maintenance and process optimization. By analyzing trends in PIST data, manufacturers can identify recurring defects or deviations from tolerance early in the production process. This data-driven approach enables quick adjustments to tooling, machinery, or assembly methods, minimizing downtime and reducing the likelihood of defects propagating through to the final product. Advanced statistical methods and machine learning algorithms can further enhance the utility of PIST scores. For example, by integrating PIST data with other manufacturing metrics, predictive models can be developed to forecast potential failures or areas of non-compliance. This proactive approach enables manufacturers to make real-time decisions that improve the quality of the BIW and reduce the need for post-production inspections.

5.3 Digitalization of PIST Inspection

The digitalization of PIST inspection is a transformative advancement in quality control and manufacturing. It involves using digital technologies, such as Coordinate Measuring Machines (CMMs) [7] integrated with computer systems and specialized software, to automate and streamline the inspection process. Digitalization not only accelerates data collection and analysis but also facilitates real-time monitoring, improving precision and overall quality assurance. This approach leverages data-driven insights to optimize truck cabin production, enhance component fit, and maintain high PIST scores, ultimately contributing to the efficiency and quality of the manufacturing process.

Digitalization has also paved the way for more collaborative and decentralized quality control processes. With data now available on digital platforms, teams across different geographical locations can access the same inspection results, facilitating better communication and faster decision-making. This level of connectivity is essential for large-scale manufacturers who operate multiple production lines or plants globally. By having centralized access to real-time data, issues related to PIST scores can be addressed swiftly, and corrective measures can be implemented before significant deviations occur.

The integration of artificial intelligence (AI) and machine learning (ML) into the digital inspection process has the potential to further enhance predictive maintenance capabilities. By analyzing historical PIST data alongside current measurements, ML algorithms can identify patterns that suggest potential defects or wear in manufacturing equipment. This predictive capability allows manufacturers to schedule maintenance at optimal times, reducing unexpected downtimes and ensuring continuous production efficiency.

A significant advantage of digitalized PIST inspections is the reduction of human errors associated with manual data entry and report generation. Automated data collection ensures that all measurements are accurately recorded and immediately available for analysis. This automation not only improves the accuracy of the inspection process but also frees up human resources to focus on higher-value tasks, such as data interpretation and process optimization.

As the automotive industry moves toward greater adoption of Industry 4.0 technologies, the role of digitalized PIST inspections is expected to expand. Future developments may include the use of augmented reality (AR) to assist operators in visualizing PIST data in real-time during the assembly process. This could provide immediate feedback on component alignments, ensuring that issues are corrected before they lead to quality defects.

5.3.1 Distribution of PIST Points on Truck Cabin

The allocation of CMM inspection points across the Truck Cabin (BIW) is determined by design specifications, manufacturing viability, and customer satisfaction criteria. The truck model (Tata Signa) used for analysis in this project has a total of 162 PIST [4] points divided into six sub-assemblies (Fig. 5.5(b)): Front Wall (71 points), Substructure (40 points), Roof (17 points), Left-Hand & Right-Hand Corner Panels (16 points each), and the Rear Wall (2 points). Each of these points plays a vital role in ensuring the overall precision of the cabin's dimensions and its visual appeal. Further details regarding the arrangement of these points in the X, Y, and Z dimensions are elaborated in Table 5.3.

The comprehensive distribution across multiple axes ensures that any deviations from tolerance can be identified early in the production process. For instance, the Front Wall sub-assembly, which comprises the majority of PIST points, is particularly critical for ensuring the structural integrity of the cabin. This part not only bears a significant load but also directly influences the cabin's aerodynamics and driver safety. In addition, the complexity of the inspection increases with the integration of different components. For example, in the corner panels, which connect the front wall to the side structures, slight misalignments can cascade into significant fitment issues during the final assembly. The allocation of inspection points in these areas is crucial for maintaining the overall cabin geometry.

The PIST distribution also takes into account factors such as thermal expansion and contraction that occur during welding processes, which can cause dimensional shifts in both the roof and substructure. Therefore, inspection points on these sub-assemblies are strategically placed to detect any post-welding distortions and ensure that corrective measures can be implemented.

Table 5.3 Sub-assembly wise distribution of PIST Points on BIW along X, Y and Z-axis

S.no	Sub-assembly	No. of PIST points	Number of X, Y and Z coordinates			Contribution to PIST (%)	Percentage contribution of X, Y and Z coordinates		
			X	Y	Z		X (%)	Y (%)	Z (%)
1	Front wall	71	15	36	20	44	9	22	12
2	Substructure	40	10	18	12	25	6	11	7
3	Corner panel—RH	16	4	11	1	10	2	7	1
4	Corner panel—LH	16	4	11	1	10	2	7	1
5	Rear wall	2	2	0	0	1.2	1	0	0
6	Roof	17	0	8	9	10	0	5	6
		162	35	84	43	100	22	52	27

5.3.2 *Development of Online Platform for PIST Digitalization*

Digitalization of CMM lab reports for truck cabin involves the process of converting traditional, paper-based reports generated by coordinate measuring machines into digital formats. This digital transformation has become increasingly important in the automotive industry, especially for truck cabin components, due to its potential benefits and the need for greater efficiency, accuracy, and data accessibility. Digitalization of CMM inspection reports through online platform enables real time monitoring & analysis of Truck Cabin PIST data. Availability of CMM reports on a digital platform along with data analytics tools significantly improves data storage, analysis and feedback mechanism and reduces time lost in data handling through excel files. In this project we developed an online platform called DataNest where data can be seamlessly collected, retrieved & available for analytics purpose (Fig. 5.6a) [8, 9].

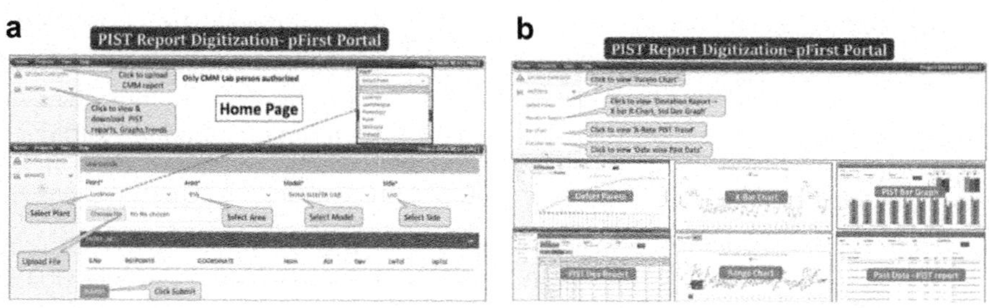

Fig. 5.6 Online platform (Data Nest) for PIST data analysis: **a** Data retrieval system **b** Data analytics tools

In addition to real-time data retrieval, the platform offers enhanced data visualization capabilities that are crucial for decision-making in manufacturing environments. The use of dashboards provides users with an intuitive interface to monitor the performance of PIST points across multiple truck cabins. These dashboards can be customized based on specific user preferences, allowing for the display of critical metrics such as average PIST score trends, sub-assembly deviations, and inspection point conformity. By presenting data in a visually accessible manner, engineers and quality control teams can quickly identify areas requiring immediate attention and intervene before production issues escalate.

5.3.3 Integration of Online PIST Digitalization Platform with CMM Lab

The integration of this online PIST Digitalization platform with a CMM machine significantly enhanced the efficiency and accuracy of PIST inspection processes. By seamlessly connecting these two systems, inspection data is automatically transferred from the CMM Lab to the online platform (DataNest), enabling real-time monitoring and analysis of PIST scores. The online platform developed has capabilities for analyzing PIST trends for selected period, retrieve PIST data stored online, generate Pareto charts of location wise PIST points, X-bar charts, Range charts, PIST deviation & various other data analytics reports (Fig. 5.6b). It also has other features like sending email notification & CMM reports to the concerned agencies for product improvement action planning, storage of CMM reports of BIW assembly fixtures & shim records of the changes made in the process for PIST score improvement [4, 8, 9].

5.4 Development of Machine Learning Model for PIST Score Prediction

Developing a mathematical equation for a machine learning model [10, 11] to predict the Truck Cabin PIST score involves creating a regression model that estimates a continuous numerical value based on input features. The first step is defining a mathematical equation that depicts the relationship between input and output variables. In this case, the output variable is the Cabin PIST, while the input variables are the measured X, Y, Z coordinates [4], Front Wall PIST, and Substructure PIST values. We defined the mathematical equation using the data in Table 5.3 to describe the relationship between Cabin PIST and its sub-assemblies, calling it the PIST equation as shown in Fig. 5.7a.

$$\text{Cabin} = \frac{\textbf{No. of OK PIST Points}}{\textbf{Total no. of PIST Points}} = \frac{\textbf{No. of OK Points}}{162} \tag{5.1}$$

$$\text{Cabin PIST} = \frac{\mathbf{FW + SS + CP + RF}}{\mathbf{162}} = \frac{\text{FW}}{162} + \frac{\text{SS}}{162} + \frac{\text{CP}}{162} + \frac{\text{RF}}{162} \tag{5.2}$$

$$\text{Cabin PIST} = \frac{71}{162} \times \frac{\text{FW}}{71} + \frac{40}{162} \times \frac{\text{SS}}{40} + \frac{34}{162} \times \frac{\text{CP}}{34} + \frac{17}{162} \times \frac{\text{CP}}{17} \tag{5.3}$$

$$\text{Cabin PIST} = 0.44\ \text{F} + 0.25\ \text{S} + 0.21\ \text{C} + 0.10\ \text{R} \tag{5.4}$$

where:

- FW = No. of OK PIST Points on Front Wall
- SS = No. of OK PIST Points on Substructure
- CP = No. of OK PIST Points on Corner Panel
- RF = No. of OK PIST Points on Roof
- F, S, C, and R represent the respective PIST scores for these sub-assemblies.

Figure 5.9 shows the mathematical calculation of the overall Cabin PIST value by substituting specific scores for each component into the equation. The formula is:

$$\text{CabinPIST} = 0.44\text{F} + 0.25\text{S} + 0.21\text{C} + 0.10\text{R} \tag{5.5}$$

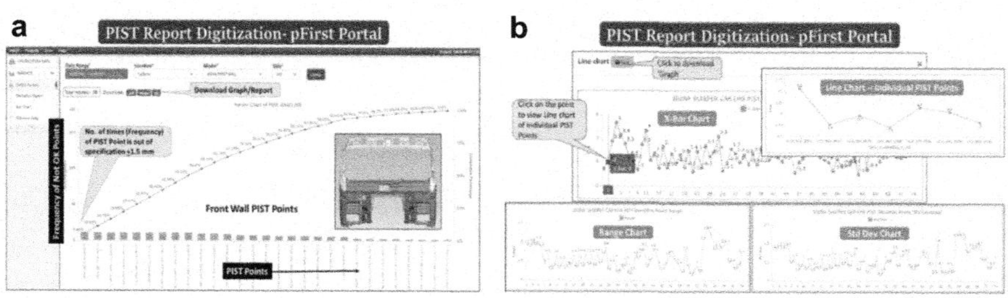

Fig. 5.7 Online platform (Data Nest) for PIST data analysis **a** Sub-assembly PIST Pareto **b** Control charts of PIST points

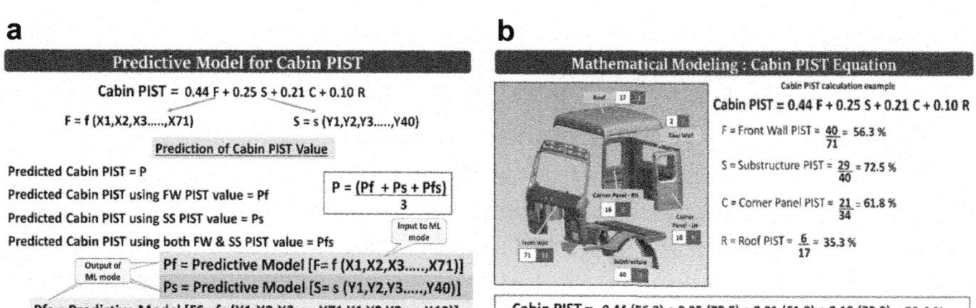

Fig. 5.8 **a** Relationship between input and out variable **b** PIST value calculation example

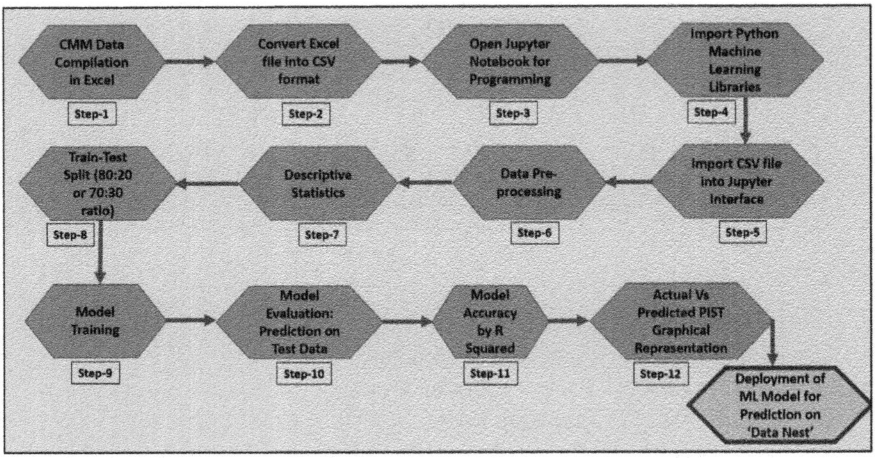

Fig. 5.9 Steps for developing machine learning model

In the example provided:

- F (Front Wall PIST): 56.3% from a total of 71 points
- S (Substructure PIST): 72.5% from a total of 40 points
- C (Corner Panel PIST): 61.8% from a total of 34 points
- R (Roof PIST): 35.3% from a total of 17 points

Substituting these values into the formula gives the overall Cabin PIST value as 59.4%. This approach effectively quantifies the overall quality of the truck cabin based on multiple component inspections and their respective importance in the total score.

To develop the machine learning models, we denote the target or output variable (Truck Cabin PIST Value) as Y, and the input features as X1, X2, X3,…, Xn. Two Python-based machine learning models—Random Forest and Gradient Boosting Regression—were used to predict Cabin PIST using these input variables [8, 9]. The key steps involved in developing the regression-based models are explained in the flowchart (Fig. 5.9).

After importing the data into the Python environment, PIST data is cleaned and pre-processed using techniques such as normalization, scaling, filtering, smoothing, outlier detection, and missing value imputation, ensuring the dataset is consistent and suitable for machine learning algorithms (Table 5.4) [12]. In the descriptive statistics shown in Table 5.5, we calculated the mean, standard deviation, and minimax values of input and output variables. The Cabin PIST data was visualized using a density plot as shown in Fig. 5.10, which indicated negative kurtosis (−0.50), suggesting a slightly flatter than normal distribution, and negative skewness (−0.18), showing the distribution was slightly left-skewed, with a mean value of 66.5 and a standard deviation of 8.09 (Fig. 5.10). The below steps provide a comprehensive framework for predicting PIST values using a machine learning model [13, 14].

Table 5.4 Data analysis using python-libraries-importing input and output variables

Input variables																Output variable
FW + SS	X1	X2	X3	X4	X5	X6	…	S36	S37	S38	S39	S40	FW PIST	SS PIST		Cab PIST
FW + SS–1	−1097.2	1373.2	−1123	889.2	−1123.4	769.2	…	1194.4	1193.9	−54.2	1183.3	499.2	56	65		59
FW + SS–2	−1096.4	1374.7	−1123	890.8	−1123.5	770.7	…	1192.8	1194.4	−55.3	1187.3	500.1	61	68		57
FW + SS–3	−1098	1376.1	−1124.6	891.3	−1125	771.3	…	1194	1194.4	−53.7	1185.6	496.7	55	65		56
FW + SS–4	−1098.9	1376.2	−1125.8	891.5	−1126.3	771.4	…	1193.9	1194.2	−53.5	1184.9	496.1	58	68		59
FW + SS–5	−1096.6	1374.5	−1122	890.2	−1122.5	770.2	…	1193.6	1195	−54.6	1186.1	496.9	58	70		54

Table 5.5 Descriptive statistics of PIST data (both input and output variables)

	X1	X2	X3	X4	X5	…	S38	S39	S40	FW PIST	SS PIST	Cab PIST
count	403	403	403	403	403	…	403	403	403	403	403	403
mean	−1097.3	1373.96	−1124.2	889.05	−1124.8	…	−50.72	1188.39	495.49	60.60	76.84	66.57
std	1.249	1.371	1.29	1.419	1.189	…	2.019	0.923	1.478	9.033	6.173	8.097
min	−1100	1371	−1127.6	885.8	−1128	…	−56.1	1185.9	492.6	39	63	44
25%	−1098	1373	−1125	888.05	−1125.5	…	−51.9	1187.7	494.5	54	73	61
50%	−1097.2	1374	−1124.2	889.2	−1124.8	…	−50.7	1188.4	495.3	61	75	68
75%	−1096.4	1374.7	−1123.5	889.7	−1124.2	…	−49.3	1189.1	496.3	68	83	72
max	−1094.7	1377.8	−1120.8	892.9	−1121.9	…	−47	1190.4	500.1	77	90	84

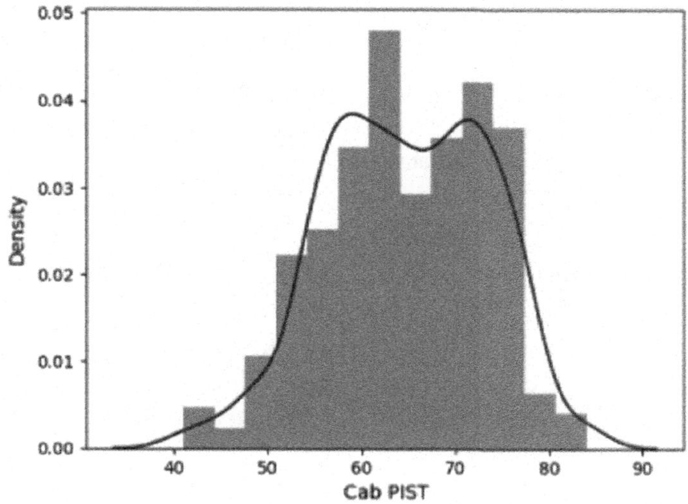

Fig. 5.10 Cab PIST data density plot

5.4.1 Importing Required Libraries

Essential libraries like numpy, pandas (for data manipulation), matplotlib, seaborn (for visualization), and sklearn (for machine learning tasks) were imported.

5.4.2 Loading the Data

The data was loaded from a CSV file, with the chardet library used for detecting encoding. The initial few rows of the dataset were displayed for structure analysis.

5.4.3 Data Preprocessing

Missing values were handled appropriately, and outliers were identified and removed using Z-scores, maintaining data integrity. The distribution of the target variable, Cabin PIST, was visualized through density plots for better insight.

5.4.4 Feature Selection and Splitting

The dataset was divided into input features (X) and the target variable (Y). Data splitting into training and testing sets was done using sklearn's train_test_split function, enabling evaluation on separate datasets.

5.4.5 Model Training

The Random Forest Regressor model was trained using the fit method, identifying patterns in the training data for predictive capabilities.

5.4.6 Model Evaluation

The model's performance was evaluated on the test set using metrics such as Mean Absolute Error (MAE), Mean Squared Error (MSE), Root Mean Squared Error (RMSE), and R-squared (R^2). These metrics provided insights into model accuracy.

5.4.7 Comparison and Visualization

Actual versus predicted PIST values were compared and visualized via scatter plots, helping to identify any prediction discrepancies.

5.5 Visualization of Predictive Model on Data Nest

Visualizing the graphical user interface (GUI) for a machine learning predictive model involves designing an intuitive, interactive platform where users can input data, select model parameters, and view predictions and accuracy results [14]. This process focuses on creating a seamless user experience, making the machine learning model accessible and functional for non-technical users. Below is a detailed explanation of how the GUI for PIST Prediction machine learning models (Fig. 5.11) have been structured to meet these goals.

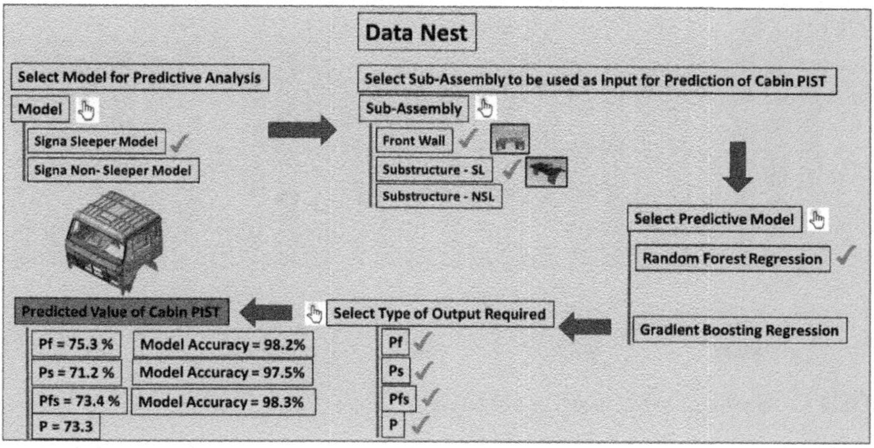

Fig. 5.11 Data Nest

5.5.1 *Selecting the Truck Model for Predictive Analysis*

The GUI features a clear, user-friendly dropdown menu or selection box allowing users to choose a specific truck model for predictive analysis. This selection defines the dataset and configurations the model will use for predictions. By pre-loading datasets based on the selected truck model, the system ensures that the relevant features and specifications are applied, tailoring the prediction process to the unique characteristics of each model. The flexibility of this option enables users to easily switch between models and perform comparative analysis.

5.5.2 *Choosing the Truck Sub-assembly as Input for Cabin PIST Prediction*

The second dropdown or selection box enables users to select the specific sub-assembly (e.g., front wall, roof, substructure) that will serve as the input for predicting the Cabin PIST score. Each sub-assembly has unique dimensions and characteristics, so selecting the correct component ensures that the machine learning model receives appropriate input data. This feature allows users to experiment with different sub-assemblies to see how they impact the overall PIST prediction, providing insights into which parts contribute most to the vehicle's quality metrics.

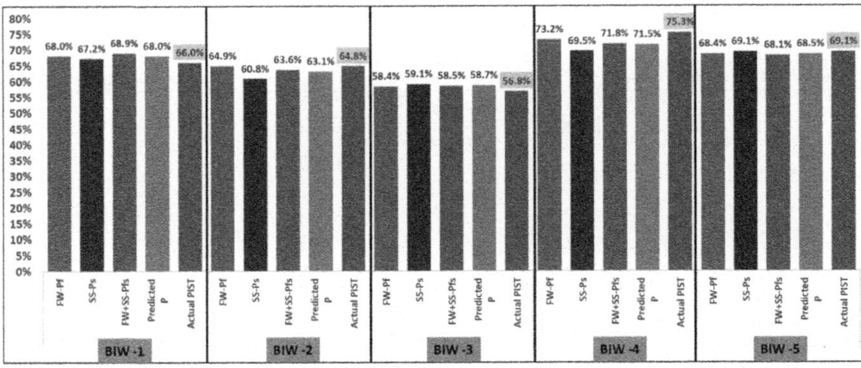

Fig. 5.12 Machine learning model on Data Nest

5.5.3 Selecting the Type of Regression Model

Users can select the type of regression model they want to apply for predictions. This can be implemented as a set of radio buttons or a dropdown menu offering several machine learning models. Providing these options in the GUI gives users the flexibility to tailor their analysis based on data type and the nature of the prediction needed, ensuring the model accommodates a wide range of tasks.

5.5.4 Selecting the Type of Output Required

Users must specify the type of output they require from the model. Options include the predicted Cabin PIST value, graphical visualizations of prediction trends over time, or comparisons of predicted versus actual values. Output options should prioritize clarity and usability, offering both numerical results and visual aids such as graphs or charts for easy interpretation as shown in Fig. 5.12. It also allows users to export data in formats like CSV or Excel which enhances usability for detailed reporting.

5.5.5 Displaying the Predicted Value of Cabin PIST and Calculating Accuracy

A key feature of the GUI is a designated section where the predicted Cabin PIST value is displayed after the user inputs data and selects the model.

Rigorous testing has been done to validate the GUI's functionality. The interface was tested under various conditions to ensure accurate predictions, clear visualizations, and an overall intuitive user experience. This ensures the GUI will be a reliable tool for users performing predictive analysis with machine learning models [15].

5.6 Results

In this section, we present a comparative analysis of the performance of the two machine learning models developed for Cabin PIST prediction: Random Forest & Gradient Boosting Regression algorithms. We have made three machine learning models for each of the regression algorithm using sub-sets of the PIST data. The three models are: (a) Cab PIST prediction using both Front Wall (FW) & Substructure (SS) data points, (b) Cab PIST prediction using Front Wall (FW) data points only & (c) Cab PIST prediction using Substructure (SS) data points only. The performance of these machine learning models is evaluated & compared using metrics like MAE, MSE, RMSEand R^2 values (Fig. 5.13).

Random Forest Regression (RF): The RF models predicted Cabin PIST with a very high accuracy in all the three cases & with very low error values of MAE, MSE, and RMSE. It achieves the R^2 value between 0.98 and 0.99, indicating that more than 98% of the variance in the Cab PIST data is explained by the model (Fig. 5.14). This high R^2 value signifies a strong linear relationship between the predicted and actual PIST values, highlighting the model's accuracy and reliability.

Gradient Boosting Regression (GBR): The GBR model also performs well with very high accuracy (above 97%). The R^2 value of above 0.97, indicates that more than 97% of the variance in the Cab PIST data is explained by the model (Fig. 5.14).

The prediction results demonstrate that rather than conducting CMM inspections on the Complete Cabin, focusing on CMM inspection of both the sub-assemblies

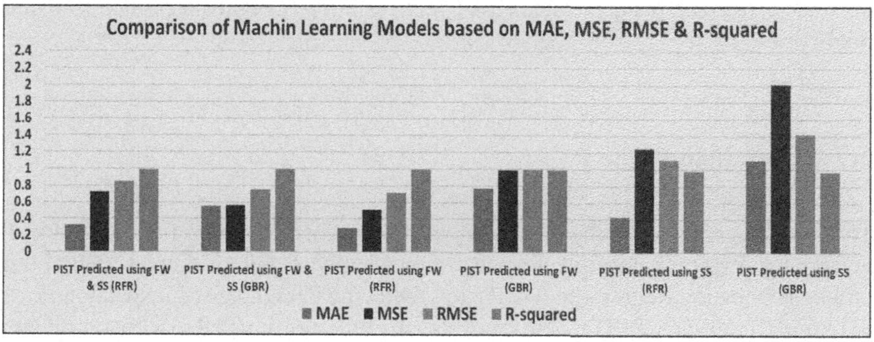

Fig. 5.13 Comparison of machine learning models (Random Forest and Gradient Boosting regression)

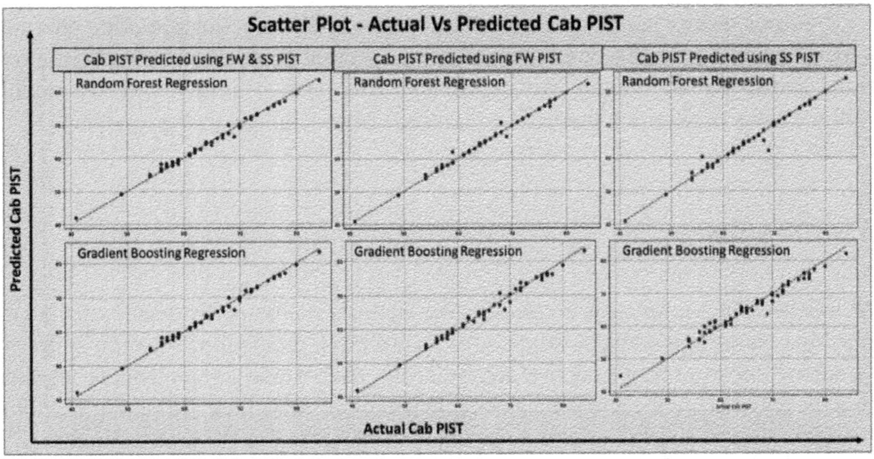

Fig. 5.14 Scatter plot between actual and predicted PIST values (Random Forest and Gradient Boosting regression

(Front Wall and Substructure) PIST data points or either of the Sub-assemblies allows for highly accurate prediction of the Cabin PIST value. This approach not only reduces CMM inspection time but also offers insights for proactive measures to enhance Cabin PIST quality before the complete Cabin assembly (BIW). Analysis of various machine learning models indicates that as the number of input variables decreases (in the case of sub-assembly-based models), the error tends to rise, with the GBR model (based on SS data points) exhibiting the highest MAE, MSE, and RMSE error values. Selection of an appropriate model can be made based on comparative charts, considering the desired accuracy level and inspection time for predicting Cabin PIST. Scatter plots comparing actual versus predicted Cab PIST values reveal tight clustering around the diagonal line through the origin, indicating strong agreement between actual and predicted PIST values with minimal deviation (Fig. 5.14). The tight clustering of data points in the scatter plot indicates that RF & GBR captures the underlying patterns in the data effectively, resulting in highly accurate PIST value predictions.

5.7 Conclusion

This project centered on the development of advanced Machine Learning models, leveraging Python's built-in libraries, specifically Random Forest (RFR) and Gradient Boosting Regression (GBR), to predict the Percentage of Inspection Points Satisfying Tolerance (PIST) for truck cabins. By utilizing detailed PIST data collected from critical sub-assemblies, including the Front Wall and Substructure of the cabin, the models demonstrated exceptional predictive accuracy, exceeding 97%. This high level of precision underscores the models' capability to effectively capture complex patterns in the inspection data.

The impressive performance of the Random Forest and Gradient Boosting Regression models can largely be attributed to their ensemble learning approach, which enhances the robustness of predictions by combining multiple decision trees. This method minimizes errors that could arise from individual trees, resulting in highly reliable and accurate predictions. The ability of these models to synthesize vast amounts of data and generate precise forecasts holds immense potential for revolutionizing quality control processes in automotive manufacturing. By providing data-driven insights, these models can help identify inefficiencies, optimize inspection procedures, and ensure higher standards of product quality.

One of the key highlights of this project is the demonstration of how regression-based machine learning models can be applied to real-world manufacturing scenarios, emphasizing the importance of continuous process improvement. The models not only reduce the reliance on manual inspections, which can be time-consuming and prone to human error, but also pave the way for more strategic, data-informed decision-making. This shift toward automated, data-driven quality control offers manufacturers a significant advantage, enabling faster, more precise identification of defects and allowing for proactive measures to improve the production process. This project represents a significant leap forward in improving both the efficiency and quality of truck cabin manufacturing. By showcasing the practical application of machine learning in this field, it highlights how data science and artificial intelligence can be harnessed to enhance operational workflows, reduce costs, and elevate the overall product quality. Moreover, the success of these models underscores the broader potential for machine learning technologies to be integrated into various stages of automotive production, fostering innovation and driving industry-wide improvements.

5.7.1 Future Scope

Looking ahead, the scope of this work can be expanded by incorporating additional vehicle components into the predictive models. This could be achieved by employing more advanced techniques, such as deep learning algorithms, which are capable of handling even larger and more complex datasets. Furthermore, the integration of real-time data collection tools, such as the Internet of Things (IoT) devices, combined with big data analytics, could further refine the accuracy of predictions and enable real-time monitoring of manufacturing processes. As the models are scaled to handle larger datasets, their ability to optimize quality control in diverse manufacturing environments will only increase, marking a new era of data-driven decision-making in the automotive industry.

References

1. Kumar, U.D., Pradhan, M.: Machine Learning using Python. Wiley, India (2019)
2. Artkin, F.: CMM machines and industrial applications. In: 1st International Conference on Engineering, Natural and Social Sciences (2023)
3. Zhang, B., Xu, S., Jin, X.: Research review of coordinate measuring machines. J. Sci. Res. Rep. 25–34 (2022) https://doi.org/10.9734/jsrr/2022/v28i430513
4. Tata Motors: BIW CMM Inspection Methodology (2012)
5. Mantel, M.R.: Coordinate Measuring Machines: A Modern Inspection Tool in Manufacturing, A Thesis New Jersey Institute of Technology, Degree of Master of Science in Manufacturing Systems Engineering (1993)
6. Mears, L., Roth, J.T., Djurdjanovic, D., Yang, X., and Kurfess, T.: Quality and inspection of machining operations: CMM integration to the machine tool. ASME. J. Manuf. Sci. Eng. 131(5): 051006 (2009)
7. Imkamp, D., et al.: Challenges and trends in manufacturing measurement technology–the "Industrie 4.0" concept. J. Sens. Sens. Syst. **5**(2), 325–335 (2016)
8. Brijith, A.: Data Preprocessing for Machine Learning (2023)
9. Francisco P.: A review of data mining, big data analytics and machine learning approaches. J. Comput. Nat. Sci. 169–18110.53759/181X/JCNS202303016
10. Carbonell, J.G., Michalski, R.S., Mitchell, T.M.: An overview of machine learning. Mach. Learn. 3–23 (1983)
11. Liu, Y., Wang, Y., Zhang, J.: New machine learning algorithm: Random Forest. In: Information Computing and Applications: Third International Conference, ICICA 2012, Chengde, China, (2012)
12. Alloghani, M., et al.: A systematic review on supervised and unsupervised machine learning algorithms for data science. Supervised Unsupervised Learn. Data Sci. 3–21 (2020)
13. Mohanty, S., Mohanta, G.K., Senapati, A.K., Rath, K.C.: Taguchi grey relational analysis for optimal process parameter of low carbon steel welded by GTAW by using ER309L filler material. Mater. Today: Proc. 44: 2556–2561 (2021)
14. Jaiswal, S.K., Mukti, S.K., Rath, K.C.: E-waste control and its recycling to build sustainable society in the global context. In Handbook of Research on Applications of AI, Digital Twin, and Internet of Things for Sustainable Development, IGI Global, pp. 200–222 (2023)
15. Rath, K.C., Khang, A., Kumar, S., Satapathy, A.K., Das, S.R., Panda, M.R.: 6 Enabling the future of manufacturing. Mach. Vis. Ind. Robot. Manuf.: Approaches Technol. Appl. 101 (2024)

Nasim Ahmed is an M.Tech scholar in Electric Vehicle Technology at the Department of Mechanical Engineering, Amity School of Engineering and Technology, Amity University, Lucknow. He holds a B.Tech in Mechanical Engineering from the National Institute of Technology, Tiruchirappalli, Tamil Nadu, India. Currently, he is working as a senior manager in Quality Operations at Tata Motors Ltd, Commercial Vehicle, Lucknow Plant, with a total experience of over 8 years. He has completed various Six Sigma projects related to quality improvements in truck cabins. His research interests include data analytics and machine learning, with a focus on utilizing these tools for quality improvement.

Vivek Verma is a part-time research scholar at the Department of Mechanical Engineering at Amity School of Engineering and Technology, Amity University, Lucknow. He holds B.E. in Industrial Engineering from IIT Roorkee (formerly University of Roorkee), India and M.Tech. in Production Engineering from MNNIT, Allahabad, India. Currently he is working as Assistant

Professor at Department of Mechanical Engineering, ASET, Amity University, Lucknow, has vast experience of teaching about 20 years. He used to hold several administrative posts at the Department and University levels. He is a member of several professional bodies such as SAE, IET (UK), and IAENG. He has supervised 10 masters and more than 50 undergraduate students. He has organized several FDP consultancy projects as well. He has research interest in mathematical modelling of mechanical systems and computational fluid dynamics.

Dr. Sachin Kumar a Ph.D. in Electronics, M.Tech. and B.E in Electronics and Communication, presently with Amity University, Lucknow Campus. He has almost 20 years of academic and research experience. His research interest includes applications of bio-medical signal processing, machine learning, artificial intelligence and applications of wireless sensor networks. He is an active member of number of professional bodies IEEE, IETE, IAENG, ISC and IET (UK), he has published more than 80 research papers, book chapters, and articles in a number reputed Journals indexed in SCOPUS and SCI and contributed 2 books with international publishers. He is an active member of editorial board/reviewer in various reputed journals. He has been instrumental in organizing number of international conferences, has given invited talks, and chaired various technical sessions in national as well as international conferences of repute.

Dr. Kheelraj Pandey Ph.D. in Production Engineering form Indian Institute of Technology Delhi, having specialization in Nano Finishing, presently working with Amity University, Lucknow Campus. He has almost 15 years of academic and research experience. His research interest includes applications nano finishing, polishing and battery thermal management system. He has published more than 15 research papers, book chapters, and articles in a number of reputed Journals indexed in SCI. He is an active member of editorial board/reviewer in various reputed SCI journals. He has two patents under his name on nano polishing of Silicon, Aluminum and Copper. He is guiding four Ph.D. scholar and guided 5 M.Tech. students.

Chapter 6
Determination of Barriers Affecting the Circular Supply Chain Adoption Towards Productivity Improvement in Automotive Sector

Rita Nagwal, Ravindra Pathak, Kumar Rohit, and Roopesh Kumar Mehra

Abstract This research systematically explored and analyzed the critical connections between the challenges of implementing circular supply chain (CSC) practices and the influence of Industry 4.0 technologies on CSC performance. A comprehensive investigation was conducted to demonstrate the impact of CSC adoption on advancing the circular economy (CE). The primary focus of this study was to identify and classify the significant barriers to CSC implementation, particularly in the context of their potential effects on the adoption of such practices within automotive companies. CSC adoption is crucial, as it aimed at sustainable growth so as to increase the overall performance while emphasizing on environmental impact by deploying practices such as, reduce, reuse, recycle, remanufacture, refurbish, etc. Finding a strategic framework for the study and analysis of CSC implementation challenges in small and medium-sized firms (SME) that impede sustainability in the Indian sector has been the target of the research effort. To identify a deeper knowledge of CSC implementation in the Indian context, obstructions to CSC implementation and expected performance outcomes have been identified and analyzed based on a thorough research of the literature. The primary focus of the research is on the performance outcomes and barriers associated with implementation of CSC in India's small

R. Nagwal · R. Pathak
Department of Mechanical Engineering, Medi-Caps University, Indore, MP, India
e-mail: enggrita@gmail.com

R. Pathak
e-mail: ravindra13479@gmail.com

K. Rohit (✉)
Department of Industrial and Production Engineering (IPE), Shri G S Institute of Technology and Science, Indore, MP, India
e-mail: rohitkumar.mechy@gmail.com

R. K. Mehra
Mammoth Equipment and Exhausts, Perth, WA, Australia
e-mail: Roopesh.Mehra@mammothequip.com.au

© The Author(s), under exclusive license to Springer Nature Switzerland AG 2025

K. C. Rath et al. (eds.), *Engineering, Technology and Management*, Information Systems Engineering and Management 54, https://doi.org/10.1007/978-3-031-93654-8_6

medium. This study focuses on identifying and prioritizing the barriers to the adoption of circular supply chains (CSC) in the automotive industry, with sustainability as a central consideration. Using the analytic hierarchy process (AHP), a prominent multi-criteria decision-making (MCDM) method, the research systematically categorized these obstacles. The findings highlight the importance of recognizing and ranking critical challenges to CSC implementation through pairwise comparisons and scoring, along with assessing their impact on CSC performance metrics. This research will identify a number of significant challenges that need to be addressed in order to successfully implement CSC practices that strive for total sustainability in automotive manufacturing companies, not only in India but also internationally.

Keywords Circular supply chain · CSC barriers · MCDM technique · Analytical hierarchy process · Productivity · Sustainability

Preface
This chapter substantially identified and investigated the core linkage between the barriers to circular supply chain deployment based on Industry 4.0 technological solutions and how they affect the circular supply chain performance in the automotive sector. To demonstrate how the implementation of circular supply chain employing Industry 4.0 technologies will impact the circular economy, one of the multi-criteria decision making (MCDM) techniques, analytic hierarchy process (AHP) was performed to deduce and highlight the key outcomes related to circular supply chain performance of the automotive sector.

6.1 About the Circular Supply Chain

Citing the contemporary world trends, the core notion of circular supply chains (CSC) is extensively researched. It is due to the challenges posed by excessive waste production, unfavorable environmental circumstances, and growing demand for natural resources. Due to a number of issues influencing its functionality, the deployment of CSC is a challenging undertaking. On the other hand, according to the extant literature in circulation, the main barriers to CSC adoption specifically for SMEs in developing countries remain unexplored. The use of Internet and information technologies in education as tools for instruction is currently growing at a rapid pace. One of the most widely used province of educational the knowledge era is virtual learning, or e-learning [1, 2]. In fact, e-learning uses computer and Web technologies to expand conventional learning paradigms into new dynamic learning models [3, 4]. Supply chain management involves overseeing and coordinating the various participants within the supply chain including distributors, manufacturers, retailers, and suppliers and [5] Customer which focuses on improving the overall performance of Industry and helps it to achieve its objectives [6, 7]. In linear supply chain the product flows from the supplier to customer, but in Circular Supply Chain once the product is completely used [8] and its has reached its end of life then it flows from the customer, and back to the manufacturer so that it can be refurbished, recycled or remanufactured in any way and made suitable for use again, so that overall waste is reduced and a cleaner and greener approach is used by the industry [9, 10]. The SMEs face significant obstacles or problems while implementing CSC techniques in their companies that offer Industry 4.0 solutions. In order to get clarity on the aforementioned research domain, these issues impacting the prospects for CSC implementation using fourth industrial revolution based on actual fashion have been carefully examined and are handled as such; Lack of Awareness: There's not much awareness of farmers or sustainable raw supplies, the organization has a poor vision, the benefits are unclear, and expertise is inadequate. Understanding the importance of Industry 4.0 is necessary for SMEs to successfully implement CSC practices [11, 17]. In a supply chain there are certain barriers which hurdles the flow of either information or product and hence reduces the overall profit and performance of an industry [10–12]. Therefore, it is very important for an industry to find these barriers and classify them so as to overcome them and increase productivity in the following step which is crucial for its growth and performance [13, 14]. Capacity and resource efficiency must be improved in order to create circular supply chains. The only way to support CSCM practices is through cross-sector collaboration [15, 16].

6.2 Implementation of Circular Supply Chains

These variables' dynamic and complicated nature makes supply chains more uncertain [17, 18]. To tackle these issues, Industry 4.0 technologies have been introduced to transform supply chains and create interconnected digital networks. Research indicates that big data analytics play a crucial role in fostering trust, cultural alignment, and collaborative behaviors within cross-industry networks and logistics chains. In the context of municipal logistics, smart collection applications leveraging Internet of Things (IoT) technology have been developed for effective waste management. Furthermore, recent studies highlight the circular economy (CE) as a strategic approach for businesses to transition from traditional linear economic models to circular systems. The CE aims to promote sustainable industrial practices that positively impact the environment while fostering a socially equitable economy. Researches that have concentrated on numerous uses of the CE have revealed its significant benefits compared to the linear system, of this economic system. Several industries have operationalized the CE concept: supply chain [26], manufacturing [27, 28], construction and services [25]. However, very few studies (24) have looked at how to apply CE in connection to supply chain operations. Regenerative methods take into account changing materials into a new condition [25] or maximizing a product's use, such as fast-moving consumer good shaving extremely little or no lifespan [26]. Activities that recover materials are known as restorative techniques parts and goods for later application. The writers stated that physically robust products could be treated with restorative techniques sectors, businesses, and organizations from a CSC standpoint, protecting the financial and environmental. It is critical to maintain the target profitable cost of "tangible durable products" in economic systems for as long as feasible. Value is maintained by extending Reabsorbing products into the life cycles of their products [27] therefore, while we work to improve the conversion of industries from linear to circular models that emphasize measurable, long-lasting goods, we have to highlight the rise of restorative practices (e.g., remanufacturing, recycling, renovating, and repairing) that extend the lifespan of tangible, long-lasting EoL goods. R-typologies are represented in the literature at the firm level by means of recycling, reusing, and remanufacturing [28, 31, 33].

A new sustainability framework has to be explored that supports circularity and all supply chain participants in order to guarantee that all of them successfully embrace the CE concept. In this context, CSC is a useful paradigm. The objective of the Circular Supply Chain (CSC) is to enhance resource circulation to meet supply chain demands. By broadening the closed-loop scope, CSC combines supply chain management (SCM) with circular economy (CE) principles to reclaim value from both primary and secondary supply chains. Additionally, to reduce the reliance on virgin materials, CSC draws on resources available within supply chain systems and extends beyond Boundaries of sustainable supply chain management (SSCM) and environmentally-friendly supply chain management (GSCM) [27, 30]. This approach enables the Reclamation of used products—Focuses on retrieving valuable components from discarded items, even those considered waste, facilitating zero waste in the CSC. In business ecosystems, CSCs promote the development of resource recovery strategies [32]. One reason for the growing interest in linking supply chains to CE is the role of CSC. Despite the advantages, challenges and bottlenecks in optimizing CSC remain. Businesses often lack the comprehensive understanding needed to assess circular initiatives or compare them with existing business models, which hampers progress toward a fully functional CSC [29, 32]. In other words, in order to implement a circular model, practitioners need to have a thorough understanding of circularity and how it affects company features. Geographical location, corporate size, and governmental, managerial, financial, and social considerations all influence the hurdles. Nonetheless, the literature enumerates a number of broad obstacles that are unspecific to any of the specific aspect. The circular supply chain concept emphasizes treating the entire product as a valuable resource that requires careful management throughout its lifecycle. This approach considers not only the supply chain but also the selection of raw materials, product usage, disposal, and other key factors [33]. Green and sustainable supply chain management seeks to minimize material consumption and lessen the environmental impact of manufacturing processes, thereby integrating environmental considerations into organizational practices for long-term sustainability [29, 33]. The circular supply chain model aims to transform raw materials into finished products while balancing economic and environmental objectives. There are numerous obstacles and difficulties in putting the circular supply chain model into practice in businesses. As such, a holistic strategy is required for the reuse, recycling, and remanufacturing of materials, power, and other resources [30, 33].

6.2.1 Coordinating the Circular Supply Network

The amalgamation of the concept of CE into the chain of supply is identified as the CS-C. The target of the CE is to oblige industrial companies in collaborating with a socially just society to create a sustainable economy [1]. Technologies of the fourth revolution of industrial are mentioned as being necessary for the economy of circular to expand in a productive way. An analysis of the pressure of a determined industry 4.0 technology (block chain) of fourth industrial revolution on the transition to a CE was conducted using a framework [2] that links the CE to Industry 4.0. Industry 4.0 techniques, such as the usage of block chain technology, increase the result of the CE by lowering carbon footprint, improving supply chain conveying medium, and minimizing anticipated cost of transaction [3, 4]. By averting dangerous working conditions, the adoption of CSC methods improved the social value of adhering to the ecological balance [9].

Reducing waste, maximizing resource efficiency, minimizing energy consumption, and lowering pollution levels were further benefits of using CSC methods [10, 11]. "Among the key factors that contribute to sustainable development within the framework of implementing CSC procedures are efforts to mitigate climate change" [12], implementing moral production practices throughout the chain of supply, creating ecologically responsible design, maximizing the manufacturing of sustainable energy, remanufacturing, recycling, and repair (3R) as well as converting trash into useful resources for greater value [13, 14]. Fourth industrial revolution is based on how modern civilizations have embraced digital transformation. Some of these civilizations show indications of heading towards ecological transition, or environmental conservation taking sustainability into account [15]. "Resource efficiency, sustainability, and environmental responsibility are given priority in the flow of material management, information's and services, and information through the innovative practice of CSC [5, 6, 33]. Circular supply chain management seeks to provide a closed cyclic process system where goods, materials, and resources are recycled, repurposed, remanufactured, and reused to prolong their life cycles [7, 8]. This is in contrast to typical linear supply chains, which operate on a "take, make, dispose" paradigm. According to previous researches, Industry 4.0 technological solutions were deployed in various industry sectors and have proven their significance to realign supply networks across the circular economy. Although there are several Industry 4.0 applications to incorporate the advancements in CSC, the current state of theory and practice overlap have to be explored. The major part of this research centralizes on theoretical paradigms and frameworks that have not been validated to be deployed as per the actual scenario of current environment.

6.2.2 Linkage Between Chain of Circular Supply Network and Industry 4.0 Solutions

Germany initiated the Industry 4.0 transition in 2011 [24]. The information and communication technologies (ICT) and digital manufacturing are linked in modern production models, which align with the Industry 4.0 approach [25]. Industry 4.0 assists producers in handling a wide range of product offers and intensifying worldwide rivalry. The recycling centre and the internet of things (IoT) have accelerated the growth of e-commerce and enabled online enterprises to thrive in the face of fierce global competition, respectively. Additionally, industry 4.0 technologies facilitate communication between organizations and customers. This makes it possible for companies to produce things that meet the needs of their customers [26]. Furthermore, Industry 4.0 techniques improve sustainable frameworks and provide opportunities for cleaner manufacturing [27]. Consequently, manufacturing processes are merged both horizontally and vertically to improve supply chain efficiency across a range of sectors [28]. Industry 4.0 technology may make it easier to implement the ideas of the CE [28]. It is advised that implementing the CE concept will increase resource efficiency and promote harmony between the environment and society [30]. The relationship between the implementation of CSC and Industry 4.0 applications with relation to CE served as the basis for the fact finding issue to be taken into consideration in this research in accordance with this given topic domain that is reachable using some specific process for S-LR. An essential part of each subject area reviewed and evaluated the pertinent literature [2]. Numerous studies showed the complementary effects of Industry 4.0 solutions and CSC practices on the CE. Due to the complexity and dynamic nature of various attributes, supply chains are now more uncertain than ever. To address these challenges, the fourth industrial revolution sector has developed solutions to re-design of framework these supply chains by building digital networks. This can lead to affordable prices and reduce emissions, which can promote environmental sustainability. Additionally, the practices can have a positive social impact and enable sustainability through firm-level and supply chain sustainability. In 2011, the industries of German implemented the Industry 4.0 revolution, which resulted in notable changes to their manufacturing processes and overall sustainability. There was significant compatibility found between industry 4.0 techniques and both traditional production models and ICT-enabled digital manufacturing [24, 28, 32]. Industry 4.0 solutions help businesses reach customer satisfaction levels by easing communication between organizations and their customers and helping them manage their extensive selection of products. They also help businesses become more competitive globally [28, 31, 33].

6.3 Identification of Barriers to Csc Adoption

The statistical software suite SPSS Statistics was developed by IBM and is resorted in data management, advanced analytics, multivariate analysis, analytical technologies, and delinquent liability. SPSS was initially created by SPSS Inc., but IBM acquired the software in 2009. From 2015 onward, it has been promoted under the name IBM SPSS Statistics. Initially developed for social science research, SPSS was originally called the "Statistical Package for the Social Sciences" before being rebranded as "Statistical Product and Service Solutions. "Today, it is widely used for statistical analysis across multiple disciplines, including social sciences, market research, healthcare, education, government operations, and data mining [30, 32] considered highly influential in sociology, allowed researchers to conduct their own statistical analyses. The base package offers tools for statistical analysis and data management, including functions for case selection, file restructuring, derived data generation, and data documentation. It also includes an integrated metadata dictionary saved within the data file. Several types of data can be altered and examined using a tool called SPSS. Virtually any source, such as Google Analytics, client databases, academic research, and server log files from websites, could be the source of this data. SPSS can open any type of file that is often used for structured data, including as spreadsheets like MS Excel or Open Office, relational (SQL) databases, state, SAS, and plain text files (.txt or.csv) [26, 29].

The statistical package for the social sciences, or SPSS, was used to analyze the relationship between the obstacles that the industry really (and frequently) experienced while implementing circular supply chain models. Out of the entire number of barriers, this program was utilized to get the list of barriers that were really confronted. First, data was entered into the SPSS program while adhering to its guidelines and requirements. The information was entered into a tabular format and categorized under categories like Name, Designation, and Questions 1, 2, 3,… 25. Subsequently, as illustrated in Figs. 6.1, 6.2, 6.3 and 6.4, the replies from the Google form that included the admirers possibilities—"Strongly not correspond, not correspond, Neutral, Accept, and Strongly Accept"—were substituted with "1, 2, 3, 4,5" in that order.

Next, the measure choice was altered from nominal to ordinal, and the data view was changed to variable view. After selecting descriptive analysis from the analyze section, the frequency table, histograms, relationship map, mean, median, mode, and sum were selected. Finally, frequency tables, histograms, mean, median, mode, and relationship maps were provided as output, as illustrated in Figs. 6.1, 6.2, 6.3 and 6.4, following the process of inputting the values, changing them, selecting the necessary output data, and selecting the output view. And as a consequence of this procedure, ten barriers were chosen (as indicated in Table 6.1) (Tables 6.2 and 6.3).

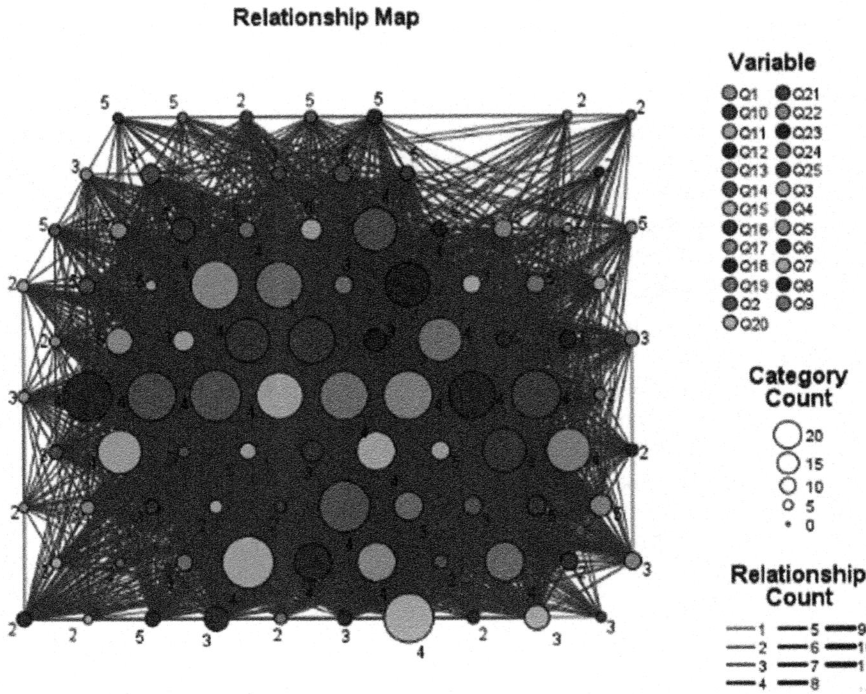

Fig. 6.1 Mapping of core linkages among all 25 identified barriers

6.4 Prioritization of Key Barriers to CSC Adoption Applying AHP Technique

It was used where the selected barriers from SPSS are further brought to be classified on the basis of their importance. For using AHP the selected barriers from SPSS were used. All the selected questions were given a code name in short form such as A, B, C, D...., J. Then a matrix of 10×10 was formed which is showed in Fig. 6.1. This matrix was formed on the basis of relative importance by using the principle from "SAATY Rating Scale" (as shown in Table 6.4). The significance of barriers was decided on the basis of following criteria as shown in Fig. 6.2. After the formation of matrix, the matrix was converted from fraction form to decimal form (Fig. 6.3). Then a 10×1 Matrix (Matrix B shown in Fig. 4.4) was formed by using the formula: (a11*a12*a13*a14*a15*a16*a17*a18*a19*a110)^(1/10);(a21*a22*a23*a24*a25*a26*a27*a28*a29*a210)^(1/10);....;(a101*a102*a103*a104*a105*a106*a107*a108*a109*a1010) ^ (1/10). Then, the sum of matrix B of order 10×1 was calculated and denoted as Sum B. Furthermore, matrix A2 (aka weights as shown in Fig. 4.4) was formed by using the formula: (a11/Sum B); (a21/Sum B); ....; (a101/Sum B). Then, matrix A3 of order

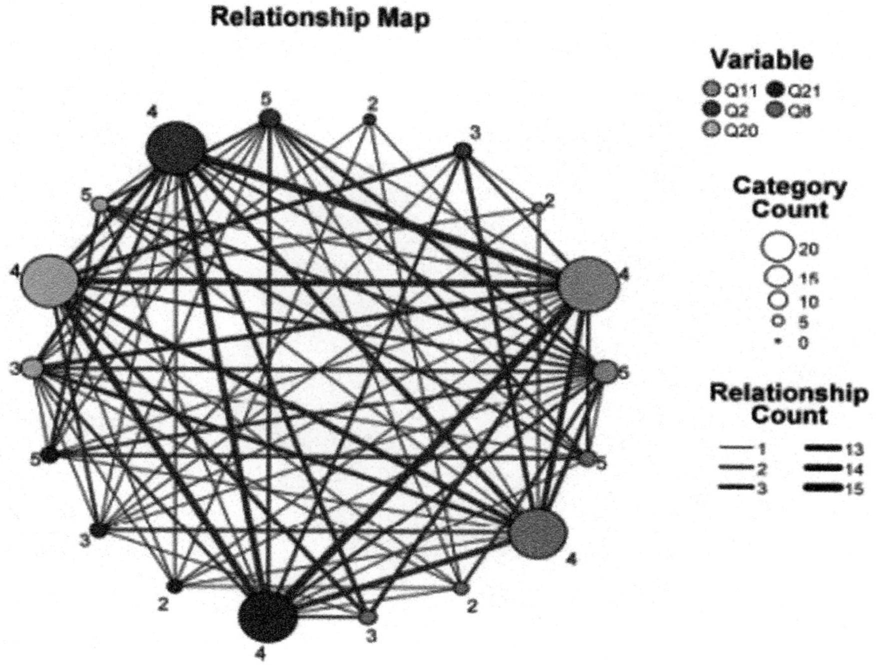

Fig. 6.2 Mapping of core linkages among barriers 11, 21, 2, 20, and 8

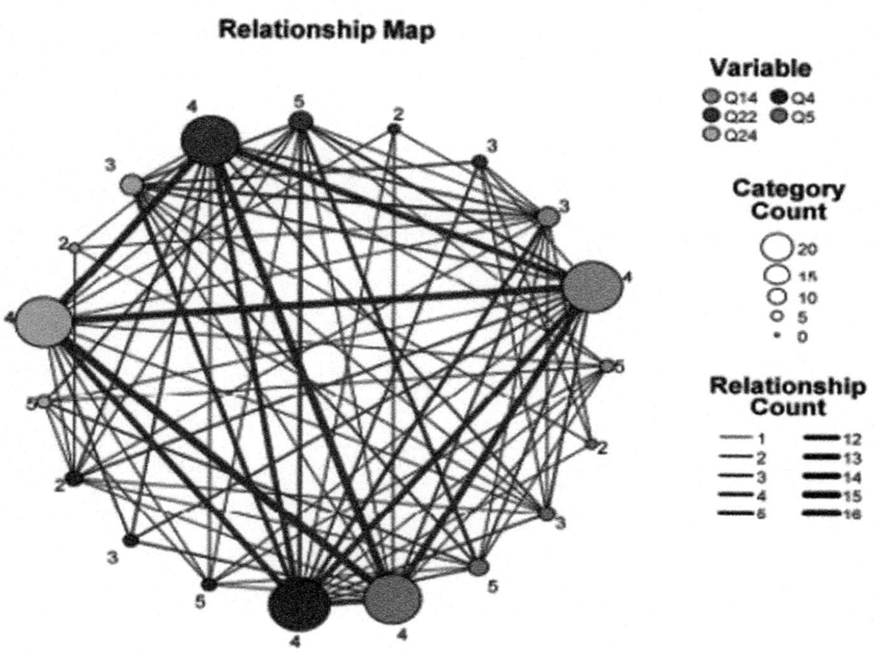

Fig. 6.3 Mapping of core linkages among barriers 14, 4, 5, 22, and 24

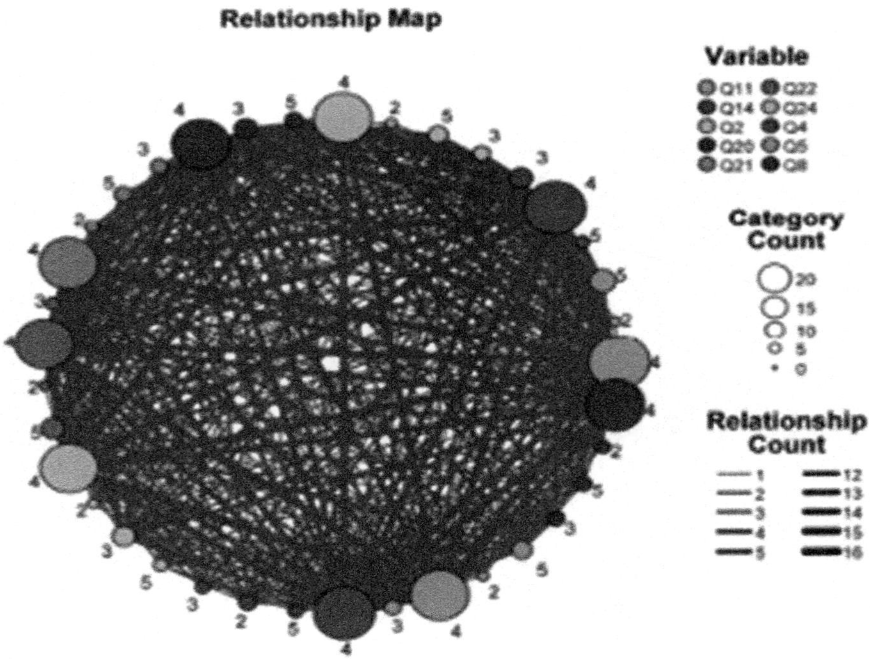

Fig. 6.4 Mapping of core linkages among barriers 11, 22, 14, 2, 4, 20, 5, 21, and 8

Table 6.1 Mean, median and range computed using SPSS (citing examples)

Q1	Periodicity	Percentage (%)	Valid percentage (%)	Cumulative percent
2	2	8.0	8.0	8.0
3	2	8.0	8.0	16.0
4	13	52.0	52.0	68.0
5	8	32.0	32.0	100.0
Total	25	100.0	100.0	100.0
Q2				
2	1	4.0	4.0	4.0
3	3	12.0	12.0	16.0
4	17	68.0	68.0	84.0
5	4	16.0	16.0	100.0
Total	25	100.0	100.0	100.0
Q3				
2	2	8.0	8.0	8.0
3	6	24.0	24.0	32.0
4	15	60.0	60.0	92.0
5	2	8.0	8.0	100.0
Total	25	100.0	100.0	100.0

Table 6.2 Computations of mean, median and range generated from SPSS (Statistical Parameters)

Statistics	Q1	Q2	Q3	Q4	Q5	Q6	Q7
N (Valid)	25	25	25	25	25	25	25
Missing	0	0	0	0	0	0	0
Mean	4.08	3.96	3.68	3.76	3.76	4.16	3.84
Median	4.00	4.00	4.00	4.00	4.00	4.00	4.00
Range	3	3	3	3	3	2	3
Minimum	2	2	2	2	2	3	2
Maximum	5	5	5	5	5	5	5
Statistics	Q8	Q9	Q10	Q11	Q12	Q13	Q14
N (Valid)	25	25	25	25	25	25	25
Missing	0	0	0	0	0	0	0
Mean	3.84	3.92	3.64	4.16	3.60	3.52	3.88
Median	4.00	4.00	4.00	4.00	4.00	4.00	4.00
Range	3	3	3	3	3	2	2
Minimum	2	2	2	2	2	3	3
Maximum	5	5	5	5	5	5	5
Statistics	Q15	Q16	Q17	Q18	Q19	Q20	Q21
N (Valid)	25	25	25	25	25	25	25
Missing	0	0	0	0	0	0	0
Mean	3.92	3.84	4.08	3.80	3.92	3.92	3.84
Median	4.00	4.00	4.00	4.00	4.00	4.00	4.00
Range	2	3	2	2	3	3	2
Minimum	3	2	3	3	2	2	3
Maximum	5	5	5	5	5	5	5
Statistics	Q22	Q23	Q24	Q25			
N (Valid)							
Valid	25	25	25	25			
Missing	0	0	0	0			
Mean	4.04	3.80	3.80	4.00			
Median	4.00	4.00	4.00	4.00			
Range	3	3	3	2			
Minimum	2	2	2	3			

10 × 1 (as shown in Fig. 4.4) was calculated by using the formula: A3 = A1*A2 (here A1 and A2 are matrix A1 & A2 which were calculated above).

In addition to this matrix A4 of the order 10 × 1 (as shown in Fig. 6.4) was formed by using the formula: A4 = A3/A2 (here A3 and A2 are matrix A3 & A2 which were calculated above). In addition to this the average of matrix A4 was calculated and was termed as LaMDA maximum. So, the value of LaMDA max = 10.33271 Later on,

Table 6.3 Outcome analysis of barriers to CSC

S. No	Barrier no	Barriers	Percentage (%) agreed values
1	14	The absence of reliable warehousing facilities and infrastructure for storing end-of-life (EOL) products within trusted networks and formal sectors poses a significant challenge to managing circular supply chains effectively	72
2	8	The lack of adequate information to establish standards hinders the ability to design recoverable products and address the necessity of recovering end-of-life (EOL) products	68
3	24	Deficiency of a stable method or infrastructure for communication between/among internal or external sectors affects take back processes, and hence affects the circularity of supply chain?	68
4	2	The absence of efficient access to tracking systems for the recovery of end-of-life (EOL) products significantly affects the management of circular supply chains	68
5	20	Lack of profit and economy in the short run also acts as a demotivating factor for implying circularity in industry?	68
6	11	The cost of embracing circularity can be categorized into four key steps: 1) implementing a recovery strategy, 2) operating the recovery process, 3) managing reverse logistics, and 4) establishing the required number of recovery facilities. Each of these steps is crucial for the successful adoption of a circular supply chain	72
7	22	Insufficiency in knowledge, information and awareness of organizational members regarding the circular supply chain model initiatives and opportunities plays a major role in affecting the circular supply chain adoption?	68

(continued)

Table 6.3 (continued)

S. No	Barrier no	Barriers	Percentage (%) agreed values
8	4	Deficiency of skills and practical knowledge for the adoption of recovery approach appears as a result of company's focus on operational aspect in spite of recovery approaches, and it affects the circular supply chain management to a great extent?	72
9	21	Lack of interest and enthusiasm by the stake holders also acts as a major barrier towards circular supply chain management?	68
10	5	The limited accessibility to tracking systems for monitoring the recovery of end-of-life (EOL) products significantly affects the efficiency of circular supply chain management	68

Table 6.4 SAATY rating scale for AHP technique

Ratings	Descriptions	Inferences
First	Identical relevance	Both factors play an equal role in achieving the objective
Third	Reasonable importance	One factor holds slightly more significance than the other.
Fifth	Pivotal importance	One factor holds significantly more importance than the other.
Seventh	Paramount importance	One factor is significantly more important
Nineth	Indispensable	One element is extremely more important than the other
Second, Fourth, Six, Eight	Balanced values	It is employed to signify trade-offs between the values mentioned above

"The Consistency Index was determined using the following formula: Consistency Index = " (LaMDA max-10)/(10–1) And the resultant value of Consistency Index is 0.03697 which lies in between 0 and 1, therefore the matrix is a Consistent in nature. Finally, Consistency Ratio was calculated by using the formula: Consistency Ratio = Consistency Index/Random Index; (here the value of consistency index is calculated above and value of random index is 1.49 for a 10 × 10 matrix). Therefore, the resultant value of Consistency Ratio was observed as 0.02481 which is less than 0.1. Hence, Consistency Ratio < 0.1 so the weights are accepted Since the matrix is consistent and weights are accepted So, Matrix A2 can be arranged in descending

Table 6.5 Criterion for assessing significance employing AHP

O	A	B	C	D	E	F	G	H	I	J
A	1	0.333	3	0.5	2	0.5	0.333	0.333	2	3
B	3	1	5	2	3	2	1	2	3	5
C	0.333	0.2	1	0.333	0.5	0.333	0.2	0.25	2	2
D	2	0.5	3	1	2	1	0.333	0.5	2	3
E	0.5	0.333	2	0.5	1	0.5	0.25	0.333	0.5	2
F	2	0.5	3	1	2	1	0.333	0.5	2	3
G	3	1	5	3	4	3	1	2	3	5
H	3	0.5	4	2	3	2	0.5	1	3	4
I	0.5	0.333	3	0.5	2	0.5	0.333	0.333	1	3
J	0.333	0.2	0.5	0.333	0.333	0.2	0.25	0.25	0.333	1

Table 6.6 Transformed matrix representation in decimal form from fraction form

B	A2 (Weights)	A3 = A1 * A2	A4 = A3 / A2
0.89569	0.073007	0.768115	10.52114
2.36178	0.192506	1.964671	10.20574
0.406422	0.033127	0.344858	10.41014
1.196112	0.097494	0.995737	10.21333
0.608243	0.049577	0.506901	10.22446
1.196112	0.097494	0.995737	10.21333
2.636037	0.214861	2.209236	10.28217
1.83463	0.149539	1.549482	10.36173
0.779743	0.063556	0.668055	10.51127
0.353811	0.028839	0.299456	10.38377

order and the final list of barriers classified onbasis of their priorities (as per the chosen criteria) can be obtained (Tables 6.5, 6.6 and 6.7).

6.5 Results and Interpretations

Circular supply chain is the new age supply chain model as it focuses on the sustainable development and growth of industries which are the very need of this hour. Linear supply chain models permit the flow of product in one direction only which is generally from the Supplier to the Customer. But, this flow of supply chain leads to excessive waste generation, waste of resources and harms the environment in numerous ways, so the circular approach of supply chain is a crucial model for keeping the environment clean and green. But when it comes to the adoption of circularity in supply chain models it brings along several obstacles along with the

Table 6.7 Computations for matrix B, A2, A3, A4, consistency index, and consistency ratio

B	A2 (Weights)	A3 = A1 * A2	A4 = A3/A2
0.89569	0.073007	0.768115	10.52114
2.36178	0.192506	1.964671	10.20574
0.406422	0.033127	0.344858	10.41014
1.196112	0.097494	0.995737	10.21333
0.608243	0.049577	0.506901	10.22446
1.196112	0.097494	0.995737	10.21333
2.636037	0.214861	2.209236	10.28217
1.83463	0.149539	1.549482	10.36173
0.779743	0.063556	0.668055	10.51127
0.353811	0.028839	0.299456	10.38377
Sum	12.26858		
LAM MAX	10.33271		
Consistency index	0.03697		
Consistency ratio	0.02481		
CR < 0.1	True		
Weights accepted	Yes		

advantages it possesses. So, in this research work, a significant effort was made to identify some of those barriers and classify them on the basis of their priority. The results as visualized in the Table 6.8 was achieved by this research and the outcome was evident as the final list of 10 barriers out of those 25 barriers that were selected in first place, was obtained from this research, as such, firstly, 10 barriers were selected from the 25 barriers employing SPSS Software.

6.6 Concluding Remarks and Future Directions of Research

Traditional supply chain models produce a lot of waste and gives harmful effects on the environment. Whereas circular supply chain models are beneficial in many ways likely, financially rewarding growth, creation of employment, business sturdiness, substantial customer commitment, reduced green-house gases emission, waste mitigation, reduced resource consumption, increased 3Rs practices, encouraged use of renewable resources, and maintaining sustainable practices and so on. But when industries approach for the implementation of circular supply chain it faces several obstacles and challenges, and in this research paper we identified and classified some of these barriers on the basis of their priority. This can further be used by the industry, as the list of classified barriers such as primary, secondary, etc. can

Table 6.8 Prioritization of key barriers to CSC adoption affecting productivity improvement deduced from AHP technique

Priority	Code	Barrier no	Barriers
1	G	20	Lack of profit and economy in the short run
2	B	11	The price of adopting circularity is divided into 4 steps because of: (1) applying a recovery approach, (2) running the recovery approach, (3) reverse logistics and (4) building the no. Of facilities for recovery
3	H	21	Lack of interest and enthusiasm by the stakeholders
4	F	8	Deficiencyof information to provide standards
5	D	2	Lack of availability of access to tracking of recovery of EOL products
6	A	4	Deficiencyof skills and practical knowledge for the adoption of recovery approach appears as a result of company's focus on operational aspect in spite of recovery approaches
7	I	22	Insufficiency in knowledge, information and awareness of organizational members regarding the circular supply chain model
8	E	5	Lack of availability of access to tracking of recovery of EOL products
9	C	14	Lack of warehousing organizations and infrastructures to store the EOL products through trustable networks and formal sectors
10	J	24	Deficiencyof a stable method or infrastructure for communication between/among internal or external sectors

provide them a direction to work on and they can focus on overcoming these barriers so as to increase their productivity in a much efficient way. The classified list of barriers which was researched provides a half way solution for overcoming barriers in adoption of circular supply chain model, as an industry can only overcome or find solution to a certain obstacle if at first it knows about what is the precise problem and what is its significance in overcoming the challenges to implementation of circularity which will lead to an effective and efficient solution. This research would place This research prioritizes the identification and classification of key barriers to the deployment of Cyber-Physical Systems (CPS) within automotive manufacturing by employing pairwise comparison and scoring methodologies to assess each obstacle's impact on CSC (Cyber-Security Control) performance metrics. Through this approach, several critical challenges are highlighted that need to be addressed to enable effective CSC implementation in the automotive manufacturing industry. The insights from this study aim to inform a future-oriented framework that comprehensively integrates the relevant factors essential for CSC practice adoption. This framework is expected to support automotive SMEs and the broader manufacturing sector in overcoming barriers to CSC deployment. Moreover, an extensive survey will utilize multi-criteria decision-making (MCDM) techniques-such as fuzzy AHP, TOPSIS, VIKOR, and DEMATEL-as well as advanced methods like Grey theory and statistical modeling approaches, including structural equation modeling with PLS and AMOS. Additionally, such study can further incorporate the R package SIMSEM (simulated-SEM) for robust analysis can validate the outcomes of this study conforming to the futuristic considerations. This will aid to identify and categorize critical barriers to CSC deployment in the automotive manufacturing set-ups. These studies would set a critical benchmark that automotive manufacturing firms adopting CSC could deploy to prosper both nationally as well as globally. This would provide these firms with a level playing baseline as they work diligently to increase their overall sustainability. The future directions of this research addressing the adoption of CSC practices while studying the barriers to CSC deployment focused on productivity improvement are stated herein. Practices in CSC implementation are in a very nascent stage in India. Every supply chain contains some level of uncertainty. They have a negative impact on supply chain performance and are connected to scenarios involving decision-making. Future studies should examine the relationship between various forms of uncertainty and the implementation of CSC techniques in the context of sustainability.

References

1. Luthra, S., Sharma, M., Kumar, A., Joshi, S., Collins, E., Mangla, S.: Overcoming barriers to cross-sector collaboration in circular supply chain management: a multi-method approach. Transp. Res. Part E: Logist. Transp. Rev. **157**, 102582 (2022)
2. Amiri, M., Hashemi-Tabatabaei, M., Ghahremanloo, M., Keshavarz-Ghorabaee, M., Zavadskas, E.K., Salimi-Zavieh, S.G.: Evaluating barriers and challenges of circular supply chains using a decision-making model based on rough sets. Int. J. Environ. Sci. Technol. 1–22 (2022)
3. Gupta, H., Kusi-Sarpong, S., Rezaei, J.: Barriers and overcoming strategies to supply chain sustainability innovation. Resour. Conserv. Recycl. **161**, 104819 (2020)
4. Farooque, M., Zhang, A., Thürer, M., Qu, T., Huisingh, D.: Circular supply chain management: a definition and structured literature review. J. Clean. Prod. **228**, 882–900 (2019)
5. Manavalan, E., Jayakrishna, K.: An analysis on sustainable supply chain for circular economy. Procedia Manuf. **33**, 477–484 (2019)
6. Wang, J.X., Burke, H., Zhang, A.: Overcoming barriers to circular product design. Int. J. Prod. Econ. **243**, 108346 (2022)
7. Saroha, M., Garg, D., Luthra, S.: Key issues and challenges in circular supply chain management implementation-a systematic review. Int. J. Appl. Eng. Res. **13**(9), 91–104 (2018)
8. Jeihoonian, M., Zanjani, M.K., Gendreau, M.: Closed-loop supply chain network design under uncertain quality status: case of durable products. Int. J. Prod. Econ. **183**, 470–486 (2017)
9. Mastos, T.D., Nizamis, A., Terzi, S., Gkortzis, D., Papadopoulos, A., Tsagkalidis, N., Ioannidis, D., Votis, K., Tzovaras, D.: Introducing an application of an industry 4.0 solution for circular supply chain management. J. Clean. Prod. 300, 126886 (2021)
10. Dwivedi, A., Paul, S.K.: A framework for digital supply chains in the era of circular economy: Implications on environmental sustainability. Bus. Strateg. Environ. **31**(4), 1249–1274 (2022)
11. Saroha, M., Garg, D., Luthra, S.: Identification and analysis of circular supply chain management practices for sustainability: a fuzzy-DEMATEL approach. Int. J. Product. Perform. Manag. **71**(3), 722–747 (2022)
12. Cao, Y., Qu, Y., Guo, L.: Identifying critical eco-innovation practices in circular supply chain management: evidence from the textile and clothing industry. Int J Log Res Appl **26**(11), 1462–1483 (2023)
13. Ghisellini, P., Cialani, C., Ulgiati, S.: A review on circular economy: the expected transition to a balanced interplay of environmental and economic systems. J. Clean. Prod. **114**, 11–32 (2016)

14. Ayati, S.M., Shekarian, E., Majava, J., Wæhrens, B.V.: Toward a circular supply chain: understanding barriers from the perspective of recovery approaches. J. Clean. Prod. **359**, 131775 (2022)
15. Mishra, R., Singh, R.K., Govindan, K.: Barriers to the adoption of circular economy practices in micro, small and medium enterprises: instrument development, measurement and validation. J. Clean. Prod. **351**, 131389 (2022)
16. Giusto, D., Iera, A., Morabito, G., Atzori, L.: (eds.) The internet of things: 20th Tyrrhenian workshop on digital communications, Springer Science and Business Media
17. Tseng, M.L., Tan, R.R., Chiu, A.S., Chien, C.F., Kuo, T.C.: Circular economy meets industry 4.0: Can big data drive industrial symbiosis? Resour. Conserv. Recycl. **131**, 146–147 (2018)
18. Navghane, S.S., Killedar, M.S., Rohokale, V.M.: IoT based smart garbage and waste collection bin. Int. J. Adv. Res. Electron. Commun. Eng. (IJARECE) **5**(5), 1576–1578 (2016)
19. Shyam, G.K., Manvi, S.S., Bharti, P.: Smart waste management using Internet-of-Things (IoT). In: 2nd International Conference on Computing and Communications Technologies. IEEE, pp 199–203
20. Merli, R., Preziosi, M., Acampora, A.: How do scholars approach the circular economy? A systematic literature review. J. Clean. Prod. **178**, 703–722 (2018)
21. Shekarian, E.: A review of factors affecting closed-loop supply chain models. J. Clean. Prod. **253**, 119823 (2020)
22. Zhu, Q., Geng, Y., Lai, K.H.: Circular economy practices among Chinese manufacturers varying in environmental-oriented supply chain cooperation and the performance implications. J. Environ. Manage. **91**(6), 1324–1331 (2010)
23. Lieder, M., Asif, F.M., Rashid, A., Mihelič, A., Kotnik, S.: Towards circular economy implementation in manufacturing systems using a multi-method simulation approach to link design and business strategy. Int. J. Adv. Manuf. Technol. **93**, 1953–1970 (2017)
24. Linder, M., Williander, M.: Circular business model innovation: inherent uncertainties. Bus. Strateg. Environ. **26**(2), 182–196 (2017)
25. Ravindra, K., Kaur, K., Mor, S.: System analysis of municipal solid waste management in Chandigarh and minimization practices for cleaner emissions. J. Clean. Prod. **89**, 251–256 (2015)
26. Smol, M., Kulczycka, J., Henclik, A., Gorazda, K., Wzorek, Z.: The possible use of sewage sludge ash (SSA) in the construction industry as a way towards a circular economy. J. Clean. Prod. **95**, 45–54 (2015)
27. Wildani, L., Yolanda, P., Laksana, S.A.: ImplementasiMasalahMatematikamelalui SPSS. Papanda J. Math. Sci. Res. **1**(2), 94–104 (2022)
28. Norman, H., Hull, C.H.: SPSS: Statistical Package for the Social Sciences. McGraw-Hill Book Company (1975)
29. Kumar, A., Mangla, S.K. and Kumar, P.: Barriers for adoption of industry 4.0 in sustainable food supply chain: a circular economy perspective. Int. J. Prod. Perform. Manag. **73**(2), 385–411
30. Lu, H., Zhao, G., Liu, S.: Integrating circular economy and industry 4.0 for sustainable supply chain management: a dynamic capability view. Prod. Plan. Control. **35**(2), 170–186
31. Kandasamy, J., Venkat, V., Mani, R.S.: Barriers to the adoption of digital technologies in a functional circular economy network. Oper. Manag. Res. **16**(3), 1541–1561 (2023)
32. Khandelwal, C., Barua, M.K.: Prioritizing circular supply chain management barriers using fuzzy AHP: case of the Indian plastic industry. Glob. Bus. Rev. **25**(1), 232–251 (2024)
33. Abdul-Hamid, A.Q., Ali, M.H., Osman, L.H., Tseng, M.L., Omar, A.R.C.: An industry 4.0 adoption in the circular economy application hierarchical model: driver, enable and barrier aspects. Ind. Manag. Data Syst. **124**(1), 386–415

Ms. Rita Nagwal She is working as an assistant professor and a PhD research scholar in the Department of Mechanical Engineering at Medi-Caps University, Indore. Her research domains are circular supply chains, Industry 4.0 applications, and MCDM techniques for industrial engineering applications.

Dr Ravindra Pathak He is the Head, Office of International Affairs, and Professor in the Department of Mechanical Engineering at Medi-Caps University, Indore. His interests have broadened beyond traditional academia, leading him to actively foster international academic collaborations.

Dr Kumar Rohit He is an assistant professor in the Department of Industrial and Production Engineering (IPE) at Shri G S Institute of Technology and Science (SGSITS), Indore. He has his doctorate from Indian Institute of Technology (IIT) Kharagpur in the research domain of new product analytics and core operations area.

Chapter 7
Enhancing Productivity and Workflows in Smart Industry Through Robotics Process Automation

Debasis Mishras, Kali Charan Rath, Santosh Kumar Tripathy, Alvaro Rocha, and Dina Darwish

Abstract A new age of digital transformation has been brought about by Industry, and robotic process automation (RPA) is a key component in increasing productivity and streamlining workflows in smart industries. This article investigates how Robotic Process Automation (RPA) may reduce manual involvement, streamline complicated industrial processes, and increase operational efficiency. Robotic Process Automation (RPA) frees up human workers to concentrate on more strategic, high-value jobs by taking over repetitive, routine duties, which boosts innovation and competitiveness. The integration of cutting-edge technologies such as Artificial Intelligence (AI), Machine Learning (ML), and the Internet of Things (IoT) enhances the capabilities of Robotic Process Automation (RPA), allowing businesses to optimize their operations in real time. Even though RPA is very good at automating repetitive, rule-based operations, the addition of AI improves its capacity to handle unstructured data, make wise decisions, and get better over time through learning. This is further refined by ML, which makes processes more flexible and efficient by enabling systems to adjust and forecast results based on historical data. The Internet of Things (IoT) gives RPA access to real-time data from linked devices, which it can utilize to automate answers or initiate tasks like predictive maintenance. When combined, these

D. Mishras (✉) · K. C. Rath · S. K. Tripathy
Department of Mechanical Engineering, Gandhi Institute of Engineering and Technology University, Gunupur, Odisha, India
e-mail: debasismishra101@gmail.com

K. C. Rath
e-mail: rathkalicharan1980@gmail.com

S. K. Tripathy
e-mail: sktripathy@giet.edu

A. Rocha
ISEG, University of Lisbon, Lisbon, Portugal
e-mail: amrrocha@gmail.com

D. Darwish
Faculty of Computer Science and IT, Ahram Canadian University, Giza, Egypt
e-mail: dina.g.darwish@gmail.com

© The Author(s), under exclusive license to Springer Nature Switzerland AG 2025

K. C. Rath et al. (eds.), *Engineering, Technology and Management*, Information Systems Engineering and Management 54, https://doi.org/10.1007/978-3-031-93654-8_7

technologies improve decision-making with data-driven insights, facilitate seamless collaboration across business divisions, and minimize downtime through predictive maintenance. Businesses can swiftly adjust to shifting customer needs and market situations because to this synergy, which increases agility and scalability. In addition, case examples of effective RPA deployment in smart sectors are presented in this chapter, with an emphasis on the advantages, difficulties, and potential developments. This paper attempts to offer insights into how RPA might be used to build more flexible, effective, and sustainable industrial settings through a thorough investigation, so making a substantial contribution to the development of contemporary smart industries.

Keywords Industrial automation · Smart manufacturing · Robotics Process Automation (RPA) · Process optimization · Digital transformation · Artificial Intelligence (AI)

Preface

This chapter explores the transformative role of Robotic Process Automation (RPA) in the evolving landscape of smart industries, where it is driving significant improvements in productivity and workflow efficiency. By automating repetitive, rule-based tasks, RPA frees up human workers to focus on higher-value, strategic initiatives, fostering innovation and enhancing competitiveness. The chapter also examines the powerful synergy between RPA and emerging technologies such as Artificial Intelligence (AI), Machine Learning (ML), and the Internet of Things (IoT), which together enable real-time operational optimization, intelligent decision-making, and predictive maintenance. Through case studies, it highlights the tangible benefits, challenges, and future opportunities of RPA, offering valuable insights into how it is shaping more agile, efficient, and sustainable industrial ecosystems.

7.1 Introduction

In the current dynamic industrial environment, firms worldwide are placing a high premium on increasing productivity and efficiency. The incessant pursuit of optimization has propelled enterprises to investigate and embrace novel technologies capable of optimizing processes, curtailing expenses, and preserving a competitive advantage. Robotics Process Automation (RPA) is a technology that has the potential to revolutionize several industries by automating repetitive, rule-based processes that were previously completed by people. This introduction lays out the importance of Robotic Process Automation (RPA) and how it may improve workflows and productivity, making it a fundamental component of contemporary industrial processes.

Technological developments and globalization drive industries to increase productivity on a continuous basis. Delivering goods and services in a timely, accurate, and economical manner is essential in markets that are highly competitive. Traditional task management, which depends on human labor, frequently faces challenges related to time constraints, accuracy, and consistency. In order to overcome these obstacles, robotic process automation (RPA) streamlines procedures, increases productivity, and automates repetitive tasks.

Robotic Process Automation (RPA) automates operations such as data entry, invoice processing, and client onboarding through the use of software bots. Businesses can automate tasks more accurately and quickly without requiring human intervention by programming bots to adhere to predetermined criteria. Scalability, accuracy, and speed are all improved by RPA. For instance, it can automate loan processing in banking, cutting down on days to hours and allowing staff to work on more complicated projects, increasing output and service quality.

7.1.1 Impact on Workflows and Operations

The impact of RPA goes beyond straightforward task automation. It can bring a new degree of consistency and efficiency that might completely change processes. Multiple employee handoffs are a common feature of traditional processes, which increases processing time and increases the chance of mistakes. Because bots manage the entire process end-to-end using RPA, these hand-offs may be reduced to almost nonexistent. Tasks are integrated seamlessly, which improves operations and lowers bottlenecks [1].

Furthermore, RPA makes it easier for businesses to grow their operations. Routine duties increase in bulk as firms expand. Without the need for more human resources, RPA offers a scalable solution that can readily accommodate an expanding workload. This scalability is especially useful in sectors where quick, precise processing is essential, including manufacturing and logistics (Fig. 7.1).

Case Studies and Applications

Manufacturing, finance, and other industries are being revolutionized by RPA. RPA automates quality assurance, inventory control, and supply chain management in manufacturing. One large automaker used RPA to improve product quality and decrease downtime. RPA in finance is saving time and money by simplifying compliance reporting [2]. An international bank, for instance, used RPA to automate its anti-money laundering (AML) processes, which led to a 60% decrease in processing time and a notable reduction in compliance costs.

RPA enhances productivity and personnel utilization in the healthcare sector by helping with tasks like patient data administration, billing, and appointment scheduling. RPA was utilized, for instance, to expedite patient intake in a large hospital system, thereby increasing admission rates [3]. One large international bank

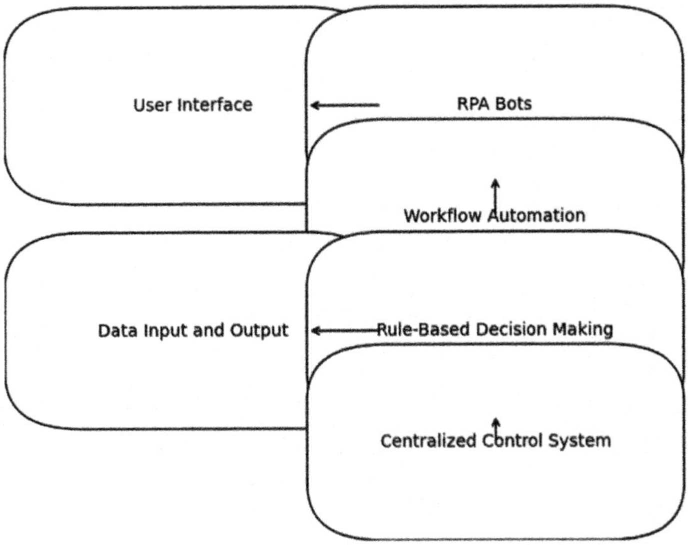

Fig. 7.1 Workflow and operations of robot

used RPA to automate data entry and verification, cutting the processing time for loan applications from 30 to 10 min.

An automotive manufacturing company reduced order processing time by 50%, increased inventory accuracy by 95%, and saved $2.5 million a year by using RPA to automate supplier contacts, order processing, and inventory management. Similar to this, a major healthcare provider enhanced customer satisfaction by 30%, cut operational expenses by 40%, and reduced processing time by 60% by automating the submission, validation, and payment of claims [4].

By automating repetitive tasks like account changes, billing inquiries, and technical support [5], a telecom industry leader was able to improve response times by 70% and increase customer satisfaction by 40%. RPA was utilized by a major retail chain to automate processes [6] such as invoice scanning, verification, and payment. This resulted in an 80% reduction in errors, a reduction in processing time from 15 min to just three minutes per invoice, and a good amount of money an annual savings (Table 7.1).

Table 7.1 Robot application and workflow in different sector

Industry	Company	Challenge	RPA implementation	Results
Banking	Leading global bank	Slow and error-prone loan processing	Automated data entry, verification, and processing	Processing time reduced by 67% (30 min to 10 min), error rate decreased by 70%, productivity increased by 200%
Manufacturing	Automotive parts manufacturer	Inefficiencies and delays in supply chain	Automated inventory management, order processing, and supplier communication	Order processing time reduced by 50%, inventory accuracy improved by 95%, annual savings of $2.5 million
Healthcare	Major healthcare provider	Slow and error-prone claims processing	Automated claims submission, validation, and payment	Claims processing time reduced by 60%, operational costs decreased by 40%, customer satisfaction increased by 30%
Telecommunications	Leading telecom operator	Overwhelming customer service requests	Automated repetitive tasks like account updates and billing inquiries	Response time improved by 70%, annual savings of $1.2 million, customer satisfaction increased by 40%
Retail	Large retail chain	Slow and error-prone invoice processing	Automated invoice scanning, validation, and payment	Processing time reduced by 80% (15 min to 3 min), errors decreased by 80%, annual savings of $500,000

7.2 Importance of Productivity and Workflow Optimization in Industry

Productivity and workflow optimization's significance in industry is based on the use of Robotics Process Automation and have become essential components for preserving efficiency and guaranteeing continuous development in today's competitive corporate scene. In this field, robotic process automation (RPA) integration has become a disruptive force with a host of advantages that meet both strategic goals and operational difficulties [7, 8]. Leveraging productivity, workflow optimization, and RPA to the fullest extent possible and attaining industry excellence requires an understanding of how these factors interact.

7.2.1 Enhancing Productivity Through RPA

For an industry to succeed, productivity is essential, and msanual, repetitive tasks can reduce output. RPA solves this by automating processes like transaction processing, report generation, and data entry, freeing up workers for more intricate work and enabling companies to handle increased workloads without compromising quality. RPA also lowers expenses by reducing manual labor and the possibility of mistakes, which can result in costly corrections. This cost-effectiveness is particularly beneficial in sectors that require stringent control over operating expenses and have small profit margins [9–11].

7.2.2 Optimizing Workflows for Greater Efficiency

Robotic Process Automation (RPA) is improved by workflow optimization, which removes inefficiencies, simplifies tasks, and facilitates transitions. RPA eliminates bottlenecks by automating repetitive tasks, guaranteeing quicker turnaround times and better service delivery [12]. Moreover, automated workflows easily interface with other tools, improving productivity, teamwork, and communication. Additionally, optimizing processes maximizes technological and human resources, freeing up staff members to concentrate on strategic roles and make full use of RPA systems. Effective resource management facilitates scalability, allowing companies to expand without requiring significant adjustments to workflow [13].

Scalability is fueled by efficient resource management, which streamlines processes and allows businesses to expand without requiring significant changes. Through process simplification and effective use of human and technological resources, companies can grow and meet rising demands without experiencing operational disruptions [14–16].

7.2.3 Real-Time Insights and Strategic Decision-Making

The ability of RPA to provide real-time data and analytics improves strategic decision-making by offering useful information about performance metrics, process efficiency, and areas for development with this data-driven approach, companies can make well-informed decisions, quickly adjust to changes in the market, and improve their strategies for greater outcomes. Furthermore, proactive problem-solving is made possible by real-time insights, which help businesses spot possible problems early and act quickly to address them. This promotes long-term resilience and operational stability [17, 18].

7.3 Overview of Robotics Process Automation

By automating repetitive tasks, RPA enhances industrial operations, as examined in "The Power of Robotics Process Automation." This increases productivity by allowing workers to concentrate on more crucial tasks. With real-world examples of successful application across industries, the article also demonstrates how RPA helps to improve accuracy, cut costs, and streamline processes. Along with insights into future trends and best practices for successful RPA adoption, it discusses implementation challenges and provides advice on how to overcome them.

Robotics Process Automation (RPA) is revolutionizing how companies increase productivity and workflow efficiency by using software robots, or "bots," to automate repetitive, rule-based tasks that were previously performed by humans. RPA improves accuracy, speeds up procedures, and lowers costs by streamlining tasks like data entry, transaction processing, and communication management. Additionally, it makes it possible to process data in real time, which helps companies react swiftly to shifts in the market and make wiser decisions. RPA promotes flexibility and scalability, which helps businesses expand without adding more physical labor. This is particularly beneficial for those going through a digital transformation. Furthermore, RPA increases worker satisfaction and productivity by automating repetitive tasks, freeing up staff members to concentrate on more valuable, creative work. However, obstacles like system integration, data security, and change management must be overcome for RPA adoption to be effective, therefore preparation, training, and continuing assistance are crucial (Fig. 7.2).

7.3.1 Evolution of RPA

Software automation advanced in the 1990s and early 2000s, moving from simple batch processing and macro scripts to tools that processed transactions and entered data automatically. The development of more sophisticated automation technologies was influenced by these early tools, which increased speed and accuracy and laid the groundwork for RPA by showcasing its potential to increase productivity and decrease errors [19, 20]. More sophisticated solutions across industries were made possible by this transition from hardware to software-based automation. Business process automation has undergone a radical change with the development of robotic process automation, or RPA. Beginning in the early 2000s with simple screen scraping and workflow automation tools, RPA has developed into a sophisticated technology that makes use of machine learning and artificial intelligence.

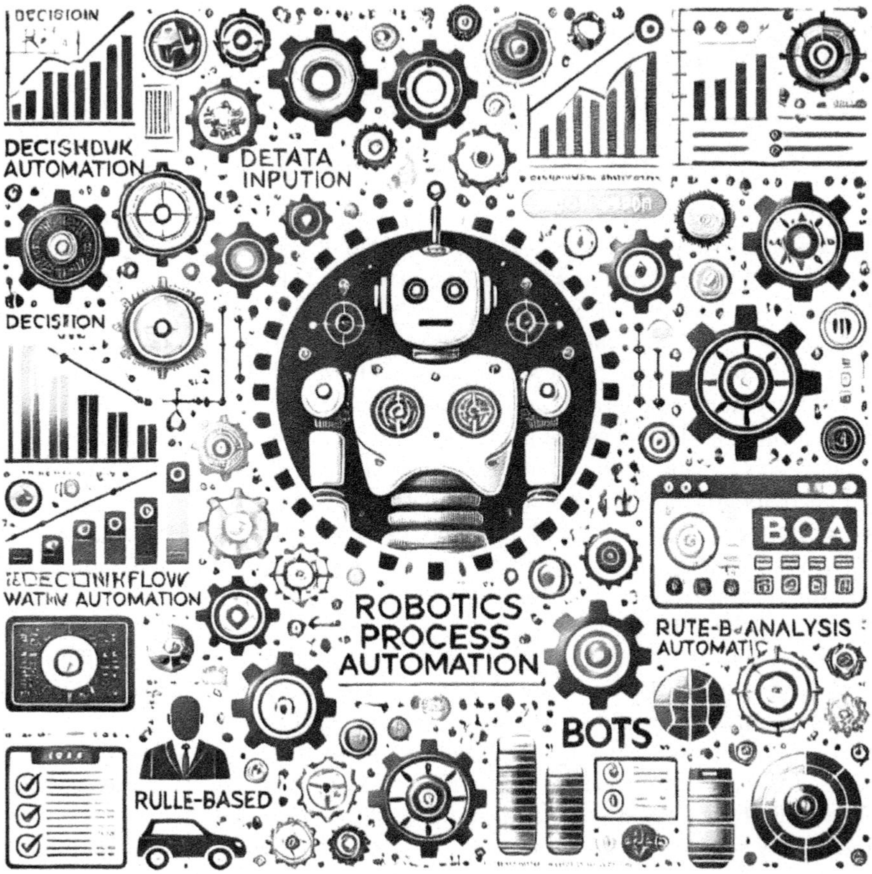

Fig. 7.2 Robot process and automation processes

The mid-2010s saw a surge in RPA adoption across industries, with software robots capable of mimicking human actions to perform repetitive tasks. Significant progress was made in robotic process automation (RPA) in the 2010s, as tools became more capable of managing intricate workflows and integrating with other systems [21, 22]. Through significant advancements, RPA progressed from basic task automation to intelligent systems. Adoption was increased by enabling non-technical users to create automations through user-friendly interfaces. Complex decision-making and unstructured data handling were made possible by integration with cognitive technologies. Intelligent automation is the result of the most recent stage, which blends RPA with analytics and AI. Hyperautomation is the result of this development, where RPA is a component of a broader ecosystem propelling digital transformation in businesses [23, 24].

It is anticipated that RPA will continue to develop in the future due to developments in machine learning (ML) and artificial intelligence (AI). These technological advancements will make it possible for increasingly complex bots to manage intricate decision-making procedures. As RPA becomes more widely used, its integration with cutting-edge technologies such as blockchain will increase its functionality and solve new business problems. Simultaneously, a stronger emphasis will be placed on ethics, data security, and governance to guarantee the responsible use of automation. RPA will ultimately contribute to more dynamic workflows, increased operational effectiveness, and new growth prospects, thereby influencing the direction of industry and work in the future.

7.3.2 Types of RPA Tools and Technologies

RPA tools and technologies encompass various categories, including proprietary solutions like Automation Anywhere and UiPath, open-source options such as Robot Framework and Selenium, AI-powered platforms like Kofax RPA and Pega, cloud-based services including Microsoft Power Automate and Google Cloud RPA, and specialized tools like process mining, OCR/IDR, and workflow management solutions, offering diverse automation capabilities for businesses.

7.3.3 Task-Based RPA

The purpose of task-based RPA solutions is to automate particular, repetitive processes that adhere to predetermined guidelines and protocols. These technologies work especially well for jobs when the procedure is clear-cut and predictable, such data input, data extraction, and basic computations [25] Task-based RPA uses pre-written scripts or procedures to simulate human interactions with software applications. When jobs are regular and don't involve complicated decision-making or adaption, this kind of RPA performs well. Organizations may realize considerable efficiency benefits, lower mistake rates, and free up human resources for more important duties by automating these repetitive operations. Examples of task-based RPA tools include UiPath Studio, which provides a user-friendly drag-and-drop interface for building automation workflows, and Automation Anywhere, which offers robust capabilities for creating and managing.

7.3.4 Process-Based RPA

Instead of automating discrete activities, process-based RPA systems concentrate on automating complete business processes. These technologies are made to handle intricate processes involving several phases and communications across various apps and systems. In contrast to task-based RPA, which works well for discrete activities, process-based RPA manages end-to-end automation by fusing together different workflow components to optimize operations. Workflow management, system integration, and scalability are important characteristics of process-based RPA solutions that allow them to manage more complex processes with many phases and data sources. In this area, Blue Prism and Pega Robotics are two noteworthy examples of tools. While Pega Robotics concentrates on integrating automation with case management and pre-existing IT systems, Blue Prism provides a full platform for automating intricate business processes [26].

7.3.5 Intelligent RPA

This kind of RPA handles more complicated tasks including unstructured data, decision-making, and condition adaption than just basic rule-based automation. AI and ML are used by intelligent RPA technologies to assess data, make defensible judgments, and modify their behavior in response to interactions and fresh information. Because of this connection, jobs requiring a higher degree of intelligence may be completed by bots that can analyze and comprehend unstructured data, including text and pictures, thanks to cognitive automation. Examples are UiPath AI Fabric, which incorporates AI and ML into RPA workflows, and Automation Anywhere IQ Bot, which uses RPA and AI to handle unstructured data [27].

7.3.6 Additional Technologies

RPA's capabilities are further enhanced by other technologies like computer vision and natural language processing (NLP). RPA technologies can detect and understand visual aspects on displays thanks to computer vision, which makes it possible to automate processes involving visual data like document processing and picture identification [28]. With the use of this technology, RPA may now operate in fields requiring visual comprehension. However, NLP improves RPA by allowing bots to comprehend and interpret human language, which allows text-based functions like email management and customer service to be automated. These innovations provide RPA a higher level of intelligence, enabling more flexible and clever automated solutions. RPA systems can perform a wider range of activities and offer more complete automation solutions by combining computer vision and NLP, addressing (Table 7.2).

Table 7.2 Different type robot process automation with example

Type of RPA	Description	Key features	Examples
Task-based RPA	Automates specific, repetitive tasks with well-defined rules. Ideal for simple, rule-based tasks	Script-based automation, high efficiency, limited intelligence	UiPath Studio, Automation Anywhere
Process-based RPA	Automates entire business processes involving multiple steps and interactions between systems	Workflow management, system integration, enhanced scalability	Blue Prism, Pega Robotics
Intelligent RPA	Combines traditional RPA with AI and ML to handle complex tasks involving unstructured data and decision-making	AI and ML integration, cognitive automation, enhanced adaptability	Automation Anywhere IQ Bot, UiPath AI Fabric
Computer vision	Allows RPA tools to recognize and interpret visual elements on screens. Useful for tasks involving visual data	Image recognition, visual data processing	Not specific to a single RPA tool but used in conjunction with RPA technologies
Natural Language Processing (NLP)	Enables RPA tools to understand and process human language, facilitating tasks involving text and communication	Text understanding, automated customer support	Not specific to a single RPA tool but integrated into various RPA

7.4 Impact of RPA on Productivity

RPA significantly boosts productivity by automating repetitive tasks, freeing up staff for strategic work. By streamlining processes, RPA reduces processing times and increases efficiency. This leads to enhanced accuracy, improved employee engagement, and accelerated business growth. Various importance are:

i. **Efficiency Gains**: RPA automates repetitive tasks, allowing employees to focus on more strategic work, thereby boosting productivity.
ii. **Mistake Reduction**: By following predefined rules, RPA minimizes human errors, reducing the need for rework and improving reliability.
iii. **Scalability**: RPA systems can easily scale operations up or down, helping businesses manage workload changes without proportional increases in costs.
iv. **Savings**: Automating manual tasks reduces labor costs, enabling organizations to reallocate savings to more strategic initiatives.

v. **Operational Speed**: RPA operates 24/7, speeding up task completion and enhancing workflow efficiency.
vi. **Standardization and Consistency**: RPA ensures tasks are performed uniformly, reducing process variability and enhancing efficiency.
vii. **Improved Data Processing**: RPA processes large volumes of data quickly and accurately, providing reliable insights for better decision-making.
viii. **Job Satisfaction**: Automating mundane tasks allows employees to focus on meaningful work, improving morale and reducing turnover.
ix. **Integration and Adaptability**: RPA integrates with existing systems, facilitating easy adaptation to new processes without major infrastructure changes.
x. **Continuous Development**: RPA provides data-driven insights into process performance, enabling ongoing improvements and increased efficiency.

Case study-I: U.S. Bank's Loan Processing Automation

Robotic Process Automation (RPA) was integrated by U.S. Bank in an effort to increase efficiency and simplify repetitive procedures related to loan processing. Prior to RPA, processing a loan application took an average of ten days. This time was cut in half, to 4 days, with the use of RPA—a noteworthy 60% decrease. After RPA was implemented, the mistake rate on loan applications—which was 5% when processed manually—was just 1%, indicating an 80% increase in accuracy. Significant reductions in operational costs were also made; for example, processing a loan application now only costs $100, as opposed to $200, which represents a 50% reduction in processing costs.

Furthermore, there was a notable increase in productivity; each employee completed 25 loan applications each day, a 150% increase from the previous 10 applications. Additionally, RPA simplified compliance handling by reducing the amount of time needed for compliance checks from four hours per application to only one hour—a 75% decrease. Due to quicker processing times and fewer mistakes, customer satisfaction also increased, going from 70 to 85%-a 21% gain. RPA has the ability to significantly increase productivity and operational processes in the financial services industry. This was demonstrated by the overall gains in efficiency, cost savings, and customer satisfaction that resulted from the introduction of RPA at U.S. Bank.

Case study-II: Vodafone's Customer Service Automation

Vodafone implemented robotic process automation (RPA) to enhance its customer support operations by automating repetitive tasks like answering questions and processing service requests. As a result, response times for customer inquiries were reduced from 48 to 12 h (a 75% improvement), and service request processing time was cut from five hours to one hour (an 80% reduction). RPA also significantly improved accuracy, reducing the error rate from 7 to 2% (a 71% improvement), and lowered costs per customer inquiry by 47%, from $15 to $8. Additionally, productivity increased by 150%, with agents handling 50 requests per day instead of 20. Complex issue resolution time was reduced by 50%, and customer satisfaction scores improved from 75 to 90% (a 20% increase). These improvements demonstrate RPA's substantial impact on efficiency, accuracy, and customer satisfaction in Vodafone's customer service operations.

7.4.1 Quantitative Benefits

Implementing RPA yields substantial quantitative benefits, including considerable cost reductions, productivity boosts, and efficiency gains. Businesses can expect notable improvements in accuracy, error reduction, and compliance adherence, ultimately leading to optimized resource allocation and sustainable growth.

7.4.1.1 Time Savings

RPA offers substantial time savings in a variety of industries, including banking, healthcare, and customer service, by automating repetitive tasks and drastically reducing processing times. For example, manually processing an invoice involves several steps and can be error-prone, taking around 8 h. With RPA, this time can be reduced by 70%, down to just 2.4 h. This allows employees to focus on more strategic tasks rather than administrative work. The cumulative effect of RPA across multiple functions, such as invoice processing, client onboarding, and report generation, can greatly enhance overall operational efficiency. A bar chart comparing processing times before and after RPA adoption visually demonstrates these time-saving benefits, making it easier to assess the value and communicate it to stakeholders (Fig. 7.3).

7.4.1.2 Error Reduction

Reducing errors is a key benefit of RPA. Manual procedures are prone to human errors like inaccurate data input or calculation, leading to operational issues such as financial discrepancies and compliance challenges. RPA addresses this by performing tasks with high precision, unaffected by fatigue or distraction. For example, automating a process with a 4% error rate can reduce mistakes to as low as 0.8%, enhancing accuracy by 80%. This reduction improves regulatory compliance, reduces rework, and maintains data integrity. A pie chart illustrating the accuracy gains from RPA compared to manual processes effectively shows how RPA enhances operational excellence and quality control [28] (Fig. 7.4).

7.4.1.3 Cost Savings

Cost reduction is a key driver for companies considering robotic process automation (RPA). By automating routine tasks, RPA reduces the need for human labor and

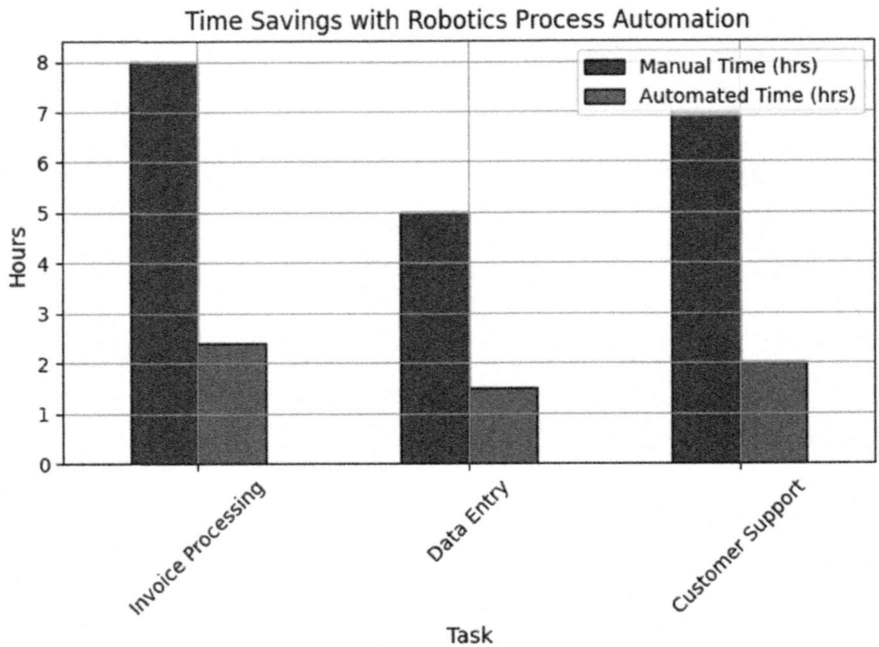

Fig. 7.3 Time saving with robotics process automation

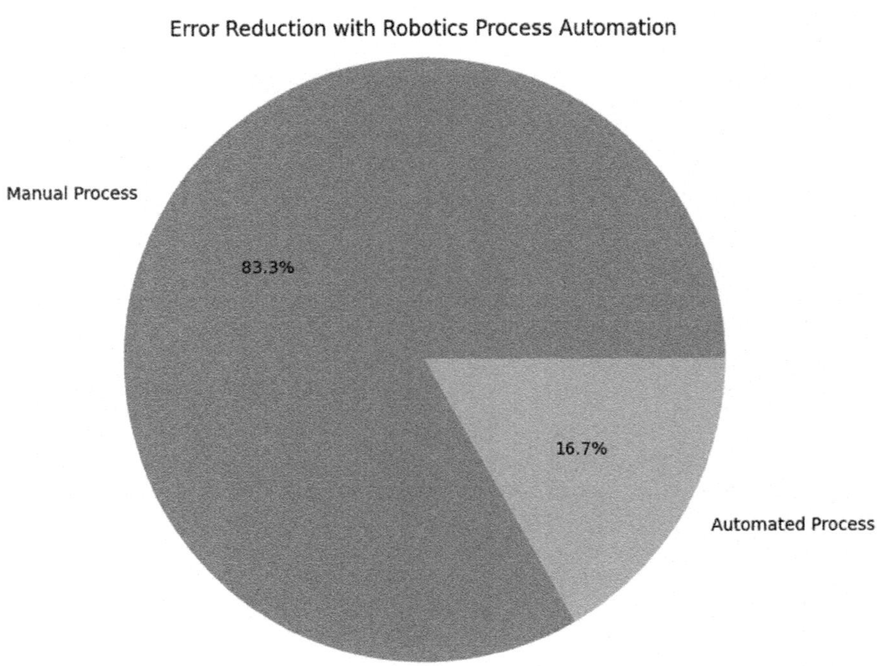

Fig. 7.4 Error reduction with robotics process automation

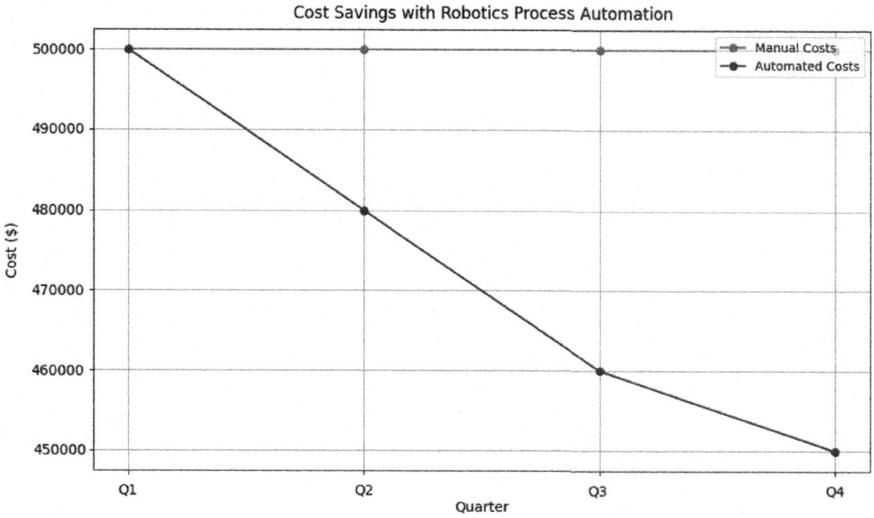

Fig. 7.5 Cost saving with robotics process automation

minimizes error-related costs, potentially lowering operating expenses by up to 30% [28, 29]. For a company spending $500,000 annually on operations, this could mean savings of $150,000. Cost savings are achieved in several ways: RPA reduces labor costs by automating manual tasks, decreases error-related expenses by improving accuracy, and optimizes resource allocation by enabling more strategic use of human resources. A line graph depicting cost trends before and after RPA implementation illustrates these financial benefits, helping organizations assess RPA's impact and make informed decisions (Fig. 7.5).

7.4.1.4 Productivity Increase

By automating repetitive tasks and allowing employees to focus on strategic work, RPA increases productivity. These efficiency gains can be seen in before-and-after stacked bar charts. For example, automating order processing can increase capacity by 35%, handling 1,350 more orders per month without extra staff (Fig. 7.6).

7.4.2 Qualitative Benefits

Automating repetitive tasks like data entry and report preparation with RPA allows workers to focus on more rewarding work that makes use of their creativity, critical thinking, and problem-solving abilities. This transforms workplace satisfaction. A TechAdvantage Solutions survey shows this impact: job satisfaction rose

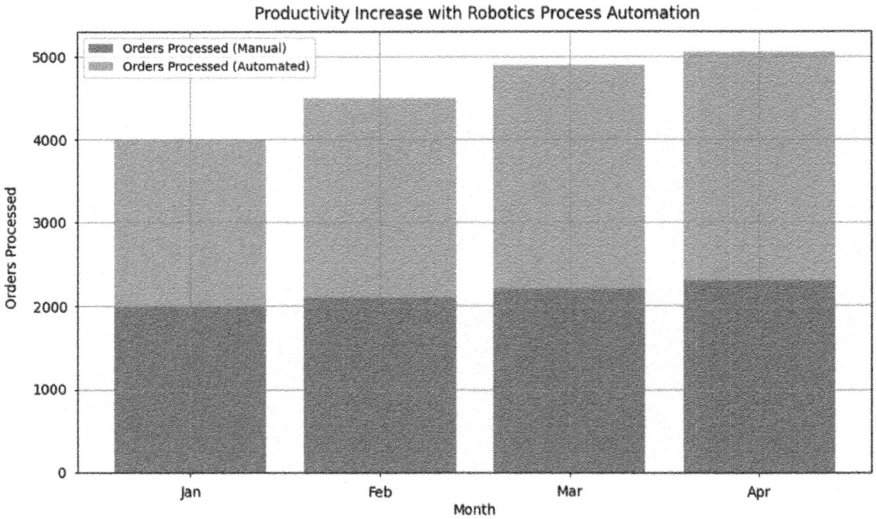

Fig. 7.6 Productivity increase with robotics process automation

from 56 to 87% after RPA implementation. This increase is due to reduced workloads and a more rewarding work environment where employees engage in tasks aligned with their interests and skills, lowering stress and boosting overall job satisfaction. Consequently, RPA improves operational efficiency and enhances the quality of the employee experience, resulting in higher retention rates and a more motivated workforce [29] (Fig. 7.7).

Fig. 7.7 Employee satisfaction before and after RPA

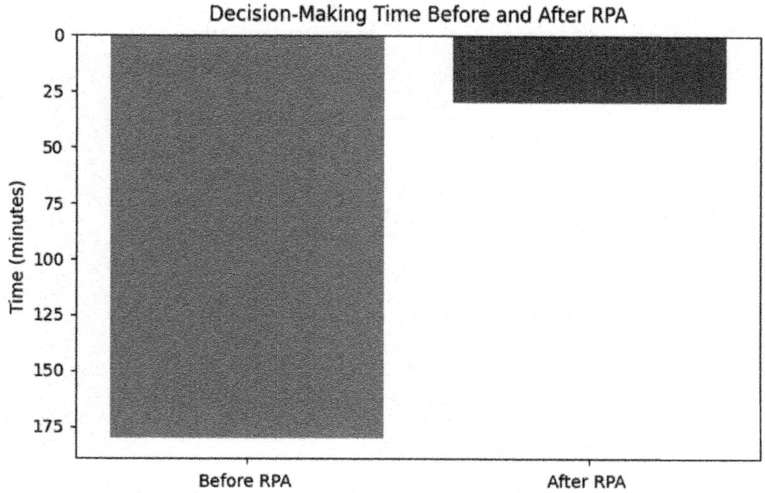

Fig. 7.8 Decision making before and after RPA

7.4.2.1 Improved Decision-Making

Robotic Process Automation (RPA) simplifies data management procedures, revolutionizing business decision-making. RPA automates data collection, processing, and analysis, giving decision-makers access to precise, up-to-date information—a vital resource in the fast-paced business world of today. Compared to pre-RPA environments, where businesses frequently struggled with out-of-date, fragmented, or siloed data, leading to delayed decisions and increased error risks, this represents a significant improvement. Businesses can react to market opportunities and changes more quickly and efficiently when RPA is used because it not only speeds up but also improves the quality of decision-making. RPA mitigates these issues by continuously gathering and analyzing data from multiple sources, enabling managers to make fact-based decisions. For instance, a financial services organization saw a 35% improvement in decision speed and accuracy after implementing RPA, saving 3 h on data retrieval and analysis, which ultimately supports strategic, data-driven decision-making for long-term success (Fig. 7.8).

7.4.2.2 Enhanced Work-Life Balance

In the age of remote work, robotic process automation, or RPA, has become a potent instrument for improving work-life balance. RPA greatly lessens employee workload by effectively automating time-consuming and repetitive tasks, releasing workers from the monotony of routine tasks. Because of this automation, professionals can now focus their time and efforts on more strategic, innovative, and satisfying facets of their work. As a result, workers' ability to pursue personal interests, spend time with

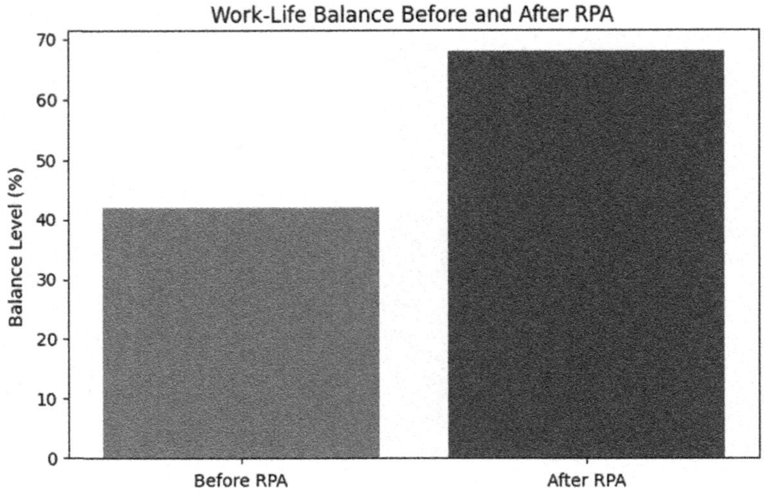

Fig. 7.9 Work-life balance before and after RPA

their families, or fund self-improvement initiatives has increased. A more harmonious blending of work and personal life is the end result, which lowers stress levels, boosts job satisfaction, and improves general wellbeing. Businesses that continue to use RPA are not only increasing productivity but also cultivating a more balanced and satisfied workforce, which eventually helps to create a more sustainable and healthy work culture. Research from the Work-Life Balance Institute shows that 68% of employees reported improved work-life balance with RPA, accompanied by a 20% decrease in stress levels. By reducing the need for long hours and enabling employees to leave work on time, RPA fosters a healthier, more productive work environment (Fig. 7.9).

7.4.2.3 Increased Collaboration

RPA not only increases productivity and reduces expenses, but it also improves teamwork by automating repetitive tasks. This frees up staff members' time for strategic, collaborative work, which in turn creates a more collaborative corporate culture. A study at a global company revealed that RPA increased team collaboration from 50 to 75%. This boost is due to more time for meetings and brainstorming, interdepartmental cooperation during RPA implementation, and improved decision-making with better data access. Ultimately, RPA fosters a collaborative environment that enhances creativity, communication, and business outcomes (Fig. 7.10).

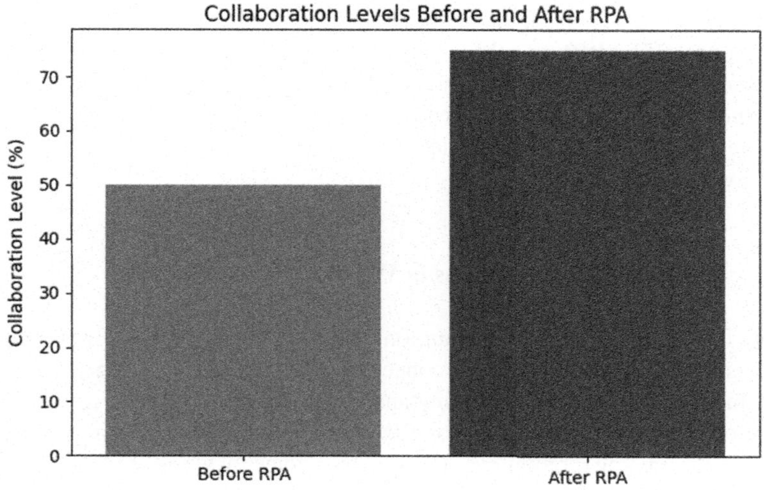

Fig. 7.10 Collaboration levels before and after RPA

7.4.2.4 Improving Workflows with RPA

RPA transforms workflow management by removing time-consuming, repetitive tasks and optimizing processes. Organizations can increase productivity and improve customer satisfaction by implementing RPA solutions, which minimize errors, increase efficiency, and free up staff time for higher-value tasks.

7.4.2.5 Workflow Automation and Streamlining Processes

RPA is vital for workflow automation and process optimization across various sectors. By automating repetitive tasks, RPA boosts productivity, reduces errors, and allows staff to focus on complex work. For instance, a multinational bank used RPA to automate 70% of its loan processing, cutting processing time from five days to hours. This led to a 40% increase in efficiency and a 25% reduction in operating expenses, alongside fewer errors and improved customer satisfaction. The integration with the existing banking system ensured a smooth transition and minimal disruption.

7.4.3 Integration with Existing Systems and Technologies

RPA's effectiveness in improving workflows largely depends on its ability to integrate with existing systems. While integrating new automation technologies can be challenging, especially with legacy systems, RPA is highly adaptable and can function across various platforms without major IT changes. A manufacturing company exemplifies this by using RPA to enhance supply chain management despite an outdated ERP system. The company automated data entry, order tracking, and inventory management, integrating smoothly with its legacy ERP system. The key to successful RPA integration is selecting a flexible solution that aligns with the business's specific needs, even with older systems.

7.4.3.1 Challenges and Solutions in Workflow Automation

Despite its benefits, RPA implementation often faces challenges such as integrating with existing systems, data security concerns, and resistance to change. These issues can be managed with careful planning and best practices. For instance, a healthcare facility overcame staff resistance to RPA by implementing a change management strategy with training, communication, and employee involvement. To address data security, it employed stringent measures like access controls and encryption. Continuous feedback and iterative testing further improved the RPA system's performance and acceptance. The result was a 35% reduction in administrative time and 99% accuracy in patient records. This case demonstrates that with strategic planning, RPA can significantly enhance accuracy, productivity, and cost savings, fostering creativity and growth in a competitive environment (Fig. 7.11).

Case study-I

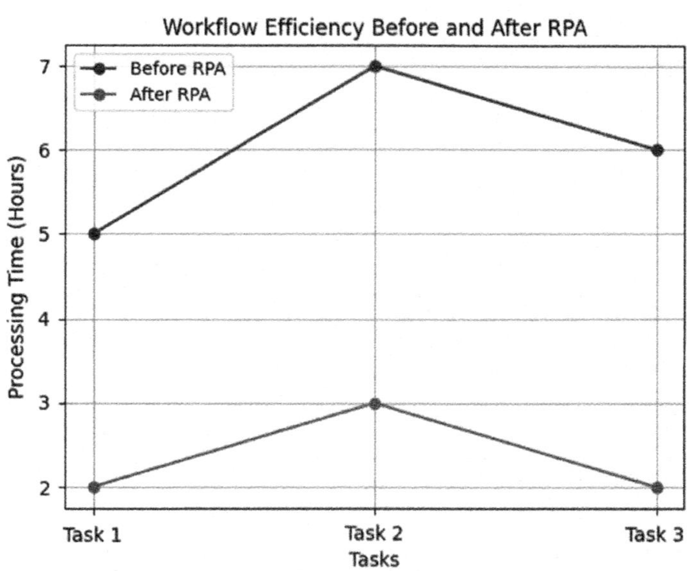

Fig. 7.11 Workflow efficiency before and after RPA

Unilever, a global consumer products company, launched a large-scale Robotic Process Automation (RPA) initiative to boost productivity and reduce costs across its HR, procurement, and finance departments. Aiming to cut operating expenses and transaction processing times by 30%, they began with a pilot project in the finance division, automating purchase order management and invoice processing, which highlighted the benefits of RPA. Unilever chose an RPA platform compatible with their ERP systems and partnered with an RPA service provider for deployment. They developed automation scripts, documented key processes, and conducted thorough testing to ensure accuracy and performance. A centralized monitoring system was established to assess performance, address issues, and support a continuous improvement program to refine RPA solutions. Unilever followed best practices by involving key stakeholders early, focusing on high-impact processes, ensuring scalability, and maintaining strong governance, while also addressing challenges like employee resistance and complex regional processes through proactive change management and clear goal-setting. Updated training programs emphasized RPA benefits for employees, contributing to a 25% reduction in operating expenses and a 40% decrease in finance transaction processing times. The initial success led to RPA expansion across other departments, showcasing how a disciplined approach and a culture of continuous learning can help a multinational company achieve significant operational improvements and maintain a competitive edge.

Case study-II

A leading industrial machinery producer implemented Robotic Process Automation (RPA) to automate labor-intensive processes such as inventory management, order processing, and equipment maintenance, integrating IoT and AI to enhance workflows and reduce costs. The RPA deployment improved productivity by 30%, cut operational costs by 20%, reduced errors by 25%, and enabled predictive maintenance, helping the company transition to a smart factory model and maintain its competitive edge.

7.4.3.2 RPA Implementation Strategies

Prior to selecting the best RPA tool based on requirements and usability, organizations must evaluate their IT infrastructure for compatibility and scalability. A well-organized deployment strategy, which includes pilot projects and a governance framework, guarantees a seamless rollout and tackles any possible obstacles.

(a) Best practices

After a successful pilot project, companies can scale RPA across other divisions using a staggered rollout strategy, which allows for close monitoring and adjustments. Continuous performance monitoring through metrics like process cycle time, accuracy rates, and cost savings is crucial to ensure RPA bots are functioning effectively and to facilitate ongoing improvements. Adopting best practices, such as starting small, involving stakeholders early, prioritizing process optimization before automation, establishing strong governance, and investing in training, enhances the chances of successful RPA implementation. Common mistakes to avoid include unclear goals, neglecting change management, automating inefficient processes, underestimating the complexity of RPA projects, and failing to maintain RPA bots, all of which can lead to suboptimal results and reduced effectiveness.

(b) Change management and training considerations

Change management is a key component of a successful RPA implementation strategy, focusing on preparing, encouraging, and supporting individuals and groups to adapt to new technology. Effective communication is essential to explain the benefits, necessity, and impact of RPA deployment, reducing resistance and uncertainty among employees. Involving employees early in the RPA decision-making process helps gain their support and address concerns. Establishing support systems and feedback mechanisms is crucial for guiding staff through the transition and making necessary adjustments. Cultivating a culture that embraces innovation and views RPA as a tool to enhance productivity, rather than replace workers, is also vital. Additionally, comprehensive training ensures employee proficiency with RPA technology, with specialized programs tailored to different roles. Hands-on experience and continuous learning programs are necessary to keep employees updated on new features, industry developments, and best practices. Regular assessment of training effectiveness and participant feedback helps improve training materials and delivery.

7.4.4 Future Trends and Innovations in RPA

Robotic process automation (RPA), which offers previously unheard-of levels of productivity, cost reduction, and efficiency, is bringing about a new era of industrial transformation. As we move further into the era of smart industries, RPA is collaborating with cutting-edge technologies like artificial intelligence (AI) and machine learning (ML) to enable cognitive automation rather than working alone. This strong integration enables bots to perform ever-more-difficult tasks, such as complex decision-making, advanced image recognition, and subtle language processing. More to the point, these systems are remarkably flexible, improving over time with little human intervention. This idea is further developed by the rise of hyperautomation, which combine RPA with AI, ML, and process mining to automate entire workflows as opposed to discrete tasks. This comprehensive approach to automation, which is being propelled by real-time analytics, is transforming industrial operations and opening the door to a more intelligent, responsive, and effective future.

While integration with the Internet of Things (IoT) improves real-time process optimization and predictive maintenance, cloud-based RPA solutions are becoming more popular due to their scalability and flexibility. Automation becomes more accessible with the emergence of code-heavy and code-free RPA platforms, allowing for wider deployment. In addition to industry-specific solutions and green automation, enhanced security and compliance features are becoming increasingly important. These developments guarantee RPA's expanding contribution to smart industry sustainability and efficiency.

7.4.5 Emerging Technologies and Their Impact on RPA

By combining AI and ML, RPA is developing quickly and enabling cognitive automation, which expands the capabilities of bots beyond conventional rule-based automation by enabling them to process unstructured data, make decisions, and recognize images. RPA is further enhanced by ML, which reduces the need for human intervention by allowing bots to learn and adapt over time. Hyperautomation automates entire business processes for increased efficiency by combining RPA with AI, ML, and other tools. RPA systems require scalability and flexibility, which cloud computing offers, while real-time, data-driven automation is supported by the Internet of Things (IoT). RPA development is made easier with low-code and no-code platforms, and security and transparency are guaranteed by Blockchain. Natural language processing (NLP) developments also improve human-bot interaction, which makes RPA a more clever and practical way to optimize processes and increase output.

7.4.6 Predictions for the Future of RPA in Industry

As robotic process automation (RPA) continues to develop in response to shifting business demands and technology breakthroughs, it is expected to have a profound impact on the sector in the future. Here are some major forecasts regarding RPA's use in business going forward. AI and new technologies are driving robotic process automation's (RPA) future. RPA will become more independent thanks to AI, allowing it to manage challenging tasks in a variety of industries. In order to automate entire processes and increase productivity, hyperautomation will combine RPA with AI and machine learning. Industries like manufacturing will gain from real-time data processing made possible by RPA's integration with IoT. Remote work will be supported by cloud-based RPA and SaaS models, which will provide more flexibility and scalability. Compliance and security will take center stage, particularly in delicate industries. RPA adoption will accelerate as low-code/no-code platforms enable non-technical users to develop RPA applications. Industry-specific RPA solutions will tackle particular problems in industries such as finance and healthcare. RPA will improve productivity by automating repetitive tasks and fostering human-robot collaboration. Through waste reduction and resource optimization, it will also promote sustainability. RPA will eventually develop into a strategic tool that promotes innovation and helps new business plans.

RPA has a promising future in business thanks to developments in cloud computing, AI, ML, IoT, and other cutting-edge technologies. RPA will change industrial workflows, boost productivity, and help firms innovate and prosper in the digital era as it develops further. It will become more intelligent, flexible, and integrated.

7.5 Conclusions

Robotic Process Automation (RPA) is a disruptive force in smart industries, reshaping workflows, increasing efficiency, and improving operations. As industries evolve under Industry 4.0, RPA provides a strategic approach for achieving operational excellence, cost efficiency, and agility. This study explores the transformative impact of RPA, particularly its ability to automate repetitive tasks, freeing human workers to engage in strategic, high-value activities that enhance job satisfaction and innovation. RPA goes beyond rule-based automation when combined with AI, IoT, and machine learning. This is because it allows for real-time data processing and cognitive functions, which are essential for applications like predictive maintenance. This results in decreased downtime and longer asset life, demonstrating the effectiveness of advanced automation in decision-making processes. Despite its benefits, RPA implementation faces challenges such as integration with existing IT infrastructure, data security, and employee resistance. Overcoming these requires a comprehensive strategy involving training, change management, and governance. The future of RPA is promising, with advancements towards Intelligent Process Automation (IPA) and hyper-automation, combining AI, ML, and IoT to create fully autonomous environments. Such innovations will enable industries to optimize operations, predict future needs, and maintain a competitive edge in increasingly complex, interconnected digital ecosystems.

7.5.1 Further Research and Development

Future research in Robotic Process Automation (RPA) for smart industries is poised to explore several key areas. Integrating advanced AI and ML algorithms into RPA systems will enhance their ability to handle complex tasks and real-time learning, improving automation's intelligence. Cloud-based RPA technologies will benefit from new scalable, secure, and cost-effective solutions to support global operations. Security and compliance frameworks tailored for RPA will address risks in regulated sectors. Enhancing user experience and human-robot collaboration will facilitate smoother integration into workflows. Combining RPA with real-time data and IoT will lead to more responsive and adaptable automation. Low-code and no-code platforms will broaden RPA adoption by enabling non-technical users to develop solutions. Industry-specific RPA solutions will improve efficiency by catering to unique sector needs. Sustainability-focused research will optimize resource use and reduce environmental impact. Integrating RPA with blockchain technology will enhance security and transparency. Lastly, developing performance standards will help assess RPA effectiveness, driving better ROI and continuous improvements.

References

1. Adams, P., Johnson, K.: RPA and machine learning: a powerful combination. J. Comput. Intell. **20**(5), 223–236 (2022). Springer, USA
2. Allen, D., Wilson, S.: Improving productivity with robotics process automation in banking. J. Financ. Serv. Technol. **14**(2), 117–130 (2023). Wiley, USA
3. Anderson, H., Thomas, J.: Leveraging RPA for efficient data management. J. Data Sci. Appl. **37**(4), 244–257 (2022). Wiley, USA
4. Brown, L., Davis, P.: Streamlining business operations with advanced robotics. Int. J. Ind. Eng. **45**(2), 143–156 (2022). Elsevier, UK
5. Carter, J., Phillips, R.: RPA and artificial intelligence in business operations. J. Bus. Manag. **22**(8), 321–334 (2023). Wiley, USA
6. Chen, M., Liu, X.: Robotics process automation in the automotive industry. J. Manuf. Syst. **21**(5), 301–314 (2023). Springer, USA
7. Chen, Y., Zhang, Q.: Robotic automation in healthcare: improving patient outcomes. J. Healthc. Eng. **18**(7), 421–434 (2023), Wiley, UK
8. Clark, S., Johnson, H.: The role of RPA in optimizing business operations. J. Bus. Optim. **33**(4), 241–254 (2021). Taylor & Francis, UK
9. Evans, M., Green, P.: RPA in the insurance industry: enhancing claims processing. J. Insur. Technol. **16**(6), 421–434, Springer, UK
10. Hall, R., Adams, J.: Innovations in robotics process automation for the energy sector. Int. J. Energy Res. **27**(6), 453–466, Wiley, UK
11. Hughes, L., Clark, E.: The future of robotics process automation in manufacturing. J. Robot. Autom. **14**(8), 321–334 (2023). IEEE, USA
12. Hughes, S., Parker, G.: The integration of RPA in corporate strategies. J. Strat. Manag. **31**(5), 271–284 (2021), Routledge, UK
13. Kumar, M., Sharma, D.: Advances in robotics process automation for business analytics. J. Bus. Intell. **25**(6), 102–115 (2022). Springer, USA
14. Lee, C., Kim, S.: Robotics process automation in financial services: challenges and opportunities. J. Financ. Technol. **15**(1), 55–69 (2023). Elsevier, USA
15. Miller, K., Nelson, D.: Adopting robotics process automation in the healthcare sector. Int. J. Healthc. Technol. **19**(2), 101–114 (2023). Elsevier, USA
16. Morgan, J., Reed, D.: The impact of RPA on operational risk management. J. Risk Manag. Financ. Inst. **18**(2), 101–114 (2021). Taylor & Francis, USA
17. Nguyen, T., Lee, H.: The impact of robotics process automation on manufacturing efficiency. IEEE Trans. Ind. Inform. **17**(5), 3023–3034 (2023). IEEE, USA
18. Patel, K., Bhattacharya, A.: Exploring the impact of RPA on financial reporting. J. Account. Financ. **40**(4), 227–240 (2022). Wiley, UK
19. Patel, S., Verma, A.: Revolutionizing supply chain management with RPA. J. Logist. Manag. **33**(3), 198–211 (2021). Taylor & Francis, USA
20. Roberts, M., Turner, L.: Improving operational efficiency with robotics process automation in retail. Int. J. Retail. Distrib. Manag. **33**(6), 423–436 (2022). Emerald, USA
21. Ross, M., Baker, L.: Robotics process automation in the public sector: opportunities and challenges. Int. J. Public Adm. **15**(6), 112–125 (2021). Taylor & Francis, USA
22. Singh, R., Gupta, N.: Impact of robotics process automation on business process reengineering. J. Eng. Technol. Manag. **21**(3), 99–112 (2023). Taylor & Francis, UK

23. Smith, J., Johnson, R.: Enhancing productivity and workflows in industry: the power of robotics process automation. J. Autom. Control. Syst. **12**(4), 254–267 (2021). Springer, USA
24. Tang, J., Lin, W.: Robotics process automation for improving supply chain resilience. J. Supply Chain. Manag. **23**(3), 185–198, Springer, USA
25. Thompson, B., Scott, T.: Enhancing compliance management with robotics process automation. J. Compliance Ethics **9**(3), 176–189 (2022). Elsevier, USA
26. Wang, Z., Sun, L.: Robotic automation in E-commerce: enhancing efficiency and customer satisfaction. J. Electron. Commer. Res. **12**(7), 421–434 (2023). Elsevier, UK
27. Williams, B., Garcia, F.: Automation and the future of work: exploring RPA in HR. Int. J. Hum. Resour. Manag. **28**(9), 523–537 (2021). Routledge, UK
28. Wilson, A., Evans, T.: Enhancing IT operations with robotics process automation. J. Inf. Technol. Softw. Eng. **35**(7), 342–355 (2022). Taylor & Francis, USA
29. Zhang, X., Li, Y.: The role of robotics in transforming traditional manufacturing. J. Adv. Manuf. Technol. **29**(3), 192–205 (2021). Springer, UK

Mr. Debasis Mishra serves as an Assistant Professor in the Department of Mechanical Engineering at Gandhi Institute of Engineering and Technology University, Gunupur, Odisha. He holds a M.Tech degree in Mechanical Engineering. With a deep passion for teaching and research, Mr. Mishra specializes in production Engineering and is committed to creating an engaging academic environment that nurtures students' intellectual growth and fosters innovation. In addition to his teaching role, he is involved in research activities. His academic and research efforts are aimed at advancing knowledge and solving real-world engineering challenges.

Dr. Kali Charan Rath is a distinguished academician and researcher, currently serving as Associate Professor in the Department of Mechanical Engineering at Gandhi Institute of Engineering and Technology University, Gunupur, Odisha. With a steadfast dedication to academic excellence, Dr. Rath has established himself as a prolific researcher, publishing extensively in esteemed national and international conferences, journals, and book chapters. His scholarly contributions have been further amplified through editorial roles in multiple books, demonstrating his expertise and leadership in the field. As a devoted PhD supervisor, Dr. Rath is guiding numerous scholars, fostering the growth of innovative research and talent. His tireless efforts, unwavering commitment, and passion for knowledge dissemination have earned him a reputation as a respected educator, researcher, and mentor in the academic community.

Santosh Kumar Tripathy is an Assistant Professor and Associate Professor in the Department of Mechanical Engineering at Gandhi Institute of Engineering and Technology University, Gunupur, Odisha. With a strong academic background and a passion for teaching, he contributes significantly to the growth of the department. His expertise spans various areas of mechanical engineering, and he is dedicated to enhancing students' learning experiences while also engaging in research and development activities. His commitment to education and innovation is reflected in his contributions to the university and the field of engineering.

Álvaro Rocha is an Honorary Professor with a Doctor of Science (DSc) in Information Science and a PhD in Information Systems and Technologies. He is currently a Professor at the University of Coimbra, as well as the President of AISTI (Iberian Association for Information Systems and Technologies) and the Chair of the IEEE SMC Portugal Section Society Chapter. With over 7,300 citations, an h-index of 41, and an i10-index of 127, his academic work and research have had a significant impact in the fields of Information Systems and Technologies.

Dina Darwish is the Vice-Dean of the Faculty of Computer Science and IT at Ahram Canadian University in Giza, Egypt. With a strong background in computer science and information technology, she plays a key leadership role in advancing academic and research initiatives at the university. Dina has also served as an editor for various academic books and has published numerous research papers in reputable journals and conferences. She is dedicated to fostering innovation, shaping the curriculum, and contributing to the academic community.

Chapter 8
Optimum Eco-friendly Refrigerant Selection Using Multi-criteria Decision-Making Methods

Pasham Nithish Reddy and Sivasakthivel Thangavel

Abstract The advancements in technology have enhanced the precision of refrigerant choice. Making the right choice involves considering its physical characteristics along with environmental, economic, and factors as well as the other pertinent variables. The article discusses multi-criteria decision-making (MCDM) tools to optimize performance with refrigerants for automobile air conditioning. Three MCDM tools were used: AHP, SAW, and GRA. The distinctiveness of this research lies in its comprehensive method of refrigerant selection which combines MCDM tools that provide an objective assessment of various aspects. In addition, the use of Spearman correlation as well as Borda count methodologies provides solid proof of relationship, stability, and the ranking of alternatives as well as ensuring the validity of the outcomes. The study provides valuable insight into the selection of sustainable refrigerants, helping to advance green and economical cooling techniques. The weightage-normalized matrix shows that R430A is the optimal refrigerant. R430A is an extremely efficient performance element with the lowest GWP, Ozone depletion potential (ODP). This research not only provides valuable insights for selecting sustainable refrigerants but also underscores the potential of MCDM frameworks in advancing green, economical cooling technologies. Ultimately, this approach supports the broader adoption of environmentally responsible practices in the automotive industry, paving the way for future studies to further refine refrigerant selection models.

Keywords Optimization · Decision making · GRA · AHP · SAW · MCDM

P. N. Reddy (✉)
R&D/Engineering Division, Retearn Technologies @ Recykal, Hyderabad, India
e-mail: nithish.pasham@recykal.com

S. Thangavel
Department of Mechanical Engineering, Global College of Engineering and Technology, Muscat, Oman

© The Author(s), under exclusive license to Springer Nature Switzerland AG 2025

K. C. Rath et al. (eds.), *Engineering, Technology and Management*, Information Systems Engineering and Management 54, https://doi.org/10.1007/978-3-031-93654-8_8

Preface
This chapter provides advancements in technology have made selecting the right refrigerant for automobile air conditioning more accurate and efficient. It explains the importance of considering not just the refrigerant's physical properties, but also its environmental impact, cost, and other factors. Using three multi-criteria decision-making (MCDM) tools-AHP, SAW, and GRA-the study offers a clear, objective method for evaluating different refrigerants. The research finds R430A to be the best option due to its strong performance, low Global Warming Potential (GWP), and minimal Ozone Depletion Potential (ODP). This chapter highlights the value of using MCDM tools to choose sustainable refrigerants and supports the adoption of greener, more economical cooling solutions in the automotive industry.

8.1 Introduction

The choice of refrigerant is important in terms of the system's performance, its environmental impact as well as economic viability. The research focuses on maximizing the choice of refrigerants that have lower GWP. The research covers the analysis of refrigerants based on properties as well as environmental and financial considerations.

8.1.1 Objectives

The main goal is to find the ideal refrigerant suitable for automobile air conditioning systems that have a balance between high performance and low environmental impacts. Particularly, this study seeks to:

1. Examine the efficiency of various refrigerants by evaluating them according to factors like effectiveness, GWP, and costs.
2. Use the MCDM tools (SAW, AHP, and GRA) to assess and classify these refrigerants.
3. Find the refrigerant that provides the highest balance of performance and environmental positives.

8.1.2 Importance of Low GWP Refrigerants

Growing concern about the effects of climate change on our environment has increased the demand to find refrigerants that have low GWP. Refrigerants that are traditional like R134a have been recognized as contributing to global warming. So, deciding on the right option that reduces environmental impacts while enhancing or maintaining the performance of the system is essential. This goal was chosen to tackle these urgent environmental problems while also ensuring automobile air conditioning systems are effective and economical.

The uniqueness of this study is its broad method of selecting refrigerants that utilize a blend of modern MCDM techniques. Through the implementation of AHP and SAW as well as GRA this study presents an efficient framework to evaluate refrigerants from a variety of perspectives. Furthermore, attention paid to refrigerants that have low GWP, like R430A, as well as applying statistical approaches such as Spearman correlation as well as Borda count to verify the results. This method not only improves the process of making decisions but aids in better and more efficient cooling technology.

8.2 Literature Review

The extensive usage of temperature refrigerants that cause global warming such as R134a in car air conditioning units has resulted in environmental guidelines that limit their use. The standards of the European Union limit GWP up to 150. Thermodynamic characteristics, flammability toxicity, and performance specifications determine the refrigerant to choose. However, the cost of manufacturing and requirements like latent heat from fusion thermal conductivity, and the specific heat's lower capacity for depletion of ozone along with low inventory costs could cause a dilemma when choosing a refrigerant. Recent studies have offered alternative options for R134a for use as an AAC refrigerant. However, the current methods are slow and cost a lot. The latest MCDM developments must deliver the most efficient solutions, which take into account the thermodynamics, environmental as well as economic aspects in a single study. Researchers identified eco-safe refrigerants for car AC systems. Their list includes hydrocarbons, hydrofluorocarbons, and carbon dioxide/ carbon dioxide as environmentally-friendly options among others. Carbon dioxide is long as a sustainable alternative to electric cars operating in low temperatures; whereas hydrofluoroolefins could be used as a short-term alternative due to the environmental impacts [1]. Few performed an energy efficiency assessment of R1234yf R1234ze(E), R600a, and lower GWP alternatives, such as R290 and R152a that serve as alternatives to R134a. The study examines alternative refrigerants with less global warming possibility (GWP) ratings as compared to R134a refrigerant, which is subject to EU guidelines because of its possible impact on the climate. The five alternatives (R290 R290, R600a R152a (R1234yf) R1234yf, and R1234a as well as R1234ze(E)) have been tested in a single facility within their context and act as substitutes that can be used in drop-in applications, which focuses on efficiency in energy use as part of the research that demonstrates the potential for them to replace R134a for cooling and air application [2]. Others enhanced the effectiveness of solar-powered milk chillers that use direct current compressors by testing HFC-134a as well as HC-600a refrigerants. They determined the optimal levels of charging with the help of exergy and efficiency coefficients. They found that HC-600a beat HFC-134a when it came to the production of ice and inefficiency as solar power reduced energy consumption

[3]. Some explored the possibility of using solar power for dairy chillers that do not require batteries, optimizing a thermostatic expansion device that incorporates a suction line exchanger for automotive AC systems. Along with the selection of refrigerants, the heat transfer aspects related to convection and diffusion are also important for efficient energy management, the fundamentals of double diffusion and heat transfer enhancement knowledge are vital [4, 5]. A study found that using an effective heat exchanger improved the efficiency of charging by 19.4%, while also increasing the cooling capacity by 8.7% and increasing the performance coefficient (COP) to 16.2%. R430A is an alternative to R134a for automotive ACs and was found to have greater COPs, and decreased energy consumption of the compressor, with reliable fluids [6].

Few compared R134a as well as R152a in ACs for passenger cars and found that R152a offered superior cooling capabilities as well as a 13.5% greater C.O.P. [7]. Recent investigations evaluated HFO-1234yf as a green refrigerant option and noted reduced cooling power and improved performance of the compressor [8]. Researchers optimized the vapor compression system using R290, R1234yf, as well as R744, determining that R290 is the most energy-efficient and environmentally sustainable [9]. Some looked at portable AC systems that incorporate thermal energy storage. They found that graphene-enhanced material improved thermal conductivity as well as less CO_2 emission [10]. Few researchers constructed models for open piston compressors with R134a, R1234yf, and R290 and emphasized the accuracy of the model in predicting the performance of compressors [11]. In recent times CO_2 heat exchangers within electric vehicles become popular because of their high efficiency at low temperatures, with good COP [12]. The use of R1234ze(E) as well as R450A in systems for vapor compression increases efficiency. IHX helped reduce energy usage [13]. Few scientists have identified R430A as a green alternative for R134a used in automobile ACs that have greater efficiency and less emission [14]. Many researchers conducted investigations of an automobile AC system that uses R152a as a refrigerant. They evaluated R-134a to highlight R152a's absence of emission of ozone and less potential for global warming. The study evaluated the efficiency of cooling in terms of power reduction, cooling efficiency, as well as the coefficient of performance (COP) in various environments. The results showed that R152a systems are more efficient than R134a and could be a feasible alternative to car air cooling [15]. Few investigated engine-powered heat pump systems using R134a as well as R152a refrigerants. They focused on the gas-source engine pump (GEHP) effectiveness. They discovered that R152a was better at heating and had higher compressor efficiency and less gas discharge pressures than R134a and R134a, which makes it a desirable choice for improving GEHP effectiveness and decreasing carbon footprints [16]. Researchers employed the TOPSIS method of selecting the best glass material to be used in solar thermal projects and compared seven glass materials using six parameters. Several studies were carried out exploring ventilation and air conditioning with numerical techniques and Taguchi together [17]. Few experiments examined the effects of R-1234yf with R-290 within open piston compressor units in different conditions. They also looked into the environmental effects and refrigerant characteristics. The researchers found that these refrigerants are acceptable alternatives to R134a and R-1234yf for mobile ACs [18]. Scientists developed an approach to fuzzy decision-making for picking small heat exchangers focused on the efficiency of industrial processes and optimization of layout. The multi-level, non-structural multi-factor design utilized fuzzy optimization to enhance design and efficiency in energy use, as well as performance [19]. Scientists created an experimental CO_2-powered vehicle AC with a swash plate compressor and fin tube cooler expansion valve that is manual, as well as an integrated heat exchanger. A thorough test revealed that the correct selection of lubricant, CO_2 charges, and adjustment of the compressor significantly increase the efficiency of the system. This highlights CO_2's capabilities as a refrigerant [20]. An extensive analysis was carried out on the methods of selection of the refrigerant and emphasized the significance of sustainability when it comes to refrigeration and heat pumps [21]. The application of MCDM instruments to select low-GWP refrigerants and highlight their efficacy in maximizing the efficiency of air conditioning has become popular [22]. New experiments were conducted on the assessment of the environmental and financial impact of refrigerants that are alternative to those used in car air conditioning systems. They identify the most important factors that influence refrigerant selection [23]. A new integrated MCDM method was introduced that evaluates refrigerants using their thermodynamic and environmental performances which highlights the importance of well-balanced criteria for decision-making [24]. Few research works focused on comprehensive reviews of the refrigerant selection methods used in HVAC systems. They focus on the use of MCDM techniques [25]. Few conducted an analysis that compares different MCDM methods of selecting refrigerants and provides insights about their efficiency in eco-friendly cooling systems [26]. A new proposal on an innovative hybrid MCDM strategy for deciding on refrigerants with low GWP in automobile air conditioning was introduced. This demonstrated greater decision-making precision [27]. In recent times different MCDM techniques for selecting refrigerants for eco-friendly cooling techniques with a focus on their contribution to sustainable development [28]. Several MCDM methods, optimization techniques, and approaches have evolved to simplify experimentation and attain the best results with the least effort [29–31]. The application of AHP as well as GRA for evaluating refrigerant choices helped demonstrate a green method of making decisions [32]. Additionally, the application of multi-objective optimization for choosing refrigerants in automotive AC systems, employing MCDM tools to achieve maximum performance and environmental sustainability was understood as effective. The results of these studies highlight the significance of MCDM methods in selecting refrigerants that are balanced between the economic, environmental, and performance requirements [33].

The novelty of this work lies in its holistic approach to refrigerant selection, integrating MCDM tools to provide a balanced assessment of multiple criteria. Additionally, the application of Spearman correlation and Borda count methods provides a robust verification of the correlation, stability, and ranking of the alternatives, ensuring the reliability of the results. This study offers valuable insights into sustainable refrigerant selection, contributing to the advancement of environmentally friendly and economically viable cooling technologies.

8.3 Multi-criteria Decision Making (MCDM)

MCDM is a process used to evaluate multiple options based on multiple criteria. The steps involved in the MCDM process, including:

I. **Define the Problem and Identify Alternatives**: Clearly identify the decision you need to make. Brainstorm and list all the potential options (alternatives) you have.
II. **Establish Evaluation Criteria**: Define the factors (criteria) that are important for evaluating your options. These criteria should be relevant to your decision and measurable.
III. **Construct the Decision Matrix**: Create a table with rows representing alternatives and columns representing criteria. Fill the table with data that reflects each alternative's performance on each criterion. This data can be ratings, scores, or any quantitative measurement relevant to the criteria.
IV. **Weighting the Criteria**: Not all criteria may have equal importance. A higher weight indicates a more critical factor.
V. **Normalization**: Depending on the MCDM method used, you might need to normalize the data in the decision matrix. This ensures all criteria are on a comparable scale for a fair evaluation. Different normalization techniques exist, like converting scores to a 0–1 range.
VI. **Weighted Normalized Decision Matrix**: Multiply the weights from Step 4 with the normalized data from Step 5. This creates a new matrix that reflects both the importance of each criterion and the performance of each alternative.
VII. **Ideal Solutions**: Some MCDM methods define positive and negative ideal solutions. These may be hypothetical constructs to serve as reference points for evaluation.
VIII. **Calculate Preference Scores**: The specific method used here depends on the chosen MCDM technique (e.g., TOPSIS, EDAS, MOORA). Each method has its algorithm for calculating a score that reflects how close each alternative is to the ideal solution (positive or both positive and negative).
IX. **Rank the Alternatives**: Based on the preference scores from step 8, rank the alternatives from the most to least preferred. The option with the highest preference score is generally considered the best choice based on the defined criteria and weights.
X. **Sensitivity Analysis**: This is a check effect of changes in the weights or the data in the decision matrix. By slightly modifying these values, you can see if the ranking drastically changes, revealing potential weaknesses in your model.

The methodology offers a general framework, and the specific steps involved might differ based on the chosen MCDM method.

8.3.1 Problem Definition

The physical properties that affect temperature, such as cooling capacity as well as efficiency coefficient (COP) the power consumed, and the rate of heat transfer contribute to AAC system effectiveness. The other important factors include toxicity, flammability asphyxiation, and physical risks in selecting refrigerants that meet safety guidelines such as the ones created by ASHRAE. Costs can also impact the system's design as well as security measures when choosing new refrigerants. In the end, choosing the best refrigerant will depend on thermodynamic characteristics along with environmental, economic, and other aspects that all work to achieve maximum efficiency.

The ladder structure of four stages was utilized to pick the AAC refrigerant suitable for use. The first stage focused on determining the most efficient refrigeration. It was then thermodynamic properties as well as environmental and financial factors. Stage three comprised nine criteria and level 4 offered a variety of choices, such as Stage 3 comprised nine subcriteria while level four provided various alternatives.

Researchers employed nine sub-criteria to choose 14 refrigerants potential subjects for studies about 14 refrigerants including cost-effectiveness, backgrounds, thermodynamic and environmental behaviors, and weightiness of the subjective matrix for the evaluation of the criteria used to make decisions The grades of these decision criteria were rated in ascending order based on the median grade, which was normalized with the score of criteria.

8.4 Methodology

The study employs a multi-criteria decision-making (MCDM) approach, using AHP, SAW, and GRA tools to evaluate refrigerant options based on environmental, economic, and performance criteria. Spearman correlation and Borda count methods are additionally applied to confirm ranking stability and validate the results.

8.4.1 Analytic Hierarchy Process (AHP)

It involves a variety of processes and mathematical formulas used to effectively evaluate and prioritize options based on a variety of parameters. These are the main elements and the steps that correspond formulas to the AHP methodology:

Step 1: Criteria Identification and Hierarchy Formation

Determine the relevant criteria to the issue of choice and arrange them in an orderly structure, that has the goal in the upper levels and the sub-criteria being placed at lower levels.

Step 2: Pair-Wise Comparison

Compare each criterion against another to establish its relative importance and use numbers on a scale to express it numerically.

To help make decisions based on criteria, create an x-sized pairwise comparison matrix in which each element (column vector) of this pairwise comparison matrix represents the preferred quality against criteria. Each element in this pairwise comparison matrix must then be assigned with specific weighting depending upon criteria weights.

$$b_{ij} = \frac{a_{ij}}{\sum_{i=1}^{n} a_{ij}} \tag{8.1}$$

and C matrix is created as

$$C = \begin{bmatrix} c_{11} & \cdots & C_{1n} \\ \vdots & \ddots & \vdots \\ c_{n1} & \cdots & c_{nn} \end{bmatrix} \tag{8.2}$$

Priority vector is given by

$$W_i = \frac{\sum_{j=1}^{n} C_{ij}}{n} \tag{8.3}$$

Step 3: Consistency Check

Determine the Index of Consistency (IC), to gauge the degree of consistency among comparison pairs, while also finding an Index of Randomness (RI), determined by the dimension of your matrix.

Calculate a consistency ratio (CR). When this exceeds an established threshold (typically 0.10), adjustments need to be made in pairwise comparisons to maximize consistency.

Step 4: Calculate the Overall Score

$$S = W\,X\,C \tag{8.4}$$

where A represents the matrix of evaluations of alternatives by criteria

8.4.2 Simple Additive Weighting (SAW)

Simple Additive Weighting (SAW) calculates the score of each option through the aggregation of a score of the attributes or the criteria. The following are the most important methods and the steps that correspond to methods for SAW:

Step 1: Criteria Identification and Weight Assignment

Find the criteria or characteristics pertinent to the issue. Affix weights to each of the criteria that reflect the relative significance. The weights are usually adjusted to make sure they add at 1.

Step 2: Alternative Evaluation

- Examine each option in relation to each of the criteria. It could mean giving scores or ratings in order to determine how the various options perform on each of the criteria.

Step 3: Normalization of Scores

- Standardize the scores for the various options for each criterion to guarantee an equal chance of comparison. This involves changing the scores to one common scale between 0 and 1.

$$\mathrm{Sij} = \frac{\mathrm{X_{ij}}}{\mathrm{maxX_{ij}}} \tag{8.5}$$

if j is the benefit attribute

$$\mathrm{Sij} = \frac{\mathrm{X_{ij}}}{\mathrm{maxX_{ij}}} \tag{8.6}$$

if j is the cost attribute

Step 4: Weighted Sum Calculation

Determine the weighted sum of every option by dividing the normalized score for every criterion by its respective weight before summing the items.

For an alternative Aj with m criteria and weights w_i and normalized scores s_{ij}, the weighted sum is calculated as:

$$SAW_j = \sum_{i=1}^{n} w_i S_{ij} \tag{8.7}$$

where

SAW_j	Weighted sum of scores for alternative j.
w_i	Weight assigned to criterion *i*.
s_{ij}	Normalized score of alternative j on criterion i.

Step 5: Ranking of Alternatives

Sort the options by their weighted sums. The option with the most weighted sum is the popular choice according to the SAW technique.

8.4.3 Grey Relational Analysis (GRA)

Analysis of Grey Relating: Analysing all influences between various variables is commonly known as factor relationship whitening.

Grey Relational Analysis (GRA) can be utilized to select an optimal condition for various input parameters to reach quality properties that most closely meet user preferences. Grey relational analysis also helps assess or judge effectiveness with minimal information provided; GRA offers one method of finding optimal conditions across multi-objective issues by assigning weights accordingly and assigning priorities according to each response—these actions and steps constitute GRA.

Step 1. Data Pre-processing: For Grey analysis to work effectively, data that will be utilized must first be converted to quantitative indexes that can normalize raw data for further use in other analyses.

Step 2. Normalizing raw data involves turning its original sequence of figures into decimal figures between 0.00–1.00 in order to facilitate comparison with future series that might use it, with those having higher sequences being normalized according to "higher-the-better." For instance, if an expected set has the shape "Higher is Better", its original one must also conform.

Step 3. Deviation Sequence: The deviation sequence of the reference sequence is given by,

$$\Delta_{0i}(k) = \|x_0^*(k) - x_i^*(k)\| \tag{8.8}$$

Step 4. Grey Relational Coefficient: The grey relationship coefficient measures the connection between theoretical and actual outcomes, providing an assessment of whether there exists any kind of gap. Essentially, this number describes its meaning by quantifying it into numbers that express correlation and causality relationships between events or actions and outcomes in general terms. It can be expressed using percentage figures.

$$\zeta_i = \frac{\Delta_{\max} + \zeta.\Delta_{\max}}{\Delta_{oi}(k) + \zeta.\Delta_{\max}} \tag{8.9}$$

where $\Delta_{oi}(k)$ is the deviation sequence of the reference sequence

Step 5. Gray Relative Grade: Once you have identified the grey relational coefficient—typically mean and median values—this value becomes known as grey relational grade and can be easily identified using its unique identification code.

$$\Upsilon_i = \frac{1}{n}\sum_{k=1}^{n} \zeta_i(k) \tag{8.10}$$

8.5 Results and Discussion

The research employs multi-objective tools for decision-making such as AHP, SAW, and GRA to determine the best refrigerant suitable for AAC applications. The values of the normalized matrix for different refrigerants were determined by using these equations in Sects. 8.4.1, 8.4.2, and 8.4.3, respectively, for AHP, SAW, and GRA in addition to the weights.

Table 8.1 presents the 14 refrigerants selected for the current study and the corresponding 13 criteria named C to C13. The first four are beneficial and C5 to C13 are considered non-beneficial criteria. The weights for each of the criteria are calculated and displayed at the bottom of the table1, whereas Tables 8.2 and 8.3 show the weights calculated by SAW and GRA Methods. The higher weights were recorded for C4, C6 and the lower for C7 to C10. Tables 8.2 and 8.3 also present the objective levels.

Table 8.4, Table 8.5, and Table 8.6 give normalized values of refrigerant fluids considering performance aspects, flammability, and temperature obtained through the AHP method, SAW, and GRA respectively. The highest weighted normalized matrix value for the GWP value is R134a which has a greater GWP (1350) over other refrigerants. The refrigerants R600a, R290, and a mix comprising R290 and R744 contain the highest toxic and flammability levels. R744 also has higher critical pressure and saturation pressure. Tables 8.7, and 8.8 represents the Gray relational coefficient and Gray relational grade obtained by the GRA method.

Table 8.9 shows AHP analysis results for refrigerant fluids, ranking R430A as the best performer. R430A has the highest assessment value of 0.786 among the 14 refrigerants. Table 8.10 presents the Ai and Ranking obtained by the SAW method. Table 8.11 represents the Grey relational grade and Rank of each alternative.

The SAW method evaluates 14 refrigerants, with R430A having the highest assessment value of 0.786. The weightage-normalized matrix shows that R430A is the optimal. However, R436A is the worst performer due to increased flammability issues.

Table 8.11 demonstrates the performance of different refrigerants, determined by GRA and GR which include R430A, R445A, and R444A ranking top and bottom in the order. R744 + R290 R1234ze(E) R1270 and R744ze(E) mix are the three most efficient. R430A is an extremely efficient performance element that has the lowest GWP, ODP, and a very low price and also an extremely high temperature latent evaporate. R744 + R290 has been deemed as the most dangerous due to the higher density of vapor and cost. R290 in combination with R744 is the most harmful due to its increased problems with flame-ability.

Table 8.1 Weights obtained using AHP methods

Criteria		C-1	C-2	C-3	C-4	C-5	C-6	C-7	C-8	C-9	C-10	C-11	C-12	C-13
Alternative (A)	Refrigerants (R)	Critical temperature (Tc)	Vapor density (rho)	Latent heat (LH)	Thermal conductivity (K)	Critical pressure (Pc)	Saturated pressure (Ps)	Liquid density (ρ)	Viscosity	Pressure (P) at 60 °C	O-D-P	G-W-P	Toxicity and flammability	Cost per kg
		°C	kg m^{-3}	kJ kg^{-1}	W m k^{-1}	bar	bar	kg m^{-3}	μPa s	bar	–	–	–	USD
		Beneficial				Non-beneficial								
A-1	R-134a	102	17	194.7	0.089	41	3.49	1278	250	16.8	1	4	1	5
A-2	R-152a	113.4	9.9	301.9	0.106	45.1	3.14	947.7	206	15	1	1	2	5.5
A-3	R-1234yf	95	21	160.02	0.074	34	3.73	1160	196	16.4	1	1	1	12.33
A-4	R-1234ze (E)	109.4	41	154.8	0.078	36.3	2.59	1112	269	12.7	1	1	1	53
A-5	R-1233zd (E)	166.6	36	188.52	0.081	36.2	0.59	1320	470	38.7	1	1	1	39.5
A-6	R-290	96.74	12	367.73	0.103	42.5	5.51	521.8	119	21.2	2	1	4	5.98
A-7	R-600a	134.6	5	349.56	0.097	36.3	1.87	574.8	187	8.69	2	1	4	10
A-8	R-744	30.98	114	214.98	0.104	73.8	39.7	896	90.8	8.69	1	1	1	10.91
A-9	R-1270	91.06	14	369.8	0.093	45.5	6.76	538.6	115	25.2	2	1	4	4.5
A-10	R744+R290	57	78	217.4	0.083	67.9	29.9	644.9	87.7	8.69	2	1	4	9
A-11	R-430A	106.9	11	295.2	0.984	40.8	3.5	802.1	180	15.7	2	1	4	7
A-12	R-436A	115.9	8.7	365.5	0.1	42.7	3.85	548	146	15.1	2	1	4	8
A-13	R-444A	101.2	29	180.5	0.9	42.3	4.47	1199	225	19.8	1	1	1	50
A-14	R-445A	104.7	29	190.4	0.95	44.9	4.67	1190	220	19.4	1	1	1	46
		166.6	114	369.8	0.984	34	0.59	521.8	87.7	8.69	1	1	1	4.5

(continued)

Table 8.1 (continued)

Criteria		C-1	C-2	C-3	C-4	C-5	C-6	C-7	C-8	C-9	C-10	C-11	C-12	C-13
Alternative (A)	Refrigerants (R)	Critical temperature (Tc)	Vapor density (rho)	Latent heat (LH)	Thermal conductivity (K)	Critical pressure (Pc)	Saturated pressure (Ps)	Liquid density (ρ)	Viscosity	Pressure (P) at 60 °C	O-D-P	G-W-P	Toxicity and flammability	Cost per kg
		°C	kg m^{-3}	kJ kg^{-1}	W m k^{-1}	bar	bar	kg m^{-3}	µPa s	bar	–	–	–	USD
		Beneficial				Non-beneficial								
	Avg	101.8	30	253.64	0.274	45	8.13	909.5	197	18	1	1	2	19.05
		Weightage calculations												
	Wj	0.018	0.1	0.0179	0.226	0.01	0.23	0.019	0.04	0.03	0	0	0	0.144

Table 8.2 Weights for the different criteria obtained by SAW method

Criteria	C-1	C-2	C-3	C-4	C-5	C-6	C-7	C-8	C-9	C-10	C-11	C-12	C-13
Objective	H	L	L	L	H	L	H	L	L	L	L	L	L
Wj	0.018	0.010	0.226	0.019	0.134	0.018	0.226	0.036	0.031	0.020	0.040	0.069	0.144

Table 8.3 Weights for the different criteria obtained by GRA

	C-1	C-2	C-3	C-4	C-5	C-6	C-7	C-8	C-9	C-10	C-11	C-12	C-13
Objective	H	L	L	L	H	L	H	L	L	L	L	L	L
Wj	0.018	0.134	0.018	0.226	0.010	0.226	0.019	0.036	0.031	0.021	0.048	0.069	0.144
MAX	166.6	114	369.8	0.984	73.8	39.7	1319.8	470	38.7	2	4	4	53
MIN	30.98	5.01	154.8	0.074	34	0.59	521.75	87.7	8.69	1	1	1	4.5
	Beneficial				Non-beneficial								

Table 8.4 Normalized matrix (AHP method)

Normalized matrix													
	C-1	C-2	C-3	C-4	C-5	C-6	C-7	C-8	C-9	C-10	C-11	C-12	C-13
	Beneficial				Non-beneficial								
A-1	0.612	0.150	0.527	0.090	0.829	0.169	0.408	0.351	0.517	1	0	1	0.900
A-2	0.681	0.087	0.816	0.108	0.754	0.188	0.551	0.426	0.579	1	1	1	0.818
A-3	0.570	0.182	0.433	0.075	1.000	0.158	0.450	0.447	0.530	1	1	1	0.365
A-4	0.657	0.356	0.419	0.079	0.937	0.228	0.469	0.326	0.684	1	1	1	0.085
A-5	1.000	0.312	0.510	0.082	0.939	1.000	0.395	0.187	0.225	1	1	1	0.114
A-6	0.581	0.104	0.994	0.105	0.800	0.107	1.000	0.737	0.410	1	1	0	0.753
A-7	0.808	0.044	0.945	0.099	0.937	0.316	0.908	0.469	1.000	1	1	0	0.450
A-8	0.186	1.000	0.581	0.106	0.461	0.015	0.582	0.966	1.000	1	1	1	0.412
A-9	0.547	0.125	1.000	0.095	0.747	0.087	0.969	0.763	0.345	1	1	0	1.000
A-10	0.342	0.687	0.588	0.084	0.501	0.020	0.809	1.000	1.000	1	1	0	0.500
A-11	0.642	0.094	0.798	1.000	0.833	0.169	0.650	0.487	0.554	1	1	0	0.643
A-12	0.696	0.076	0.988	0.102	0.796	0.153	0.952	0.601	0.575	1	1	0	0.563
A-13	0.607	0.250	0.488	0.915	0.804	0.132	0.435	0.390	0.439	1	1	1	0.090
A-14	0.628	0.258	0.515	0.965	0.757	0.126	0.438	0.399	0.448	1	1	1	0.098

This study reviews the various refrigerants using AHP, SAW, and GRA techniques and the comparison of the ranks is shown in Fig. 8.1. R430A is the most efficient refrigerant due to its larger latent heat, low viscosity, low GWP and lower T and F, and low OZP as well as its lower cost. AHP, SAW, and GRA are the most similar coefficients. This means that R430A can meet all specifications for performance. R-436A, as well as R1233zd are among the most hazardous refrigerants for SAW as well as GRA in both instances due to their cost and high OZP as well as their high GHP.

Spearman's Correlation Coefficient

Spearman correlation coefficient is a statistical dependence of ranking between two variables.

$$R = 1 - \frac{6\sum D^2}{N^3 - N} \tag{8.11}$$

where N = no. of data points

D = ranks difference

For the current study when the rank is compared, the Spearman correlation coefficient between the AHP and SAW is 0.9948, and for GRA with AHP/SAW is about 0.899. The interpretation for Spearman coefficients suggests [33] that any value above 0.8 has a very strong correlation and the results are stable.

Table 8.5 Sample normalized matrix obtained by SAW method

Normalized matrix

	C-1	C-2	C-3	C-4	C-5	C-6	C-7	C-8	C-9	C-10	C-11	C-12	C-13
A-1	0.6120	0.8290	0.1690	0.4080	0.1500	0.5270	0.0900	0.3510	0.5170	1.0000	0.2500	1.0000	0.9000
A-9	0.5470	0.7470	0.0870	0.9690	0.1250	1.0000	0.0950	0.7630	0.3450	0.5000	1.0000	0.2500	1.0000
A-10	0.3420	0.5010	0.0200	0.8090	0.6870	0.5880	0.0840	1.0000	1.0000	0.5000	1.0000	0.2500	0.5000
A-13	0.6070	0.8040	0.1320	0.4350	0.2500	0.4880	0.9150	0.3900	0.4390	1.0000	1.0000	1.0000	0.0900
A-14	0.6280	0.7570	0.1260	0.4380	0.2580	0.5150	0.9650	0.3990	0.4480	1.0000	1.0000	1.0000	0.0980

Table 8.6 Normalized matrix obtained by GRA

	C-1	C-2	C-3	C-4	C-5	C-6	C-7	C-8	C-9	C-10	C-11	C-12	C-13
A-1	0.5237	0.1109	0.1856	0.0165	0.8241	0.9259	0.0523	0.5755	0.7298	1.0000	0.0000	1.0000	0.9897
A-2	0.6077	0.0448	0.6842	0.0352	0.7211	0.9348	0.4663	0.6906	0.7897	1.0000	1.0000	0.6667	0.9794
A-3	0.4721	0.1440	0.0243	0.0000	1.0000	0.9197	0.1997	0.7167	0.7431	1.0000	1.0000	1.0000	0.8386
A-4	0.5782	0.3265	0.0000	0.0044	0.9422	0.9489	0.2610	0.5258	0.8664	1.0000	1.0000	1.0000	0.0000
A-5	1.0000	0.2807	0.1568	0.0077	0.9447	1.0000	0.0000	0.0000	0.0000	1.0000	1.0000	1.0000	0.2784
A-6	0.4849	0.0632	0.9904	0.0319	0.7864	0.8742	1.0000	0.9181	0.5831	0.0000	1.0000	0.0000	0.9695
A-7	0.7640	0.0000	0.9059	0.0253	0.9422	0.9673	0.9335	0.7403	1.0000	0.0000	1.0000	0.0000	0.8866
A-8	0.0000	1.0000	0.2799	0.0330	0.0000	0.0000	0.5310	0.9919	1.0000	1.0000	1.0000	1.0000	0.8678
A-9	0.4430	0.0843	1.0000	0.0209	0.7111	0.8422	0.9789	0.9286	0.4499	0.0000	1.0000	0.0000	1.0000
A-10	0.1919	0.6724	0.2912	0.0099	0.1482	0.2506	0.8457	1.0000	1.0000	0.0000	1.0000	0.0000	0.9072
A-11	0.5598	0.0522	0.6530	1.0000	0.8291	0.9256	0.6487	0.7586	0.7664	0.0000	1.0000	0.0000	0.9485
A-12	0.6262	0.0335	0.9800	0.0286	0.7814	0.9166	0.9671	0.8475	0.7864	0.0000	1.0000	0.0000	0.9278
A-13	0.5178	0.2155	0.1195	0.9077	0.7915	0.9008	0.1512	0.6409	0.6298	1.0000	1.0000	1.0000	0.0619
A-14	0.5436	0.2238	0.1656	0.9626	0.7261	0.8957	0.1626	0.6539	0.6431	1.0000	1.0000	1.0000	0.1443
MAX	1	1	1	1	1	1	1	1	1	1	1	1	1
MIN	0	0	0	0	0	0	0	0	0	0	0	0	0
	Beneficial				Non-beneficial								

Table 8.7 Grey relational coefficients calculations

Grey relational coefficient													
	C-1	C-2	C-3	C-4	C-5	C-6	C-7	C-8	C-9	C-10	C-11	C-12	C-13
A-1	0.5121	0.36	0.3804	0.33704	0.7398	0.871	0.3454	0.5408	0.6491	1	0.3333	1	0.9798
A-2	0.5604	0.3436	0.61288	0.34134	0.6419	0.885	0.4837	0.6177	0.704	1	1	0.6	0.9604
A-3	0.4864	0.3687	0.33882	0.33333	1	0.862	0.3845	0.6383	0.6606	1	1	1	0.75592
A-4	0.5424	0.4261	0.33333	0.33431	0.8964	0.907	0.4036	0.5132	0.7891	1	1	1	0.33333
A-5	1	0.4101	0.37226	0.33505	0.9005	1	0.3333	0.3333	0.3333	1	1	1	0.40928
A-6	0.4926	0.348	0.98111	0.34057	0.7007	0.799	1	0.8593	0.5453	0.3333	1	0.3333	0.94248
A-7	0.6794	0.3333	0.84155	0.33905	0.8964	0.939	0.8827	0.6581	1	0.3333	1	0.3333	0.81513
A-8	0.3333	1	0.4098	0.34082	0.3333	0.333	0.516	0.984	1	1	1	1	0.79093
A-9	0.473	0.3532	1	0.33804	0.6338	0.76	0.9595	0.875	0.4761	0.3333	1	0.3333	1
A-10	0.3822	0.6042	0.41362	0.33555	0.3699	0.4	0.7642	1	1	0.3333	1	0.3333	0.84348
A-11	0.5318	0.3454	0.59033	1	0.7453	0.87	0.5873	0.6744	0.6816	0.3333	1	0.3333	0.90654
A-12	0.5722	0.3409	0.96154	0.33981	0.6958	0.857	0.9383	0.7663	0.7007	0.3333	1	0.3333	0.87387
A-13	0.509	0.3893	0.3622	0.84416	0.7057	0.834	0.3707	0.582	0.5746	1	1	1	0.34767
A-14	0.5228	0.3918	0.3747	0.93047	0.6461	0.827	0.3739	0.591	0.5835	1	1	1	0.36882
Wj	0.0179	0.1344	0.01789	0.22559	0.01	0.226	0.019	0.0363	0.0313	0.0213	0.0475	0.0694	0.14351
	Beneficial				Non-beneficial								

Table 8.8 Sample grey relational grade calculations

Grey relational grade													
	C-1	C-2	C-3	C-4	C-5	C-6	C-7	C-8	C-9	C-10	C-11	C-12	C-13
A-1	0.0092	0.0484	0.0068	0.0760	0.0074	0.1967	0.0066	0.0196	0.0203	0.0213	0.0158	0.0694	0.1406
A-2	0.0100	0.0462	0.0110	0.0770	0.0064	0.1998	0.0092	0.0224	0.0220	0.0213	0.0475	0.0417	0.1378
A-14	0.0093	0.0526	0.0067	0.2099	0.0065	0.1869	0.0071	0.0214	0.0182	0.0213	0.0475	0.0694	0.0529
	Beneficial				Non-beneficial								

Table 8.9 Ranking of alternatives by AHP method

	C-1	C-2	C-3	C-4	C-5	C-6	C-7	C-8	C-9	C-10	C-11	C-12	C-13	Sum	Rank
	Beneficial				Non-beneficial										
A-1	0.011	0.020	0.009	0.020	0.008	0.038	0.008	0.013	0.016	0.000	0.000	0.000	0.129	0.376	7
A-2	0.012	0.012	0.015	0.024	0.008	0.042	0.010	0.015	0.018	0.000	0.000	0.000	0.117	0.378	6
A-3	0.010	0.024	0.008	0.017	0.010	0.036	0.009	0.016	0.017	0.000	0.000	0.000	0.052	0.337	13
A-4	0.012	0.048	0.008	0.018	0.009	0.051	0.009	0.012	0.021	0.000	0.000	0.000	0.012	0.338	12
A-5	0.018	0.042	0.009	0.019	0.009	0.226	0.008	0.007	0.007	0.000	0.000	0.000	0.016	0.499	2
A-6	0.010	0.014	0.018	0.024	0.008	0.024	0.019	0.027	0.013	0.000	0.000	0.000	0.108	0.340	11
A-7	0.014	0.006	0.017	0.022	0.009	0.071	0.017	0.017	0.031	0.000	0.000	0.000	0.065	0.346	10
A-8	0.003	0.134	0.010	0.024	0.005	0.003	0.011	0.035	0.031	0.000	0.000	0.000	0.059	0.455	5
A-9	0.010	0.017	0.018	0.021	0.007	0.020	0.018	0.028	0.011	0.000	0.000	0.000	0.144	0.369	8
A-10	0.006	0.092	0.011	0.019	0.005	0.004	0.015	0.036	0.031	0.000	0.000	0.000	0.072	0.368	9
A-11	0.011	0.013	0.014	0.226	0.008	0.038	0.012	0.018	0.017	0.000	0.000	0.000	0.092	0.526	1
A-12	0.012	0.010	0.018	0.023	0.008	0.035	0.018	0.022	0.018	0.000	0.000	0.000	0.081	0.320	14
A-13	0.011	0.034	0.009	0.206	0.008	0.030	0.008	0.014	0.014	0.000	0.000	0.000	0.013	0.485	4
A-14	0.011	0.035	0.009	0.218	0.008	0.029	0.008	0.014	0.014	0.000	0.000	0.000	0.014	0.498	3

Table 8.10 Rank obtained by SAW method

Alternative	Refrigerant	Ai	Rank
A-1	R-134a	0.375887	7
A-2	R-152a	0.377768	6
A-3	R-1234yf	0.337083	13
A-4	R-1234ze (E)	0.338445	12
A-5	R-1233zd (E)	0.498795	2
A-6	R-290	0.340154	11
A-7	R-600a	0.345852	10
A-8	R-744	0.454765	5
A-9	R-1270	0.36888	8
A-10	R-744 + R-290	0.367669	9
A-11	R-430A	0.525563	1
A-12	R-436A	0.32004	14
A-13	R-444A	0.484739	4
A-14	R-445A	0.498164	3

Table 8.11 Ranking by grey relational analysis

Alternative	Sum	Grade	Rank
A-1	0.638183	0.045585	6
A-2	0.652388	0.046599	4
A-3	0.642064	0.045862	5
A-4	0.599347	0.042811	13
A-5	0.616026	0.044002	11
A-6	0.61775	0.044125	10
A-7	0.636181	0.045441	7
A-8	0.631747	0.045125	8
A-9	0.614314	0.04388	12
A-10	0.546128	0.039009	14
A-11	0.761001	0.054357	1
A-12	0.621247	0.044375	9
A-13	0.688197	0.049157	3
A-14	0.709984	0.050713	2

Borda Count Method

The Borda count method provides three options for modifications to ensure that R430A is ranked first among AAC refrigerants when used in conjunction with three multi-criteria decision-making techniques (MCDMs). Assuming equal voting for AHP, SAW, and GRA values are 33, 34, and 33 respectively (ref. Table 8.12). This

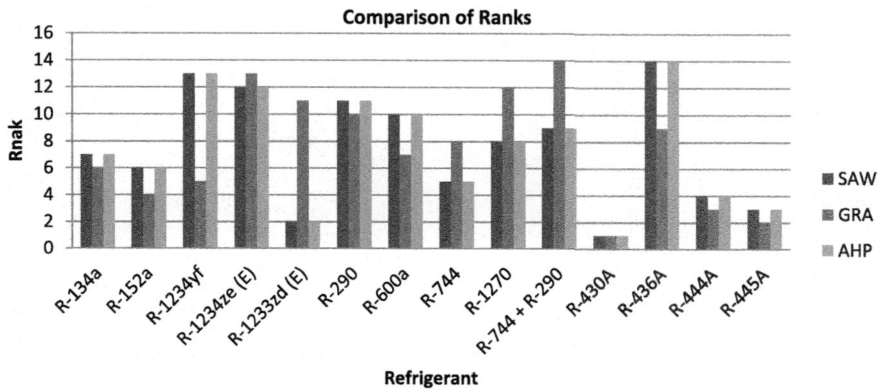

Fig. 8.1 Comparison of ranks obtained by different methods

method is where points are awarded to alternatives based on their rankings one point for the last option, two points for the second choice, and so on as shown in Fig. 8.2. The points for each ballot are added up to decide who gets the most points and wins declared the winner as shown in Table 8.12. By the three MCDM methods and the results gathered by Borda counting, R430A ranked first with 1400 points and R436A was less effective with only 265 points.

Table 8.12 Total Borda count and ranking for each alternative

Alternative	AHP	SAW	GRA	B. count	Rank
A-1	264	272	297	833	7
A-2	297	306	363	966	5
A-3	66	68	330	464	10
A-4	99	102	66	267	13
A-5	429	442	132	1003	4
A-6	132	136	165	433	12
A-7	165	170	264	599	8
A-8	330	340	231	901	6
A-9	231	238	99	568	9
A-10	198	204	33	435	11
A-11	462	476	462	1400	1
A-12	33	34	198	265	14
A-13	363	374	396	1133	3
A-14	396	408	429	1233	2

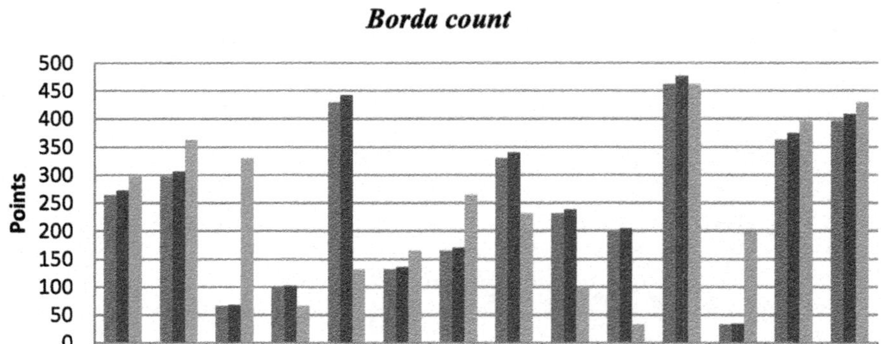

	A-1	A-2	A-3	A-4	A-5	A-6	A-7	A-8	A-9	A-10	A-11	A-12	A-13	A-14
AHP	264	297	66	99	429	132	165	330	231	198	462	33	363	396
SAW	272	306	68	102	442	136	170	340	238	204	476	34	374	408
GRA	297	363	330	66	132	165	264	231	99	33	462	198	396	429

Fig. 8.2 Borda points based on the rank of the refrigerant

8.6 Conclusions

There are three main factors to take into consideration to consider when choosing a refrigerant right one: environmental impact; economic factors as well as thermal properties. AHP, SAW, and GRA methods were utilized to pick the most efficient refrigerant. The following conclusions were drawn:

- R-430A was selected because it most closely matched all requirements for performance by MCDM analysis.
- A thorough analysis has concluded that R430A is an ideal refrigerant option for replacing R134a and supplies decision-makers with detailed guidelines to make their preferred refrigerant.
- R-436A, as well as R1233zd are among the most hazardous refrigerants for SAW as well as GRA in both instances due to their cost and high OZP as well as their high GHP.
- Following the three MCDM methods and the results gathered by Borda counting, R-430A ranked first with 14 points each. R744 and R290 were less effective.
- In addition, alternative options such as R445A and R744 are proven to be suitable for certain MCDM techniques.
- R744 + R290 has been deemed as the most dangerous due to the higher density of vapor and cost. R-290 in combination with R-744 is the most harmful due to its increased problems with flame-ability.
- Results of the Spearman correlation coefficient suggest a very strong correlation between the results obtained by the three methods.
- By Borda counting, R-430A is the best alternative with 1400 points and R-436A is the worst with 265 points.

Decision-makers need to have an impartial and practical approach to evaluating refrigerants. Therefore, this study is an essential resource for companies that are involved in AAC manufacturing, or research and development.

References

1. Mohanraj, M., Abraham, J.D.A.: Environment friendly refrigerant options for automobile air conditioners: a review. J. Therm. Anal. Calorim. 1–26 (2020)
2. Sanchez, D., Cabello, R., Llopis, R., Arauzo, I., Catalan-Gil, J., Torrella, E.: Energy performance evaluation of R1234yf, R1234ze(E), R600a, R290 and R152a as low-GWP R134a alternatives. Int. J. Refrig. **74**, 269–282 (2017)
3. Sidney, S., Prabakaran, R., Dhasan, M.L.: Charge optimisation of a solar milk chiller with direct current compressors. Proc. IME E J. Process Mech. Eng. **235**(3), 679–693 (2021)
4. Reddy, N., Murugesan, K.: Numerical investigations on the advantage of nanofluids under DDMC in a lid-driven cavity. Heat Transf. Asian Res. **46**(7), 1065–1086 (2017)
5. Reddy, N., Murugesan, K.: Numerical study of double diffusive convection in a lid-driven cavity with linearly salted side walls. J. Appl. Fluid Mech. **10**(1), 69–79 (2017)
6. Rajendran, P., Narayanaswamy, G.R., Dhasan, M.L.: Tuning thermostatic expansion valve for implementing suction line heat exchanger in mobile air conditioning system. J. Braz. Soc. Mech. Sci. Eng. **41**, 191 (2019)
7. Sumeru, K., Sunardi, C., Abdul, A.A., Henry, N., Moshood, A.A., Muhamad, S.M.F.: Comparative performance between R134a and R152a in an air conditioning system of a passenger car. J. Teknologi **78**, 10–12 (2016)
8. Rajendran, P., Sidney, S., Ramakrishnan, I., Dhasan, M.L.: Experimental studies on the performance of mobile air conditioning system using environmental friendly HFO-1234yf as a refrigerant. Proc. IME E J. Process Mech. Eng. **235**(3), 731–737 (2021)
9. De Paula, C.H., Duarte, W.M., Rocha, T.T.M., de Oliveira, R.N., Maia, A.A.T.: Optimal design and environmental, energy and exergy analysis of a vapor compression refrigeration system using R290, R1234yf, and R744 as alternatives to replace R134a. Int. J. Refrig. **113**, 10–20 (2020)
10. Prabakaran, R., Sidney, S., Lal, D.M., Harish, S., Kim, S.C.: Experimental performance of a mobile air conditioning unit with small thermal energy storage for idle stop/start vehicles. J. Therm. Anal. Calorim. (2021)
11. Sotomayor, P.O., Parise, J.A.R.: Characterization and simulation of an open piston compressor for application on automotive air conditioning systems operating with R134a, R1234yf and R290. Int. J. Refrig. **61**, 100–116 (2016)
12. Wang, D., Yu, B., Hu, J., Chen, L., Shi, J., Chen, J.: Heating performance characteristics of CO_2 heat pump for electrical vehicle in a cold climate. Int. J. Refrig. **85**, 27–41 (2018)
13. Mota-Babiloni, A., et al.: Drop-in analysis of an internal heat exchanger in a vapour compression system using R1234ze (E) and R450A as alternatives for R134a. Energy **90**, 1636–1644 (2015)
14. Andrew Pon Abraham, J.D., Mohanraj, M., Selvakumar, M., Raj, A.K.: Experimental assessments on R430A as an environment-friendly replacement to R134a in vehicle air conditioners. J. Braz. Soc. Mech. Sci. Eng. **43**(3), 1–18 (2021)
15. Yoo, S.Y., Lee, D.W.: Experimental study on performance of automotive air conditioning system using R152a refrigerant. Int. J. Automot. Technol. **10**(3), 313–320 (2009)
16. Ma, Z., Liu, F., Tian, C., Jia, L., Wu, W.: Experimental comparisons on a gas engine heat pump using R134a and low-GWP refrigerant R152a. Int. J. Refrig. **115**, 73–82 (2020)
17. Chandra, A.S., Reddy, P.N., Harish, R.: Natural ventilation in a large space with heat source: CFD visualization and Taguchi optimization. J. Therm. Eng. **8**(5), 642–655 (2022)
18. Zhou, G.Y., Wu, E., Tu, S.T.: Optimum selection of compact heat exchangers using non-structural fuzzy decision method. Appl. Energy **113**, 1801–1809 (2014)

19. Navarro, E., Martınez-Galvan, I.O., Nohales, J., Gonzalvez-Macia, J.: Comparative experimental study of an open piston compressor working with R-1234yf, R-134a and R-290. Int. J. Refrig. **36**, 768–775 (2013)
20. Liu, H., Chen, J., Chen, Z.: Experimental investigation of a CO_2 automotive air conditioner. Int. J. Refrig. **28**, 1293–1301 (2005)
21. Chua, K.J., Toh, H.K.: A review of refrigerant selection methodologies for sustainable heat pump and refrigeration systems. Int. J. Refrig. **135**, 70–84 (2022)
22. Hou, J., Wang, L.: Multi-criteria decision-making for low-GWP refrigerant selection in air conditioning systems. Energy Rep. **9**, 456–470 (2023)
23. Wang, T., Li, X., Liu, Y.: Environmental and economic assessment of alternative refrigerants in automotive air conditioning systems. Appl. Therm. Eng. **212**, 118549 (2023)
24. Yazdani, M., Zavadskas, E.K., Shadkam, E.: An integrated MCDM approach for evaluating refrigerants based on environmental and thermodynamic performance. Sustainability **14**(15), 9043 (2022)
25. Khatri, S.K., Chaudhary, R.: A comprehensive review of refrigerant selection techniques for HVAC systems: emphasis on MCDM approaches. J. Build. Eng. **58**, 104781 (2022)
26. Zhang, H., Chen, L., Wu, Y.: Comparative analysis of MCDM methods for refrigerant selection in sustainable cooling systems. J. Clean. Prod. **395**, 136490 (2023)
27. Patel, M., Sharma, S.: A hybrid MCDM approach for selecting low-GWP refrigerants in automotive air conditioning. J. Environ. Manag. **340**, 117847 (2023)
28. Singh, P., Singh, A.: Exploring MCDM methods for refrigerant selection in eco-friendly cooling technologies. Therm. Sci. Eng. Prog. **31**, 101361 (2022)
29. Mohanty, S., Mohanta, G.K., Senapati, A.K., Rath, K.C.: Taguchi grey relational analysis for optimal process parameter of low carbon steel welded by GTAW by using ER309L filler material. Mater. Today Proc. **44**, 2556–2561 (2021)
30. Muduli, K., Kusi-Sarpong, S., Yadav, D.K., Gupta, H., Jabbour, C.J.C.: An original assessment of the influence of soft dimensions on implementation of sustainability practices: implications for the thermal energy sector in fast-growing economies. Oper. Manag. Res. **14**, 337–358 (2021)
31. Jaiswal, S.K., Mukti, S.K., Rath, K.C.: E-waste control and its recycling to build sustainable society in the global context. In: Handbook of Research on Applications of AI, Digital Twin, and Internet of Things for Sustainable Development, pp. 200–222. IGI Global (2023)
32. Gupta, R., Kumar, P.: Evaluation of refrigerant alternatives using AHP and GRA: a sustainable approach. Sustain. Prod. Consum. **38**, 234–246 (2023)
33. Liu, Z., Zhang, X.: Multi-objective optimization of refrigerant selection for automotive AC systems using MCDM tools. Renew. Sustain. Energy Rev. **159**, 112204 (2022)

Dr. P. Nithish Reddy obtained his doctorate from IIT Roorkee and Masters from IIT Hyderabad. His area of specialization includes Energy, CFD, Hypersonic flows, Numerical Heat transfer, Nano fluids, Magnetohydrodynamic (MHD), DDC. He has submitted his Ph.D thesis at the age of 26 and having several SCI international research publications on his name. He has expertise in Fluid dynamics related solvers, FVM&FDM and guides students in Heat transfer and Aerospace related projects. He has experience of working on several CFD&HT related projects. Dr. Reddy is highly resourceful in executing projects, domain related inputs and Skills and has several achievements at national and international level.

Dr. Sivasakthivel Thangavel has been working as a Senior Lecturer in the department of Mechanical Engineering at Global College of Engineering and Technology since May 2018. Dr. Siva received the B.E. degree from Anna University, Chennai, India, in 2008, the M.Tech. degree from the National Institute of Technology Silchar, Assam, India, in 2011. He received his Ph.D.

degree from the Indian Institute of Technology Roorkee, Uttarakhand, India in Mechanical Engineering in 2015. From 2016 to 2018, he was a JSPS Postdoctoral fellow in Japan, where he worked on Pyrolysis, Gasification and Combustion processes. He has published more than 30 papers in refereed journals and conference proceedings.

Chapter 9
The Study of Quartz Fabrication Using ECDM

Amar Prakash and Raj Ballav

Abstract Electrochemical Discharge Machining is one of the most efficient techniques for micro-machining non-metallic brittle materials. ECDM works well to create microchannels and holes in brittle materials like glass, quartz, composites and ceramics. ECDM is one of the technologies used to manufacture micro features on complicated and fragile materials. This chapter discusses the procedure involved in material elimination and the working principle of the process. This chapter examines the effect of varying important parameters of ECDM and their effect on material removal in machining. It suggests the range of parameters to be employed in the experiment. The different workpieces, such as glass, quartz, ceramics, and composites, are fabricated by electrochemical discharge machining. The primary attention is given to quartz machining because a trial experiment has to be performed. ECDM, one of the non-traditional machining tools, made fabricating features possible in a second plentiful material in the earth's crust, i.e., quartz. The quartz is used in the MEMS, microfluidic devices and lab on a chip. The through-hole of <1 mm (micro) diameter in the experimental trial was fabricated on the quartz plate using the ECDM process.

Keywords ECDM · Micro machining · Quartz · Ceramics · Composites · Non-traditional machining · Hybrid machining

Preface
This chapter helps to decide various parameters during the fabrication of the quartz. A trial run was performed on the quartz plate, and the micro hole was machined on the quartz plate. The chapter gives information about the different non-conductive materials machined by the ECDM process, like glass, quartz, ceramics and composite.

A. Prakash (✉) · R. Ballav
Department of Production and Industrial Engineering, National Institute of Technology Jamshedpur, Jamshedpur, Jharkhand, India
e-mail: 2020rspi003@nitjsr.ac.in

R. Ballav
e-mail: rballav.prod@nitjsr.ac.in

© The Author(s), under exclusive license to Springer Nature Switzerland AG 2025
K. C. Rath et al. (eds.), *Engineering, Technology and Management*, Information Systems Engineering and Management 54, https://doi.org/10.1007/978-3-031-93654-8_9

The chapter deals with the important process parameters and their range employed during the machining of insulating materials. This chapter discusses the mechanism involved in the ECDM process. Thus, it gives a complete overview of the fabrication of the quartz using the ECDM process.

9.1 Introduction

ECDM is an emerging non-traditional process for fabricating nonconductive and brittle materials like glass, quartz, ceramics, and composites. The machining of insulating and hard materials is arduous due to their high brittleness compared to conventional machining [1]. Some nontraditional techniques for machining these materials are Laser Beam Machining (LBM), Ultrasonic Machining (USM), and Electrochemical discharge Machining ECDM. The LBM produces high-energy processing, which leads to micro-sputters created on the surface of the workpiece, and the USM has a slow machining process. The ECDM is a cost-effective method with a portable setup. It is a nontraditional technique that produces high temperatures by the electric discharge at the tool's tip, which melts and vaporises the workpiece accompanied by the chemical etching of the machined area. Studies conducted on ECDM in recent years have indicated the potential for further development of this approach to establish micro-features at high speed and with improved surface quality. ECDM is primarily used for creating microchannels, micro holes, and engraving on difficult-to-process materials. Furthermore, in contrast to most other micromachining techniques, this technology may allow control over the surface finish and structure during a single cutting stage. The tool electrode is placed just above the material to be machined in the tank during this process. The tank is filled with electrolytes, and the workpiece, supporting electrode, and tooltip are submerged. The area of the supporting electrode is 100 times that of the tool, allowing the process to run more smoothly. As the potential difference rises, the density of the current rises, and there is advancement in the bubble numbers, resulting in the evolution of bubble coalescence and a gas die electric layer around the tool. The dielectric film has very high resistance; it serves as an obstacle between the tool and the electrolyte. As the potential difference applied elevates and reaches the ultimate potential difference, the film around the tool collapses, generating sparks. The material to be fabricated is to hold on close to the tool electrode when the spark strikes, which leads to melting and vaporisation, and more heat is generated near the tool, enhancing the chemical etching of the piece. The current of the circuit abruptly drops to nought, causing discharge to occur along the bubble bridge's places, and the process is then repeated. The ECDM is a hybrid machining technique that combines the concepts of ECM and EDM in machining material. The gas film wettability and locations of nucleation on the tool electrode surface are influenced by various parameters such as surface tension, buoyancy force, tool immersion depth, the radius of the tool, electrolyte resistivity, current density, and thermocapillary flow caused by the gradient of temperature. The pace at which bubbles form is dependent on the current density. The surface tension opposes the upward motion of the bubbles, forcing them to flow in that direction. Bubble coalescence causes the gas film formation to begin, and the applied potential difference excels the ultimate value. The vicinity of the liquid and, consequently, the nearby heat impacts the wettability. The potential difference further elevates, reaching the critical voltage and producing the sparks. When a workpiece is brought into proximity to the sparks, it is responsible for removing material from it. Through chemical etching, melting, and evaporation, the material is removed.

The ECDM is a micromachining tool fabricating micro-holes and slots on brittle and insulating material. The procedure is classified as micro-machining if the material is removed in the micron range. The ECDM technique produces smooth surface features because the gas film remains stable when micro-holes are created at lower voltage and depth. Thick gas coating surrounds the tool as voltage rises, resulting from a sequence of chemical reactions that start the process and initiate electrochemical machining. In addition to serving as a dielectric, this thick gas film generates a plasma channel at a critical voltage that resembles a spark due to the rapid velocity of ion transport. When a workpiece is brought into proximity to this spark, it is responsible for removing material from it. Through chemical etching, melting, and evaporation, the substance is removed. When using the ECDM technique to create microchannels, the tool travels velocity and applies voltage to control the smoothness of the manufactured surface. The material removal mechanism in this fabricating causes the spark to melt and vaporise the workpiece, and the electrolyte present causes chemical etching. Scientists investigated the effect of variables like DC power supply, electrolyte concentration, stand-off distance, and gap between two electrodes. The MRR of this machining rises as the applied potential difference rises because the density of hydrogen gas bubbles is created, which forms a thin layer around the tool and boosts the discharge energy at the sparking zone. The pulse voltage in this non-traditional machining process reduces the machined workpiece's radial overcut and heat-affected zone. Some researchers supplied ultrasonic support to the tool electrode in this fabricating process. High-frequency vibrations eliminated unwanted bubbles. The film's thickness decreased as a result. In these trials, the tool electrode surface was cleaned of stray bubbles to reduce layer thickness. The high-frequency, low-intensity discharges generated by this enhanced the quality of the machined hole. Another significant element influencing the model's accuracy is the distribution of discharge locations on the tool electrode. The discharges were simplified in earlier ECDM drilling models using cylindrical tool electrodes, and they were dispersed at the electrode's bottom surface or edge. However, the actual ECDM drilling material removal process is dynamic, meaning that the discharge sites vary as the material is removed, particularly for gravity-feed micro-drilling. Thus, to create an improved FEM model for the ECDM process, the dynamics of the discharge location. The spark generated between the electrolyte and the tool electrode material was removed using the discharges near the material to be machined.

9.2 Parameters of ECDM

The parameters that affect the machining features in the ECDM. The significant parameters include power source, stand-off distance, inter-electrode gap, and electrolyte. With the change in parameter ranges, one can regulate various outputs (MRR), radial overcut, and tool wear rate (TWR). In the electrolyte the type of electrolytes and their concentration play an essential role in the workpiece fabrication. The fishbone diagram of the ECDM process with its important parameters is shown in Fig. 9.1. The outcome using different process parameters, including voltage, electrolyte concentration, and current, on the ECDM method performance. The terms "MRR," "fabricating quality," and "desired component dimensions" allude to the ECDM process's machining performance. However, the machined surface's dimensions and quality still matter more than MRR regarding the micro-machining of precise items.

9.2.1 *Power Source*

The Direct Current power source has been used in the ECDM. It provides a stable spark, which strikes and machines the non-conductive material. The potential difference range and current type affect the fabrication of the workpiece in this machining type. The polarity power supply allows the tool to connect to the anode or cathode terminals. Direct polarity refers to connecting the tool to the negative terminal (cathode), whereas reverse polarity refers to connecting it to the positive terminal (anode). Reverse polarity leads to faster material removal, overcutting, and penetration rates than straight polarity. In ECDM, tool wear is less in direct polarity compared to reverse polarity while machining.

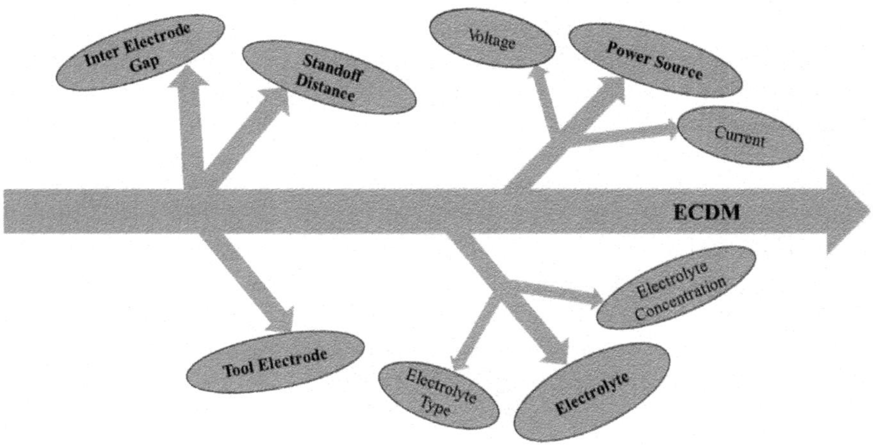

Fig. 9.1 Fish bone diagram for ECDM

9.2.1.1 Voltage

In the ECDM process, voltage plays an essential role. When the voltage rises and reaches the critical voltage, the first spark appears on the tool in the machining process. Increasing the applied potential difference raises the MRR in the ECDM. A gas layer surrounds the tool terminal, and the potential difference is raised over the crucial voltage, causing bubbles to congregate. At greater temperatures, the gas sheet thins, becomes more homogenous and vibrates. The electrolyte's bulk temperature determines the thickness and homogeneity of the gas sheet. A little area on the active electrode is the source of the electrochemical discharges. This results from the electrolyte's decreased electrical conductivity. Workpiece fabricating occurs due to these discharges through high-temperature chemical etching, melting, and vaporisation. This creates a layer of bubbles, increasing discharge energy at the sparking zone. The machining zone experiences a higher frequency of sidewall discharge and sparking violations at elevated voltages, leading to higher MRR, HAZ, roughness of the surface, and fabricating depth. The MRR of the workpiece is enhanced by the voltage enhancement up to a limit; when the voltage surpasses this point, debris is trapped in the machining gap, and the MRR starts to decline. Consequently, flushing the waste material with an electrolyte becomes difficult, resulting in a lower MRR. A small amount of microcracks will also form in the fabricating area at high voltage due to excessive heat accumulation. These microcracks could spread and cause the workpiece to rupture if the applied voltage rises further.

9.2.1.2 Current

The workpiece is fabricated using the DC (Direct Current) power source. The characteristics demonstrate how the current rises with time as the electrolytic solution heats up and becomes more conductive, and the inter-electrode resistance lowers. The gas generated if the current exceeds a critical value triggers gas bubbles' nucleation. The inter-electrode resistance value at the beginning of the potential difference input determines the film-forming time. The current in the ECDM with the rise in voltage till it reaches the critical peak value of current. Subsequently, there was an abrupt current downfall with the voltage increase. After that, the spark gets generated. This current behaviour is observed in the current–voltage characteristics curve. The majority of ECDM experiments are carried out with a DC power supply. Using pulse instead of direct current (DC) reduces surface damage and HAZ (heat-affected zone) in glass cutting. The sparks are produced only during the pulse-on duration in the pulse current power source. The heating zone in ECDM can be controlled, leading to the enhancement of the surface finish of the fabricated features. The power supply and its variable characteristics significantly impact the machining process. Voltage, pulse-on time, pulse-off time, pulse frequency, current, and electrode polarity are the electrical parameters that fluctuate with power supply and affect machine efficiency. Although the pulsed DC power supply allows for the discharge of focused spark energy and higher spark stability, it is frequently employed to perform machining operations by μ-ECDM. Additionally, the intermittent sparking and cooling in pulse DC power supplies results in less heat being evolved than in DC power supplies, and it tends to reduce the fabricated area microcracks and heat damage.

9.2.2 Electrolyte

Electrolytes play a vital role in this micro-machining process. The material to be machined, supplementary electrode, and tooltip are dipped in the electrolyte, which is stored in a tank. The hydrogen bubble evolves near the tool when the DC power is turned on. The density of the hydrogen bubble rises and forms a layer around the tool, interrupting the current flow. Sparks form in the electrolyte when the current density rises and exceeds the critical value.

9.2.2.1 Electrolyte Type and Concentration

Electrolytes can be alkaline, neutral, or acidic in composition. Alkaline electrolytes (NaOH and KOH) are generally utilised as electrolytes in this hybrid process due to their minimal hazardous nature. A comparison was made between the NaOH, HCL, and NaCl. The sodium- hydroxide creates the least hazardous gases and toxic chemicals. A basic NaOH solution that is relatively concentrated is used. The irregular and disordered arc in the acidic electrolyte (H_2SO_4) solution causes severe wear of the tool and damage to the workpiece. The natural salt (NaCl) causes significant tool wear but minimal workpiece damage. The primary electrolyte (KOH) reduces tool wear and sparks, resulting in a smoother surface finish for the workpiece. The preheating impact of electrolytes is explored in machining. A combination of NaOH and KOH electrolytes. The material to be removed primarily has an effect on applied potential difference and electrolyte concentration. When base electrolytes dissolve in water, they generate negatively charged hydroxides (OH^- radicals), which are required for chemical etching. In contrast, acid electrolytes dissolve protons (H^+) in water. It might be decided to investigate these topics more in the future. Further research into the potential effects of mixing multiple electrolytes might be pretty fascinating. Another difficult task in ECDM research is improving high aspect ratio drilling efficiency since the electrolyte flow to the machining zone gets more complex. Another critical area of research is increasing the typically achievable material to be removed to make the ECDM process more practical. The non-Newtonian fluid electrolyte increases the depth of machining and reduces the HAZ. The tool wear can be minimised by maintaining the distance between the tool and the workpiece. Some ECDM experiments used a mixed electrolyte, including conductive particles. Using 1.0 wt.% graphite powders with 30% NaOH minimises microcracks and roughness. They conducted the study using two different electrolytes, Newtonian fluid electrolytes (KOH) and non-Newtonian fluid electrolytes (a combination of polyacrylamide and KOH), in varying quantities. The experiment found that using a non-Newtonian fluid electrolyte resulted in a more permanent and thin gas layer. In some experiments, fresh electrolytes are injected into the fabricating area. The electrolyte in the tank contaminates after some machining period and disrupts spark production. Therefore, another stream of electrolytes is pumped into the machining zone. This is mainly employed to achieve higher depths in manufacturing workpieces. The electrolyte's conductivity rises with electrolyte temperature, speeding up the electrolysis process.

9.2.3 Standoff Distance

In ECDM, the distance separating the material to be machined and the tooltip is referred to as the stand-off distance (SOD). It isn't easy to control the tool-workpiece spacing length in ECDM. According to the study, proper spacing is necessary to remove material remarkably and reduce tool wear. According to an experiment, overcut and tool wear rates drop as SOD increases the rate of material elimination. The reduction of the sparking phenomenon was caused by the decrease in the momentum energy of the electrochemical reaction caused by the rise in SOD. The SOD is also widely known as the working gap. The experiments were performed on different non-metallic substrates with various working gap values. Studies reveal that increasing the work gap leads to shallow, narrow microchannels with poor surface smoothness.

9.2.4 Inter-Electrode Gap

The separation between two electrodes is called an inter-electrode gap. The electrodes in the ECDM are submerged in the electrolyte. Sarkar et al. [5] research study indicates that the MRR reduces when the inter-electrode distance expands during silicon nitride ceramic machining. The potential difference, amount of electrolytes, and electrode distance are the parameters that the experiment looks at. Using an L9 orthogonal array, the ECDM experiment was investigated. The experiment was designed according to the Taguchi technique. The results suggest that the inter-electrode spacing is the least impacting parameter. In another experiment, the inter-electrode spacing enhances MRR and decreases hole dapperness. In work conducted by Saini et al., the e-glass fibre-reinforced polymer composite was machined. When the gap size is less than the ratio of the working voltage to the dielectric resistance, electrical sparks will erupt at a certain point when the distance between the electrodes is narrow. The workpiece material undergoes anodic disintegration due to the presence of the electrical field. In addition to increasing the material removal rate, the workpiece material's anodic dissolution causes an increase in surface roughness at higher work voltages.

9.2.5 Tool Electrode

The tool electrode in this hybrid machining is essential in fabricating the workpiece. The geometry, surface roughness, and conductivity of the tool affect the precise fabricated part. The cylindrical structure of the tool electrodes is generally used in the ECDM. The tool has strong wettability, a high melting temperature, and good electric conductivity. The needle shape performs better at lower voltages and higher drilling rates than cylindrical tool-electrode combinations. The average drilling velocity achieved with the needle tool is approximately ten times quicker than the cylindrical geometrical shape. According to Wuthrich et al. [13], the tool end submerged in the electrolyte tip is where most of the discharges in the needle-structures tool occur. The discharge rate is often more significant with smaller cross-sectional tool electrodes, resulting in a higher MRR. This effect is noteworthy only in the discharge regime at low voltages. The drilling velocity develops practically independent of potential difference as soon as the tool-electrode enters the hydrodynamic phase. The dissimilarity between the cylindrical and needle-structured tools is barely noticeable within the hydrodynamic regime due to the elevated spinning rotating tool velocity in ECDM. The investigation determined that when the tool electrode rotating speed increases, the overcut and HAZ in ECDM decrease. Tool-electrode feed rate selection is essential since it influences machining time and surface quality. When the feeding rate of the drill operation surpasses the rate of material elimination, either the material to be machined or the tool fractures. However, low feed rates cause drilling to take longer, which raises the risk of a (HAZ) forming surrounding the hole. According to Cook et al. [2], the feed rate of 3 m/s was appropriate for micro-structuring glass, resulting in minor roughness (Ra = 0.099 m) for a 25 mm depth of cut. Conversely, fissures were noted along the channel's walls at a 5 m/s feed rate. Comparing a flat sidewall tool to a cylindrical tool during the drilling process improved the quality of the thin layer of hydrogen gas. As a result, fabricating precision and efficiency rise. It is discovered that the gas film's quality is stable and dense at changeover potential difference. This transition potential difference is referenced at certain levels above the ultimate potential difference. The quality of the gas film influences surface roughness and workpiece geometric accuracy. The gas film quality indexes used were machined contours and current indications. Elevated electrolyte concentration accelerates electrolysis, reducing transition potential difference in enhanced chemical etching and superior surface quality. Compared to a standard cylindrical tool (100 mm diameter), Yang et al. [2] a spherical end tool (sphere of 150 mm diameter) reduces the hole entrance diameter by 65% and requires 83% less time to machine a 500 mm depth through the hole. This occurs because the spherical tool's curving surface reduces the region where the workpiece and electrode come into contact. This makes it easier for electrolytes to flow to the electrode end and speeds up the development of the gas layer. The curve surface reduces the overabundance of current density concentration, which raises the discharge frequency and produces a more uniform bubble production speed. Therefore, with better shape precision and machining performance, this kind of tool electrode drills the micro holes effectively. A novel tool was created by

Zheng et al. [3]; it features a flat front and sidewalls. Along with reducing hole taper and improving machining precision, these sidewalls also lessen discharge around the tool. The centre of the Pyrex glass influenced by heat is measured using a circular and flat-front instrument. The front flat tool's cross-section area and rectangular pulse voltage at their ideal values improve the accuracy of the machining process. There was an increase in process performance with the least amount of overcut when the tool rotational rate reached a high. Three electrodes, namely stainless steel (SS), tungsten carbide (WC), and tungsten, were used by Yang et al. [4] in their article. These tools have different surface finishes even though they are made using the same procedure because of the physical characteristics of the material (WEDG). Electrolyte wettability is impacted by surface roughness. Because more and more gas bubbles agglomerate and raise surface tension, poor wettability causes a thicker gas layer and bigger hole diameters to be machined. Voltage, which varies depending on the type of tool material, also affects gas film development. As a result, the tool material affects energy discharge. The minor hole diameters that could be achieved and the least amount of tool wear were tungsten carbide, which performed the best during the machining process. Side-insulated tools focus the spark action at the tooltip, minimising electrolytic stray and minimising overcut and taper. Because of the cooling effect, the intermittent voltage in pulses lessens the thermal damage to the workpiece. Side insulation and pulse voltage effects show reduced overcut and taper using electrolyte ultrasonic vibrations to reduce taper and overcut and enhance machining depth.

It was found that tool-shaped process variables influence structural correctness and process efficiency. The appropriate cross-section design at the tool point is referred to as the tool geometry. The application of a side-insulated tool improves geometrical accuracy. The tool motion may provide a surface of higher quality and more efficiency. This means that each case warrants an investigation into the impact of tool motion. ECDM has been linked to drawbacks such as low precision and buildings with low aspect ratios. The ECDM has utilised some external energies to overcome these constraints. As a result, various ECDM process versions were developed. The ECDM process has been concurrently affected by these outside energies. The tool's ultrasonic vibration allows for increased depth and geometric accuracy of the hole. Side tool inflation and pulse voltage are used to pronounce this effect. An essential factor is the tool's surface roughness in this unconventional type. Surface roughness process parameters impact MRR, tool wear rate, and workpiece surface finish. The ideal surface roughness can, therefore, improve machining performance. Textured tools are thus recommended for improved surface smoothness and machining efficiency. The two most widely utilised modes of machining are gravity feed and constant-velocity feed. In constant-velocity feed machining, the tool is continuously moved in the direction of the work. In contrast, the gravity-feed mechanism keeps the tool in touch with machining materials. Some other machining modes used in ECDM are force feedback and current feedback mechanism, which are designed to upgrade fabricating performance by analysing the current signal and contact forces.

9.3 Workpieces Fabricated by the ECDM Process

Both metal and non-metal can be machined using the ECDM process. Limited nonconventional machining is used to fabricate electrically nonconducting workpieces. Table 9.1 shows the latest work done on ECDM. Insulating, hard, and fragile materials can be created using the ECDM technique. The workpieces fabricated by the researchers are mainly glass, quartz, composites, and ceramics, which is the quantity of study of different materials taken from the Scopus database. The percentage of workpieces machined by ECDM in various studies is displayed in Fig. 9.2.

Table 9.1 Recent work in ECDM

Workpiece	Parameters	Inference
Glass	Voltage level: 45, 50 and 55 V Electrolyte (NaOH) concentration level: 20, 25 and 30 (wt.%) Duty ratio level: 0.45, 50 and 0.55 Pulse frequency (Hz): 50, 75 and 100	The desirable conditions for micro-channelling performances on the glass during the micro-machining process are 50 V applied potential difference, 25% concentration of electrolyte, 0.5 duty ratio, and 50 Hz pulse frequency [5]
Glass	Applied voltage: 54 V Electrolyte (NaOH) concentration: 20(wt.%) Electrolyte flow rate: 8 ml/h Air flow rate: 0.6 LPM	A through micro hole was fabricated on a 3500 μm thick borosilicate glass using an electrolyte-air mixture injected into the workpiece [6]
Quartz	Voltage level: 50, 55 and 60 V Electrolyte (KOH) concentration level: 20, 25 and 30 (wt.%) Feed rate level: 0.8,1 and 1.2 (mm/min)	With a 30% EC and a 60 V voltage, the feed rate was 1.2 mm/min to maximise the MRR of 0.9301 mg/min [7]

(continued)

Table 9.1 (continued)

Workpiece	Parameters	Inference
Silica (Quartz)	Applied voltage level: 35, 45 and 55 V Electrolyte (NaOH) concentration level: 15, 20 and 25 (wt.%) Inter electrode gap: 25, 35, 45	For the maximum (MRR) and minimum (ROC and HAZ). The parameter combination for GRA-based multi-response optimisation is 35 V applied voltage, 15% electrolyte concentration, and 25 mm IEG [8]
Alumina ceramics	Applied voltage: 50–60 V Frequency level: 1,3,5,7,9 Duty cycle level: 50, 70 and 90 Electrolyte (NaOH) concentration: 20 (wt.%)	The ECDM study was broadened by emphasising the gas film evolution law, the alumina ceramic surface heat

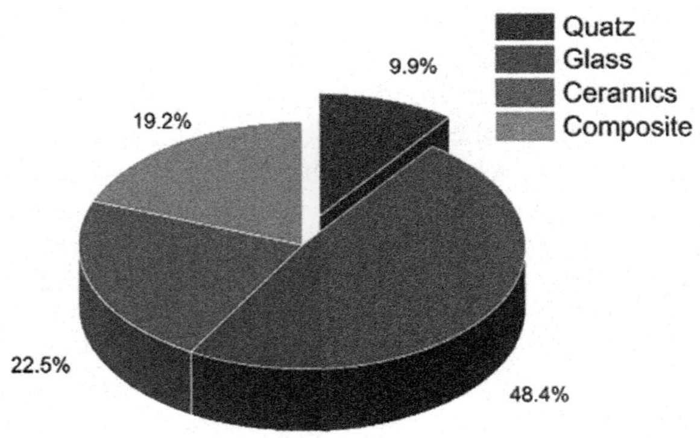

Fig. 9.2 Workpieces fabricated by ECDM process

9.3.1 Glass

The glass is a fragile, hard, and non-conductive material. It is transparent and chemically inert and has its application in the lab on chips and microelectronics mechanical systems (MEMS). A steady discharge zone produces a smooth surface when a hole is machined at low voltage and depth in glass. Altering the electrolyte's concentration can regulate TWR. The time needed to build up gas film is increased due to the rise in inter-electrode resistance, resulting in low drilling speed. Variations in the machining depth and voltage impact the surface properties of the glass fabricated zone. The taperness of the hole fabricated in the glass is reduced using the rotating tool during the ECDM process. The Microchannel surface properties can be adjusted by varying the voltage and tool speed. The terminal voltage and electrode temperature can influence the gas film phenomenon. The variation in applied voltage and tool microchannels with various features were created, such as a smooth surface, a rough outline, and HAZ ends. Charak et al. [9] they experimented with borosilicate glass with electrolyte stirring. In the experiment, they observed that workpiece machining increased above the applied voltage of 53 V while the electrolyte was being stirred. The stirring action lowers the possibility of the recasting layer forming and enhances the surface polish by 20–79%. The creation of 3 × 3 array micro-holes in glass with a 3 × 3 multi-tip array tool while taking "through glass" into consideration. High-quality micro-holes were generated with KOH electrolyte at reduced input heat energy for duty ratios ranging from 60 to 80%. At the potential difference of 56v and the duty ratio of 70%, the MRR increased. Precision features were obtained for 60 and 70% duty ratios in the frequency gap between 0.5 and 100 kHz. To achieve high aspect ratios and enhance surface integrity and quality, it is advised to reduce the levels of electrolyte content. While beneficial in increasing machining quality, this technology presents a difficulty in terms of an inefficient machining pace. The creation of microchannels on glass can be studied using computational and experimental methods with this fabricating process. The experiment demonstrated that, at low electrolyte concentrations, a magnetic field is used to smooth the fabricated zone of the glass, and the MRR of the glass rises with a potential difference. The High-Speed Steel tool is utilised to cut glass at 30–35 V, while the electrolyte is a 15–30% aqueous solution of NaOH [10]. The experiment was studied to create 3D microstructure, micro grooving, and micro holes generated on glass. Increasing the voltage during the micro-hole production process elevates the energy of the spark discharge, which improves the machining and deepens the hole. It will, however, widen the entry and change the shape of the entrance hole. The applied voltage determines the surface coarseness, which is inversely related to the power source frequency, tool electrode feeding speed, and groove.

9.3.2 Ceramics and Composites

Ceramics and composites are hard, brittle, chemically inert, and difficult to machine. Sarkar et al. described the micromachining of silicon nitride ceramics [11]. In an experiment, they determined the best settings for the highest MRR, or 1.20 mg/h, in his microfabrication at the applied potential difference of 70 V, the amount of (NaOH) electrolyte at 18%, and the interelectrode gap at 27 mm to generate the highest machining rate. While drilling a hole in silicon nitride ceramics, the least amount of radial overcut is 50 V, 24 percentage of electrolyte intensity, and 30 mm between electrodes. The ECDM, pressurised feeding, and electrolyte flow create micro holes on Yttria-stabilized zirconia. When using an ECDM electrolyte flow, the micro-hole depth increases by 42% and the hole width decreases by 33% during the same operation [12]. An experiment using ECDM to micromachine CRPF (carbon fiber-reinforced polymer). The Entropy-VIKOR technique is used to determine various weights for output quality characteristics. The approach identified the finest hole in the CRPF composite [13]. Antil et al. examined the ECDM method of drilling micro holes in SiC-reinforced polymer matrix composites. This work optimised output quality parameters like MRR, overcut and taper using Taguchi's grey relational analysis (GRA). The drilled holes showed that the composite matrix had cracks and a Hazard on the wall surface [14]. The (G-ECDM) technique added the grinding with the ECDM, which was attempted to machine MMC. The ECDM process's process capability was improved in this study due to increased grinding activity in the fabricating area. Because the recast layer was removed and the surface roughness was decreased, the machined surface's surface properties were significantly better with G-ECDM than with conventional ECDM. Researchers investigated the ECDM technique for machining fibre-reinforced polymer matrix composites and metal matrix composites Yan et al. [15] have investigated the superalloy surface properties and MRR, which were affected by the inner diameter of a tubular electrode. Because of improved flushing action, MRR in this study increases as the internal diameter of the tubular-shaped tool rises. The ECDM process used for manufacturing zirconia was investigated [16].

9.3.3 Quartz

The ECDM is efficiently capable of micromachining the hard and brittle material. Quartz is an insulating material with high hardness and brittleness in nature. It is a piezoelectric material with low thermal expansion and is chemically inert. Quartz is used in different mechanical systems, electronics, and biomedical industries. Some researchers used the ECDM to witness the quartz being machined. Madhavi et al. [17] have experimented with creating channels and texture over the quartz by this unconventional machining process. The optimal variables for the highest yield are a voltage of 60 V, an amount of electrolyte of 30%, and a duty ratio of 60%. The micro-hole created over the quartz with different diameter tool electrodes, for the minor entrance diameter and the least overcut, the ideal feed rate was 0.8 μm/s [18]. The fabrication of the quartz plate by Jain et al. [19] with different polarity conditioning of the power supply. They apply a heat transfer search method to optimise process parameters during ECM quartz machining. TWR and MRR are the two outputs. The maximum quartz removal rate was achieved at a voltage of 60 V, a feed velocity of 1.2 mm/min, and an electrolyte content of 30% [20]. In their experiment, Yang et al. [4] machined quartz using various tool material electrodes, and the results are analysed. The tool material comprises stainless steel, tungsten carbide, and tungsten. They experimented, and the findings showed that tungsten carbide, tungsten, and stainless-steel exhibit the best machining stability during fabricating of quartz. The least amount of tool wear and the best machining performance are displayed by tungsten carbide. A mixture of 5% weight SiC and 5 mg KOH was added to the electrolyte while magnetic stirrers running at 150 rpm were in operation. The titrated electrolyte flow clears out the machined particles produced at the fabricating area. The suggested technique yielded improvements in surface finish and 7.5% mean slit width. The most effective process parameters to achieve results were a supplied potential difference voltage of 35 V, a flow rate of 4 mL/min, a machining time of 210 s, a feed rate of 5 μm/s, and a 5 wt.% Sic powder concentration [21]. Machining of quartz by ECDM. The MRR elevated from 21.47 to 75.82%, and the fabricated work roughness decreased from 7.93 to 22.25% using MHDs [22]. Few studies have been conducted on analysing the Quartz material removal rate using finite element models. Comparing the simulated MRR to the findings of other studies revealed a 16.66% increase. Additionally, it was noted that the current experimental results error is 16.66% greater in magnitude than the experimental results error by 4.16%, with the spark region measured by a high-speed camera at 400 μm [23].

9.4 Trial Experiment Performed on Quartz

The least amount of experiment work is performed over the quartz, as shown in Fig. 9.1. A trial run of the ECDM process was conducted on the quartz plate. The through-hole is created over the quartz plate of dimension $25 \times 25 \times 1\ mm^3$. Table 9.2 shows different parameters of the ECDM process hole machined across a depth of 1 mm. A copper tool electrode was employed, with a rotation of 30 rpm provided to the tool by the steeper motor. The rotating tool enhances the depth of the machining. The rotating tool dipped in the electrolyte adds centrifugal force to the electrolyte during machining, resulting in the dielectric's film shrinkage. The machining was carried out at 95 V for 240 s. The stand-off distance is zero, which means the tool touches the quartz plate, and the inter-electrode gap is 60 mm. The NaOH electrolyte was used at 15 per cent of the weight concentration. The optical image of the through hole fabricated on the quartz plate is shown in Fig. 9.3a, b.

The through hole diameter of 865 μm by ECDM on the Quartz plate. The figures above show the holes created on the quartz plate and the heat-affected zone surrounding the hole.

Table 9.2 Parameters of the trial experiment

Parameters	
Voltage (V)	95
Electrolyte (NaOH) concentration (wt.%)	15
Tool material	Copper
Tool rotational speed (rpm)	30
Inter electrode gap (mm)	60
Standoff distance (mm)	0

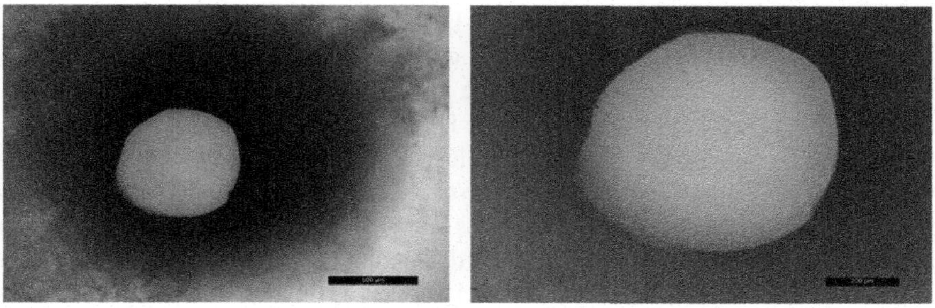

a b

Fig. 9.3 Optical image of the hole fabricated on quartz plate at, **a** 5X magnification and **b** 10X magnification

9.5 Summary and Future Scope

The ECDM phenomena involved in the material removal include thermal erosion and chemical etching in this hybrid machining. The setup cost for the ECDM is less than that of other nontraditional machining. It consists of a DC power source and an aqueous electrolyte solution submerging the subsidiary electrode, workpiece, and tool. The creation of gas film and spark stabilisation are critical to determining the parameters. The performance of machining is purely dependent on specific process parameters. After reviewing the literature and conducting trial experiments, we modify these leads to effective and precise machining. The through hole created on the quartz plate has a diameter of 865 μm. We can achieve high precision and accuracy from this micro-machining process.

9.5.1 Future Scope

In the future, ECDM can be used for industrial purposes if its repeatability, precision, and surface finish are enhanced. Research is ongoing to improve the accuracy of ECDM technology. The existing research determined that the disintegration of hydrogen bubbles on the top surface is the cause of sparking. Unfortunately, the mechanics underlying the disintegration of the gas film or hydrogen bubble layer are still unclear. Using a high-speed camera in the ECDM process will allow the various stages of hydrogen bubble layer production and disintegration to be studied in the future. The literature showed that drilling holes in the hydrodynamic zone results in numerous irregularities; as a result, it is necessary to investigate ways to improve ECDM performance in the hydrodynamic regime. The irregularities include low fresh electrolyte circulation, uneven machining surface, and unstable gas film. These issues can be resolved using rotary tool electrodes, powder-mixed electrolytes, changing tool electrode shapes, and direct and indirect vibration to the tool.

Simple machining can be done with a DC power supply. However, pulsed mode power supplies are necessary for more complex machining techniques. Surface roughness and machining over-cut can be minimised by selecting small duty ratios and high-frequency voltage pulses. Large HAZs and thermal fractures, on the other hand, are the outcome of machining at elevated potential differences and low tool velocity. Adding abrasives to the electrolyte or utilising a pulsed voltage source are two ways to manage the surface roughness and over-cut. The ECDM can be used on hard and brittle materials to create highly detailed micro features with a large aspect ratio. Surface polish and dimensional precision can be improved further by changing the power supply, electrolyte, and tool designs. Modern machining methods should make use of pulse-based sources of power. Low-duty ratios and high-frequency voltage pulses can reduce the machined surface's over-cut and surface roughness. Conversely, when cutting at low tool travel speeds and high voltage, heat fractures and large HAZs are generated. The electrolyte is treated with abrasives to reduce surface roughness. Enhancing the tool electrode's wettability can be achieved by coating it, adding texture, and using the proper manufacturing process. Wettability affects the rates at which gas and bubble films form. The tool and its tip can be manufactured with a proper geometric cross-section, resulting in minimum taperness and overcutting of the machined features. Several investigations have revealed several combinations for the tool movements. The process might use hybrid techniques like ECDM and ultrasonic machining to achieve better surface quality. It is possible to diminish the taper and overcut in ECDM by designing the tool and tooltip cross-section with the correct shape. A more advanced tool design with ideal process settings must address substandard machining results and workpiece thermal damage.

References

1. Prakash, A., Kumar, A., Ballav, R.: Process capability of electrochemical discharge machining: a review. Proc. Inst. Mech. Eng. Part L J. Mater. Des. Appl. **238**, 282–300 (2024). https://doi.org/10.1177/14644207231218386
2. Yang, C.K., Wu, K.L., Hung, J.C., Lee, S.M., Lin, J.C., Yan, B.H.: Enhancement of ECDM efficiency and accuracy by spherical tool electrode. Int. J. Mach. Tools Manuf. **51**, 528–535 (2011). https://doi.org/10.1016/j.ijmachtools.2011.03.001
3. Zheng, Z., Su, H., Huang, F.: The tool geometrical shape and pulse-off time of pulse voltage effects in a Pyrex glass electrochemical discharge microdrilling process, 265. https://doi.org/10.1088/0960-1317/17/2/012
4. Yang, C.-K., Cheng, C.-P., Mai, C.-C., Cheng Wang, A., Hung, J.-C., Yan, B.-H.: Effect of surface roughness of tool electrode materials in ECDM performance. Int. J. Mach. Tools Manuf. **50**, 1088–1096 (2010). https://doi.org/10.1016/j.ijmachtools.2010.08.006
5. Mondal, K., Mallick, B., Kumar Sarkar, M., Kumar Biswas, S., Kumar Mahato, J.: Optimization of micro-ECDM performances during micro-channeling on glass using Taguchi technique. Mater. Today Proc. 1–7 (2024). https://doi.org/10.1016/j.matpr.2024.05.014
6. Arya, R.K., Dvivedi, A.: Improving the electrochemical discharge machining (ECDM) process for deep-micro-hole drilling on glass by application of the electrolyte-air injection. Ceram. Int. **49**, 8916–8935 (2023). https://doi.org/10.1016/j.ceramint.2022.11.047

7. Nema, A., Patel, V., Kumar, A., Oza, A., Jagani, A.B.: Use of heat transfer search algorithm while optimizing the process parameters of ECDM to machine quartz material. Mater. Today Proc. **62**, 7148–7155 (2022). https://doi.org/10.1016/j.matpr.2022.02.362
8. Rajput, V., Pundir, S.S., Goud, M., Suri, N.M.: Multi-response optimization of ECDM parameters for silica (Quartz) using grey relational analysis. Silicon **13**, 1619–1640 (2021). https://doi.org/10.1007/s12633-020-00538-7
9. Charak, A., Jawalkar, C.S.: Experimental investigation and analysis on borosilicate glass using electrochemical discharge machining process (2022). https://doi.org/10.1007/s12633-021-00980-1
10. Hajian, M., Razfar, M.R., Movahed, S.: An experimental study on the effect of magnetic field orientations and electrolyte concentrations on ECDM milling performance of glass. **45**, 322–331 (2016)
11. Sarkar, B.R., Doloi, B., Bhattacharyya, B.: Parametric analysis on electrochemical discharge machining of silicon nitride ceramics. Int. J. Adv. Manuf. Technol. **28**, 873–881 (2006). https://doi.org/10.1007/s00170-004-2448-1
12. Singh, T., Dvivedi, A.: Fabrication of micro holes in Yttria-stabilized zirconia (Y-SZ) by hybrid process of electrochemical discharge machining (ECDM). Ceram. Int. **47**, 23677–23681 (2021). https://doi.org/10.1016/j.ceramint.2021.05.017
13. Priti, Singh, M., Singh, S.: Micro-Machining of CFRP composite using electrochemical discharge machining and process optimization by Entropy-VIKOR method. Mater. Today Proc. **44**, 260–265 (2021). https://doi.org/10.1016/j.matpr.2020.09.463
14. Antil, P., Singh, S., Manna, A.: Glass fibers/SiCp reinforced epoxy composites: effect of environmental conditions. J. Compos. Mater. **52**, 1253–1264 (2017). https://doi.org/10.1177/0021998317723448
15. Yan, Z., Zhengyang, X., Jun, X., Di, Z.: Effect of tube-electrode inner diameter on electrochemical discharge machining of nickel-based superalloy. Chin. J. Aeronaut. **29**, 1103–1110 (2016). https://doi.org/10.1016/j.cja.2015.12.016
16. Kumar, M., Vaishya, R.O., Suri, N.M.: Machinability study of zirconia material by micro-ECDM. In: Sharma, V.S., Dixit, U.S., Sørby, K., Bhardwaj, A., and Trehan, R. (eds.) Manufacturing Engineering, pp. 195–209. Springer Singapore, Singapore (2020)
17. Madhavi, J.B., Hiremath, S.S.: Machining and characterization of channels and textures on quartz glass using μ-ECDM process (2019). https://doi.org/10.1007/s12633-019-0083-6
18. Saranya, S., Sankar, A.R.: Fabrication of precise micro-holes on quartz substrates with improved aspect ratio using a constant velocity-feed drilling technique of an ECDM process
19. Jain, V.K., Adhikary, S.: On the mechanism of material removal in electrochemical spark machining of quartz. **0**, 460–470 (2007). https://doi.org/10.1016/j.jmatprotec.2007.08.071
20. Nema, A., Patel, V., Kumar, A., Oza, A., Jagani, A.B.: Materials today: proceedings use of heat transfer search algorithm while optimizing the process parameters of ECDM to machine quartz material. Mater. Today Proc. (2022). https://doi.org/10.1016/j.matpr.2022.02.362
21. Kuo, K., Wu, K., Yang, C., Yan, B.: Wire electrochemical discharge machining (WECDM) of quartz glass with titrated electrolyte flow. Int. J. Mach. Tools Manuf. **72**, 50–57 (2013). https://doi.org/10.1016/j.ijmachtools.2013.06.003
22. Oza, A.D., Kumar, A., Badheka, V.: Improving quartz micro-machining performance by magnetohydrodynamic and zinc-coated assisted traveling wire-electrochemical discharge machining process. Mater. Today Proc. **28**, 970–976 (2020). https://doi.org/10.1016/j.matpr.2019.12.334
23. Vaishya, R., Sharma, V., Gupta, A., Pathania, J., Oza, A., Dixit, A.K., Bhole, K.S., Makwana, M., Patel, A.: Finite element modeling of quartz material for analyzing material removal rate in ECDM process. Int. J. Interact. Des. Manuf. **17**, 2453–2459 (2023). https://doi.org/10.1007/s12008-022-01037-2

Amar Prakash is a dedicated Research Scholar in the Department of Production and Industrial Engineering at the National Institute of Technology (NIT), Jamshedpur. He holds a Master of Technology in Production Engineering from NIT Tiruchirappalli (2016–2018) and a Bachelor of Engineering in Mechanical Engineering (Manufacturing) from Annamalai University (2008–2012). With a year of teaching experience each at Godavari Institute of Engineering and Technology, Rajahmundry (2018–2019), and Veltech University, Avadi, Chennai (2019–2020), Amar's research interests lie in polymers, composites, and non-traditional machining methods, showcasing his commitment to advancing the field of production engineering.

Raj Ballav is working as a Professor in the Department of Production and Industrial Engineering at the National Institute of Technology (NIT), Jamshedpur, where he has been serving since February 25, 1997. With over 25 years of teaching and research experience, he has made significant contributions to areas such as Welding, Reverse Engineering, Rapid Prototyping, CAD/CAM, Production, and Non-Traditional Manufacturing. Dr. Ballav holds a Ph.D. in Production and Industrial Engineering from Ranchi University, an M.E. in Production Engineering from BIT Sindri (1999), and a B.E. in Production Engineering from Ranchi University (1990). His academic journey and professional dedication highlight his commitment to advancing knowledge and innovation in production and industrial engineering.

Chapter 10
Impact of Technologically Vigilant Leadership on Smart Sustainable Circular Supply Chain Management in the Context of Industry 6.0

Pham Quang Huy and Vu Kien Phuc

Abstract This manuscript targets on setting forth as well as developing an impeccable paradigm for procuring the influences of technologically vigilant leadership (TVL) as a potentially facilitator of Industry 6.0 implementation (IRI) to reach the highest performance of smart sustainable circular supply chain management (SSCSCM). From a methodological standpoint, this model has been established conceptually and substantiated by structural equation modeling instrument rested on the consideration of a cross-sectional sample of 483 valid informants retrieved from a well-structured questionnaire survey strategy in small and medium enterprises (SMEs). The outputs demonstrated the existence of strong interconnections between the two constructs of hypothesized model. More concretely, TVL was corroborated to be the greatest contribution to IRI. Additionally, the IRI was authenticated statistically to be the triggering component for amelioration and enhancement of SSCSCM. Apart from the theorical contribution to the movement of research trends targeted at Industry 6.0, this research was valuable to deepening the knowledge and understanding of SMEs as they strived to realize SSCSCM through TVL and IRI, most notably, these in-depth insights would offer guidelines for policymakers in formulating better TVL mechanism to harness the benefits of IRI and therefore intensify SSCSCM.

Keywords Circular economy · Industry 6.0 · Leadership · Sustainable development · Technological vigilance

P. Q. Huy (✉)
Accounting School, College of Business, University of Economics Ho Chi Minh City, Ho Chi Minh City, Vietnam
e-mail: pquanghuy@ueh.edu.vn

V. K. Phuc
Accounting School, UEH Vinh Long Campus, University of Economics Ho Chi Minh City, Vinh Long City, Vietnam
e-mail: phucvk@ueh.edu.vn

© The Author(s), under exclusive license to Springer Nature Switzerland AG 2025
K. C. Rath et al. (eds.), *Engineering, Technology and Management*, Information Systems Engineering and Management 54, https://doi.org/10.1007/978-3-031-93654-8_10

Preface
This chapter delivers the practical applications and implications of the model developed to explore the role of Technologically Vigilant Leadership (TVL) as a key enabler of Industry 6.0 (IRI) and its impact on achieving Smart Sustainable Circular Supply Chain Management (SSCSCM). Building on a robust methodology, the chapter presents empirical findings from structural equation modeling, based on data from 483 small and medium enterprises (SMEs). It highlights the critical contributions of TVL in facilitating IRI and underscores the importance of IRI in enhancing SSCSCM performance. The chapter concludes with actionable insights for SMEs and policymakers, aiming to optimize TVL mechanisms and leverage IRI for sustainable, high-performance supply chain practices.

10.1 Introduction

In a linear economy, the conventional supply chain process involves using raw materials for industrial manufacture and then discarding them as trash once they are used. Nevertheless, the linear economy model lacks the ability to effectively regulate the equilibrium between supply and demand when it comes to the usage of natural resources. Furthermore, the implementation of inventive manufacturing techniques, models, and services as a corporate strategy to safeguard natural resources and the environment has resulted in heightened global recognition of climate change and ecosystem degradation [1]. Based on the perspectives Ref. [2], the imminent sustainability challenge can be tackled through the implementation of circular economy. The circular economy is an economic framework that prioritizes efficiency in resource utilization to reduce waste, preserve long-term value, eliminate primary resource consumption, and establish closed-loop systems for products, components, and materials with socio-economic and environmental benefits. This model also emphasizes the restriction of waste that circulates into the atmosphere through the effective utilization of energy, water as well as resources. Circular supply chain is of utmost importance [3]. It became imperative to devise alternatives for integrating circular economy principles with sustainable supply chain collaboration. By integrating the principles of sustainability and circular economy, a sustainable circular supply chain emerges as an ecological strategy. Furthermore, they must deliver sustainable outcomes [4] and minimize human–machine interaction to facilitate the implementation of sustainable practices and the promotion of circular economy principles. To achieve this objective, Industry 4.0 technologies implement the "smart concept" and produce circular supply chains that are sustainable, flexible, and adaptable, enabling them to be transparent and collaborative in a sustainable manner [5].

Unfortunately, the extent to which Industry 4.0 incorporates digital connectivity as well as human–machine collaboration may be limited. Industry 6.0 surpasses both Industry 4.0 and Industry 5.0 by one level: in this stage, automated machines carry out all operations under the control of human minds, transcending the boundaries of the planet [6]. Industry 6.0 is anticipated to feature hyper-connectivity between industries, mass customization and personalization of services and products on a massive scale, coupled with the concept of dynamic supply chain management and a mindset centered on the one-lot size of highly customized goods and services, and the ability for information to flow across national borders [6]. More importantly, a deficiency in assurance appears to exist regarding the commencement or successful completion of strategic digital transformation endeavors [7]. The primary concern in integrating people and digital technology is the risk of digital transformation failure. Therefore, the authors argue that digital transformation should be founded upon a paradigm shift in thinking and a comprehensive strategy, as opposed to technological advancements. In light of the ongoing worldwide information revolution, it is imperative to adopt a vigilant technological stance [8]. The line manager assumes a pivotal role as a vigilant executive, owing to their obligation to uphold integrity and honesty throughout the organization. Nevertheless, active participation and the assumption of the role of vigilance executive by every leader within the organization are imperative for the success and efficacy of vigilance initiatives. A technologically vigilant leader conscientiously monitors the scientific and technological milieu, ensuring a comprehensive understanding of all contemporary technological methodologies.

Until recently, there has been limited research on the combined study of technology vigilance, sustainable circular supply chain, and Industry 6.0, primarily because their relationships is a recent development. Furthermore, the literature has not yet explored the correlations between supply chain, sustainability, circular economy, and Industry 6.0. All of these factors indicate a deficiency in research. The target of this manuscript is to examine how technologically vigilant leadership (TVL) may promote the implementation of the Industry 6.0 (IRI) in order to achieve optimal performance in smart sustainable circular supply chain management (SSCSCM), addressing the existing research gap. Therefore, the primary objective of this study is to broaden the current boundaries of comprehension regarding SSCSCM in the context of Industry 6.0. Furthermore, this may be the first academic work to offer innovative and essential viewpoints on the highly unique impacts of TVL on Industry 6.0. This paper tries to address the existing gap by offering proactive and reactive answers and contributions to this topic. Research questions are systematically developed based on the stated reasons to achieve the analysis of the research's main purposes.

RQ1. *Does TVL influence IRI? To what extent does TVL influence IRI*?
RQ2. *Does IRI influence SSCSCM? To what extent does IRI influence SSCSCM*?

This manuscript is structured as follows. Section 10.2 establishes the literature review. In Sect. 10.3, a summary of the conceptual model and the rationale for the hypotheses are presented. The subsequent section focuses on the research methodology utilized in the present manuscript. In Sect. 10.5 presented the principal findings. In conclusion, Sect. 10.6 contains a list of the most significant insights derived from the findings for practice, as well as inherent drawbacks and orientations for future research.

10.2 Literature Review

A literature review on technologically vigilant leadership's impact on smart sustainable circular supply chain management in Industry 6.0 examines the intersection of leadership, technology, and sustainable supply chain practices.

10.2.1 Theorical Background

Information processing theory: Information processing theory is expanding upon the viewpoints put forth by [9], information processing theory effectively elucidates the interplay between the concepts of information processing and organizational behavior. Numerous elements are contributory to uncertainty. The aforementioned factors encompass the degree of interdepartmental communication, the susceptibility of the supply chain environment to change, standardized operational procedures, and stratified Ref. [10]. Information processing theory posits that in the event of unfavorable circumstances, a company that possesses the capability to analyze data originating from its supply chain will exhibit enhanced capability in fulfilling the information requirements of its supply chain partners.

Stakeholder theory: The notion of stakeholder interests is fundamental to stakeholder theory [11] emphasized the interdependence of interests and contended that they are inherently merged. In order to establish shared sustainability interests, stakeholders must engage in negotiations that capitalize on the respective values and priorities of each stakeholder group. As such, when formulating future strategies and decisions, consideration should be given to the expectations of stakeholders due to their significance. The literature has previously addressed stakeholder interests in relation to the implementation of circular economy. The theoretical underpinnings of the effects of circular economy actions are established by stakeholder theory [12].

10.2.2 Conceptual Framework

New convergence ideas from Industry 4.0 and Industry 5.0 have been combined to create Industry 6.0 [6]. One of the characteristics of Industry 5.0 is the synergistic integration of machine and human labor. Alternatively, Industry 6.0 is expected to include interplanetary resource production and collection, fully autonomous machinery, and the development of renewable energy sources [13]. Industry 6.0 is anticipated to bring about hyper-interfacing of ventures with businesses, extreme mass customization, and fundamental mass personalization of services in the future [6].

The principles of earlier research serve as the foundation for SSCSCM, which builds on the body of knowledge to develop a supply chain architecture that is more robust and ecologically conscientious [14]. Circular economy is the optimization of natural resource utilization through the deployment of circular material flows during manufacturing processes, with the aim of reducing waste and the demand for additional raw materials. Circular supply networks make it easier to switch from a linear to a circular model by combining supply chain operations and adding return procedures that capture more value [3]. The sustainable development goals aim to facilitate the collaboration of different stakeholders in order to synchronize their actions towards a shared objective. According to this research, SSCSCM is the integration of Industry 6.0 initiatives and the advantages of the circular economy to help achieve sustainable development goals.

Technologically vigilant leadership. Building on the perspectives of [15], technical vigilance entails the ongoing surveillance of technology to identify and mitigate potential concerns prior to their escalation into significant impediments. This involved the upkeep of existing gear and software, knowledge of developing risks and weaknesses, and the efficient enforcement of security measures. In this research, TVL refers to the leader's practice of actively monitoring and ensuring that all individuals under their authority comply with the established policies and objectives using information technology.

10.3 Hypothesis Development

In light of the continuous information revolution that is taking place all over the world, it is absolutely necessary to implement a policy of technical vigilance [8]. Businesses that are in possession of advanced technological knowledge and abilities are more likely to achieve optimal performance and to give effective support for their employees. Every executive in the organization must take responsibility for becoming a watchful executive in order for the vigilance programs to be effective and successful. This requires the active engagement of each executive in the organization. The leader who is technologically vigilant pays close attention to the most recent scientific and technological breakthroughs and ensures that they have a comprehensive understanding of all contemporary technical ways on the market. In view of this, the first hypothetical statement guiding our investigation is considered as follows.

Hypothesis 1 (H1). *TVL significantly and positively influences IRI.*

Digital transformation impacts on nearly every aspect of an organization's functioning has been recognized, with particular emphasis on the supply chain [16]. Digital transformation specifically aids in cost reduction and performance enhancement through the facilitation of collaborations within the supply chain. The organization offers real-time supply chain management solutions. The extant body of literature also highlights the significant advantages that emerging technologies, including blockchain technology [17] and artificial intelligence, provide to the supply chain. Industry 6.0 aims to facilitate interconnected sectors, establish dynamic value networks, establish a flexible supply chain, and promote information flow

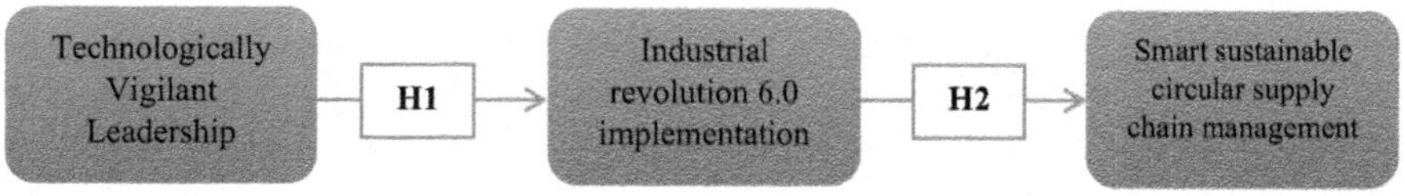

Fig. 10.1 Hypothesized model

transparency across all levels of government, both domestically and internationally [6]. Over the forthcoming years, this revolution is anticipated to engender inventive and original concepts in manufacturing and various other domains, while also augmenting the fundamental attributes of both humans and machines [6]. Industry 4.0 was founded upon the principles of mass production and intelligent production, while Industry 5.0 aimed to foster collaborations between humans and machines to achieve sustainability resilience. Promising antifragile manufacturing, ubiquitous human digital twins, and customer-centricity, Industry 6.0 envisions hyper-connected manufacturing sectors where customer needs take precedence over the utilization of sustainable resources [6]. In view of this, the second hypothetical statement guiding our investigation is considered as follows.

Hypothesis 2 (H2). *IRI significantly and positively influences SSCSCM.*

Therefore, all the aforementioned hypotheses and variables were illustrated in Fig. 10.1 as shown below.

10.4 Methods and Materials

A quantitative method was used to examine the effects of TVL on SSCSCM. A thorough literature review will be carried out in order to pinpoint important concepts and connections. After that, a survey questionnaire were created. The suggested research model will be validated and the quantitative data analyzed using structural equation modeling (SEM).

10.4.1 Measurement Scale

Utilizing a standardized survey questionnaire, statistical data for the validation of the aforementioned hypotheses were gathered. In order to guarantee uniform responses and make data processing and analysis easier, the study used a standardized questionnaire. The primary survey was originally drafted in English before being translated into Vietnamese using a rigorous translation and back-translation process to ensure precision and consistency. Prior to conducting a comprehensive survey, a panel of ten experts comprised of lecturers from public university accounting faculties and a subset of individuals with direct or indirect experience in information technology or accounting in SMEs was assembled for the jury because of their credibility, and reliability. As a result, the expert panel's fresh perspective would be beneficial in conducting a diagnostic analysis of the draft questionnaire, as the developers often became too intimately acquainted with the subject to identify issues during the screening process. The insights gained from these specialists were subsequently considered and incorporated into the complete iteration of the survey in order to enhance its lucidity and comprehensibility. This study used a measurement instrument that has been modified from prior studies to assess the variables included in the investigation. A five-point Likert scale, with 1 denoting strong disagreement and 5 denoting strong agreement, was used in the instrument.

(a) ***Smart sustainable circular supply chain management***: *The SSCSCM construct was identified in this study as an advanced combination of four core constructs: smart end-of-life product and waste management, smart circular procurement, cleaner production, and smart circular product design. More instrumentally, the criteria employed to evaluate Smart circular product design were inherited from the recommendations of* [18]. *The criteria employed to evaluate Smart circular procurement were inherited from the recommendations of* [19]. *The criteria employed to evaluate Smarter and cleaner production were inherited from the recommendations of* [20]. *The criteria employed to evaluate Smart end-of-life product and waste management were inherited from the recommendations of* [21].
(b) ***Industrial revolution 6.0 implementation***: *This investigation determined the IRI construct as an elevated amalgamation of two fundamental constructs-Green sustainability; Dynamic supply chain-which were inherited from the recommendations of* [6].
(c) ***Technologically vigilant leadership***: *This investigation utilized three items that were derived from those proposed by* [8] *in order to evaluate TVL.*

10.4.2 Sampling and Data Collection

The respondents were representatives of SMEs in Vietnam, spanning across many industries. Each SME had a minimum of two accountants and two information technology staff were involved in this investigation. There was an agreement that the collected data could be shared in an anonymized format and published for research purposes. Due to the challenges in collecting statistical data from businesses in a growing economy, as well as time constraints, the sampling approach used in this study was non-probability convenience and snowball sampling techniques. Ensuring the reliability and accuracy of study findings is contingent upon determining an adequate sample size [22]. A key component of the PLS-SEM method is figuring out the minimum sample size. Based on the idea that a reasonable sample size should be at least ten times the number of variables included in the study, the "10-times rule" approach is frequently used to estimate the minimum sample size required to conduct PLS-SEM as recommended by [23]. The period of data collecting was September

Table 10.1 Demographic information

Demographic profile	Variables	Usable responses	Weight (%)
Gender of respondent			
	Male	294	60.87
	Female	189	39.13
Age of respondent			
	30–under 40	85	17.60
	40–under 50	358	74.12
	Over 50	40	8.28
Experience of respondent (years)			
	Under 10	16	3.31
	10–under 20	265	54.87
	20–under 30	194	40.17
	30–under 40	8	1.66
Education			
	Undergraduate	465	96.27
	Postgraduate	18	3.73

2023–February 2024. There were 483 instances left after the questionnaires were screened and evaluated, yielding a data loss rate of 25.69%. All the demographic information that was collected for this study was laid out in Table 10.1. The SPSS version 29.0 facilitated the utilization of numerous statistical procedures and techniques for the analysis of the completed surveys. The measurement and structural models were subsequently evaluated with SmartPLS 4.1.0.3.

10.5 Inferential Statistics

Inferential statistical techniques will be applied to draw conclusions about the population based on the sample data. Specifically, Structural Equation Modeling (SEM) will be used to test the hypothesized relationships between technologically vigilant leadership, smart sustainable circular supply chain management, and Industry 6.0 implementation.

10.5.1 Assessment on Measurement Model

Construct reliability and Convergent validity. The findings presented in Table 10.2 validated that the Cronbach's alpha values and Composite reliability (rho_c) for all latent variables exceeded 0.7, signifying a high degree of concordance among the indicators [24]. As stated by [25], the attainment of internal consistency reliability was deemed to be indicated by Composite reliability (rho_a) values exceeding 0.70. Building on the perspectives of [24], all outer loadings in the measurement models surpassed the critical threshold of 0.707. All variables exhibited an average variance extracted (AVE) that exceeded 0.5 [24]. The result in Table 10.2 validated convergent validity and internal consistency.

In this correlations and discriminant validity study, an evaluation was conducted on Fornell–Larcker criterion and heterotrait–monotrait ratio (HTMT) for discriminant validity. The square root of AVE had to be greater than the highest correlation between the constructs [26] for the discriminant validity criterion to be satisfied. The result in Table 10.3 validated discriminant validity.

In addition, the results of the discriminant validity analysis were displayed in Table 10.4. This table comprised the HTMT ratios, which were compared to a threshold value of 0.85, as recommended by [26, 27]. The results in Table 10.4 validated discriminant validity [27].

Table 10.2 Results summary of measurement model assessment

Constructs and operationalization	Items (Abbreviation)	Convergent validity		Construct reliability			Inference
		Factor loadings	AVE	Cronbach's alpha	Composite reliability (rho_c)	Composite reliability (rho_a)	
Technologically vigilant leadership	**EL**	0.792–0.860	0.681	0.766	0.865	0.772	Retained
Industrial revolution 6.0 implementation	**IRI**						
Green sustainability	GS	0.835–0.850	0.706	0.792	0.878	0.792	Retained
Dynamic supply chain	DSC	0.804–0.885	0.708	0.793	0.879	0.794	Retained
Smart sustainable circular supply chain management	**SSCSCM**						
Smart circular product design	SCPD	0.805–0.891	0.737	0.821	0.894	0.824	Retained
Smart circular procurement	SCIP	0.802–0.872	0.690	0.775	0.869	0.780	Retained
Smarter and cleaner production	SCLP	0.805–0.871	0.711	0.797	0.881	0.796	Retained
Smart end-of-life product and waste management	SEPW	0.793–0.851	0.683	0.774	0.866	0.800	Retained

Table 10.3 Results summary for discriminant validity on Fornell–Larker criterion

	DSC	EL	GS	SCIP	SCLP	SCPD	SEPW
DSC	0.842						
EL	0.315	0.826					
GS	0.397	0.332	0.840				
SCIP	0.213	0.306	0.288	0.830			
SCLP	0.127	0.099	0.091	0.220	0.843		
SCPD	0.270	0.190	0.268	0.268	0.316	0.859	
SEPW	0.033	0.022	0.111	0.072	0.010	0.017	0.827

Table 10.4 Results summary for discriminant validity on Heterotrait–Monotrait ratio

	DSC	EL	GS	SCIP	SCLP	SCPD	SEPW
DSC							
EL	0.403						
GS	0.501	0.425					
SCIP	0.272	0.395	0.364				
SCLP	0.158	0.123	0.113	0.277			
SCPD	0.333	0.240	0.331	0.334	0.389		
SEPW	0.056	0.056	0.150	0.089	0.062	0.039	

10.5.2 Full Structural Equation Modeling and Hypotheses Testing

Table 10.5 and Fig. 10.2 provided examples of the statistical findings. There were no significant collinearity issues because all inner variance inflation factor (VIF) values were far below the crucial threshold value of 3 [24]. Hinged on the bootstrap outcomes, TVL was validated to induce a substantial as well as positive influence on IRI ($\beta = 0.387$; t-value = 10.270; p-value = 0.000). In the same vein, IRI was corroborated to induce a substantial as well as positive influence on SSCSCM ($\beta = 0.360$; t-value = 8.898; p-value = 0.000). To that end, H1 and H2 were supported.

The R^2 was 0.150 for IRI while R^2 for SSCSCM was 0.129. As shown in Table 10.5, TVL had a medium effect size on IRI (0.177) and IRI had a small effect on SSCSCM (0.149). The Q^2 values were 0.073 for IRI and 0.035 for SSCSCM which was above zero [24].

Table 10.5 Results summary of hypotheses acceptance

Relevant path	Path coefficient	SE	95% confidence interval	VIF	t-value	p-value	Result
TVL → IRI	0.387	0.038	[0.311–0.458]	1.000	10.270	0.000	Accepted
IRI → SSCSCM	0.360	0.040	[0.275–0.433]	1.000	8.898	0.000	Accepted
R^2	$R^2_{IRI} = 0.150$; $R^2_{SSCSCM} = 0.129$						
f^2	$f^2_{TVL \Rightarrow IRI} = 0.177$; $f^2_{IRI = SSCSCM} = 0.149$						
Q^2	$Q^2_{IRI} = 0.073$; $Q^2_{SSCSCM} = 0.035$						

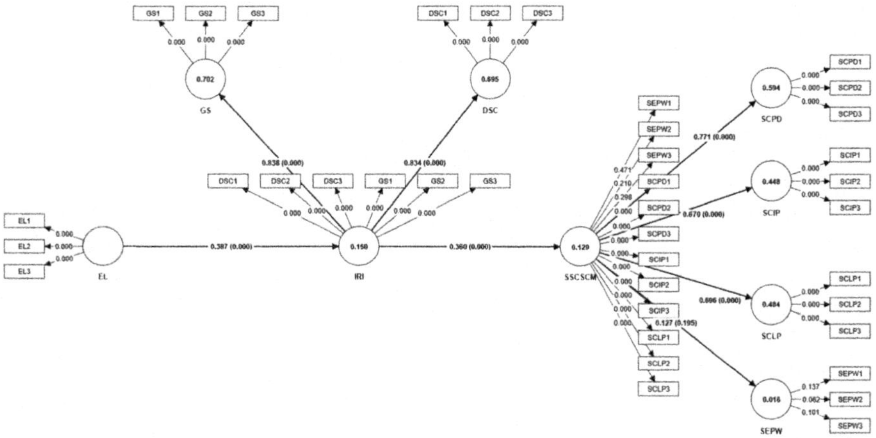

Fig. 10.2 Structural model

10.6 Conclusion

In the framework of Industry 6.0, this study seeks to offer insightful information about the role of TVL as a key enabler of IRI and its impact on achieving SSCSCM. It is anticipated that the results of this study will add to the corpus of knowledge already in existence and offer useful recommendations for supply chain experts, executives, and legislators. In order for firms to be competitive in the age of Industry 6.0, the study's findings will also emphasize the necessity of implementing TVL. The correlations among the constructs analyzed in this study will be further explored by identifying future research objectives.

(a) **Practical implication**

The implications for management will result from the findings of this study, which will establish a basis for practitioners to expand their strategies for implementing an effective SSCSCM by highlighting the interdependencies between the IRI and

SSCSCM. By employing the framework suggested in this research, coupled with an accurate assessment of investments, one can enable the establishment of a streamlined transition procedure within the supply chain. Furthermore, this research furnished managers with more comprehensive and precise understandings pertaining to the potential and execution of Industry 6.0. Consequently, they will possess an enhanced capacity to understand the most feasible enabling technologies and prerequisites associated with the implementation of Industry 6.0. The statistical empirical data regarding the influence of TVL indicates that it operates as a mechanism that allows enterprises to efficiently and effectively leverage the advantages of their technological investments. Consequently, it became imperative that executives of SMEs possess digital competencies, which include skills in technology, leadership, and organization. As a result, there is a motivation for all leaders of SMEs to participate in development courses that enhance their technological proficiency. This promotes a feeling of empowerment in their professional pursuits and facilitates the sharing of knowledge.

(b) **Limitation and agenda for further research**

There are some inherent limitations in this manuscript. Firstly, even though this study met the minimal sample size requirement, it is suggested to increase the sample size to potentially mitigate the presence of multivariate non-normality in the data. Additionally, selection bias may have occurred in this study due to the use of convenience and snowball sampling methods. This thus led to the fact that the obtained findings may not be applicable to the broader population. Hence, it is recommended that future study employ a comprehensive examination using a random sampling technique. On the other hand, it is presumed that the method variance is heightened using self-report scales in this study, as they are solely completed by the participants. Therefore, future research endeavors should aim to overcome these limitations by gathering secondary data. Last but not least, the investigation was conducted as a descriptive and cross-sectional study. Longitudinal studies are more appropriate for establishing causal links in the suggested paradigm.

Acknowledgements This work was funded by University of Economics Ho Chi Minh City (UEH) in Vietnam.

This study is the academic product of Ministerial-level Technology and Science Project sponsored by the Ministry of Education and Training "Integrating Inclusive Green Growth model into building a measurement framework for the level of Circular Economy implementation in Public Entities" with Grant Code B2025-KSA-09.

References

1. Bassetti, T., Blasi, S., Sedita, S.R.: The management of sustainable development: a longitudinal analysis of the effects of environmental performance on economic performance. Bus. Strategy Environ. **30**(1), 21–37 (2021)
2. Marjamaa, M., Salminen, H., Kujala, J., Tapaninaho, R., Heikkinen, A.: A sustainable circular economy: exploring stakeholder interests in Finland. South Asian J. Bus. Manag. Cases. **10**(1) 50–62 (2021)
3. Taddei, E., Sassanelli, C., Rosa, P., Terzi, S.: Circular supply chains in the era of industry 4.0: a systematic literature review. Comput. Ind. Eng. **170**, 1–24 (2022)
4. Zouari, D., Ruel, S., Viale, L.: Does digitalising the supply chain contribute to its resilience? Int. J. Phys. Distrib. Logist. Manag. **51**(2), 149–180 (2021)
5. Duan, Y., Aloysius, J.A., Mollenkopf, D.A.: Communicating supply chain sustainability: transparency and framing effects. Int. J. Phys. Distrib. Logist. Manag. **52**(1), 68–87 (2021)
6. Chourasia, S., Tyagi, A., Pandey, S.M., Walia, R.S., Murtaza, Q.: Sustainability of industry 6.0 in global perspective: benefits and challenges. MAPAN **37**, 443–452 (2022)
7. Ghosh, S., Hughes, M., Hodgkinson, I., Hughes, P.: Digital transformation of industrial businesses: a dynamic capability approach. Technovation **113**, 1–18 (2022)
8. Amayreh, K.T.: The role of knowledge management in delivering the organization to the state of performance excellence: mediating role of technological vigilance. Manag. Sci. Lett. **11**, 1277–1286 (2021)
9. Egelhoff, W.G.: Information-processing theory and the multinational enterprise. J. Int. Bus. Stud. **22**(3), 341–368 (1991)
10. Albhirat, M.M., Rashid, A., Rasheed, R., Rasool, S., Zulkiffli, S.N.A., Zia-Ul-Haq, H.M., Mohammad, A.M.: The PRISMA statement in enviropreneurship study: a systematic literature and a research agenda. Clean. Eng. Technol. **18**, 1–12 (2024)
11. Freeman, R.E., Harrison, J., Wicks, A., Parmar, B., De Colle, S.: Stakeholder Theory: the State of the Art. Cambridge University Press (2010)
12. Gu, J., Wang, J., Shi, X., Xu, X.: Investigating supply chain participants' circular economy action effects on firm financial performance from a stakeholder theory perspective. J. Purch. Supply Manag. 1–16 (2024)
13. Duggal, A.S., Malik, P.K., Gehlot, A., Singh, R., Gaba, G.S., Masud, M., Al-Amri, J.F.: A sequential roadmap to industry 6.0: exploring future manufacturing trends. IET Commun. **16**(5), 521–531 (2022)
14. Mahdiraji, H.A., Govindan, K., Yaftiyan, F., Garza-Reyes, J.A., Hajiagha, S.H.R.: Unveiling coordination contracts' roles considering circular economy and eco-innovation toward pharmaceutical supply chain resiliency: evidence of an emerging economy. J. Clean. Prod. **382**, 1–17 (2023)
15. Dyer, B.: A pragmatic approach to resolving technological unfairness: the case of Nike's Vaporfly and Alphafly Running Footwear. Sports Med. Open **6**(1), 1–10 (2020)
16. Lang, L.D., Behl, A., Phuong, N.N.D., Gaur, J., Dzung, N.T.: Toward SME digital transformation in the supply chain context: the role of structural social and human capital. Int. J. Phys. Distrib. Logist. Manag. **53**(4), 448–466 (2023)
17. Venkatesh, V.G., Kang, K., Wang, B., Zhong, R.Y., Zhang, A.: System architecture for blockchain based transparency of supply chain social sustainability. Robot. Comput. Integr. Manuf. **63**, 1–9 (2020)
18. Den Hollander, M.C., Bakker, C.A., Hultink, E.J.: Product design in a circular economy: development of a typology of key concepts and terms. J. Ind. Ecol. **21**(3), 517–525 (2017)
19. Zhu, Q., Geng, Y., Sarkis, J.: Motivating green public procurement in China: an individual level perspective. J. Environ. Manag. **126**, 85–95 (2013)
20. Sousa-Zomer, T.T., Magalhães, L., Zancul, E., Campos, L.M.S., Cauchick-Miguel, P.A.: Cleaner production as an antecedent for circular economy paradigm shift at the micro-level: evidence from a home appliance manufacturer. J. Clean. Prod. **185**, 740–748 (2018)

21. Hsu, C., Choon Tan, K., Hanim Mohamad Zailani, S., Jayaraman, V.: Supply chain drivers that foster the development of green initiatives in an emerging economy. Int. J. Oper. Prod. Manag. **33**(6), 656–688 (2013)
22. Islam, M.F., Mofiz Uddin, M.M., Rahman, M.M.: Factors affecting retailer social responsibility: a PLS-SEM approach in the context of Bangladesh. Soc. Responsib. J. 1–21 (2024)
23. Hair, J.F., Jr., Matthews, L.M., Matthews, R.L., Sarstedt, M.: PLS-SEM or CB-SEM: updated guidelines on which method to use. Int. J. Multivar. Anal. **1**(2), 107–123 (2017)
24. Hair, J.F., Sarstedt, M., Ringle, C.M., Gudergan, S.P.: Advanced Issues in Partial Least Squares Structural Equation Modeling (PLS-SEM), 2nd edn. Sage, Thousand Oaks (2024)
25. Dijkstra, T.K., Henseler, J.: Consistent partial least squares path modeling. MIS Q. **39**, 297–316 (2015)
26. Fornell, C., Larcker, D.F.: Evaluating structural equation models with unobservable variables and measurement error. J. Mark. Res. **18**(1), 39–50 (1981)
27. Henseler, J., Ringle, C.M., Sarstedt, M.: A new criterion for assessing discriminant validity in variance-based structural equation modeling. J. Acad. Mark. Sci. **43**(1), 115–135 (2015)

Pham Quang Huy is an Accounting Associate Professor at University of Economics Ho Chi Minh City, Vietnam. His research interests encompass accounting-to-finance, digital era and innovation with sustainable environment management. His research has been published in *Cogent Business and Management, Asian Journal of Law and Economics, Public Organization Review*, and elsewhere.

Vu Kien Phuc is an experienced accounting lecturer of at University of Economics Ho Chi Minh City in Vinh Long Campus. She has been currently a Ph.D. candidate. Her research interests focus on public accounting with innovation, sustainability and cyber security. Her publications have been published in *Cogent Education, Journal of Innovation & Entrepreneurship*, and various reputed journals.

Chapter 11
Recent Trends in Natural Fiber Reinforced HDPE Composites: Progress and Challenges

Rakesh Upadhyay, Sujeet Kumar Gautam, Rajat Upadhyaya, Bipin Kumar Singh, Siddharth Yadav, Kamlesh Kumar Singh, and Sunil Kumar Sharma

Abstract Nowadays, environmentally friendly composites are encouraged by the worldwide community due to their easy recyclability and degradability. So, natural fibre-based composites are capturing the market exponentially. The vast application opportunities in the manufacturing sector, such as automobiles, marine, piping, and grocery items, are leaders in applying natural fibre components. Several usual fibres like rice, oil palm, coconut, and kenaf have been successfully implemented and have better mechanical properties. Additionally, reinforcing natural fibre inside high-density polyethene (HDPE) plays a key role among many polymers as it acquires numerous benefits, such as strength and excellent chemical stability. HDPE has several other advantages, such as being durable, lightweight, and machine-friendly,

R. Upadhyay
AcSIR, Ghaziabad, Uttar Pradesh, India

CSIR - IIP, Dehradun, India

S. K. Gautam (✉) · R. Upadhyaya · K. K. Singh
FFT, NIAMT, Ranchi, India
e-mail: sujeetkumargautam@gmail.com

R. Upadhyaya
e-mail: rupadhyaya@niamt.ac.in

K. K. Singh
e-mail: kksingh@niamt.ac.in

B. K. Singh
CAID, Department of Mechanical Engineering, SECE, Coimbatore, Tamil Nadu, India
e-mail: bipinmech2008@gmail.com

S. Yadav
CAS, AKTU, Lucknow, India
e-mail: drsiddharth@cas.res.in

S. K. Sharma
Chair of Rail Vehicle Technology Technische Universitat Dresden, Dresden, Germany
e-mail: sunil_kumar.sharma@tu-dreden.de

© The Author(s), under exclusive license to Springer Nature Switzerland AG 2025
K. C. Rath et al. (eds.), *Engineering, Technology and Management*, Information Systems Engineering and Management 54, https://doi.org/10.1007/978-3-031-93654-8_11

making it usable in medical, construction, maritime, and food. As a result, this review thoroughly discusses the mechanical characteristics of this reinforced in HDPE and various parameters affecting the structural features of polymer composite, such as loading, fiber dimensions, and fiber structure. A critical observation is also done to understand the broader effect of the mechanical properties on varying the fibre wt.% and loading. This review article also outlined the survey and collected earlier works demonstrating natural fibres' mechanical properties and their composites. Thus, there is an extensive availability of knowledge for additional investigations into the characteristics of synthetic fibres.

Keywords HDPE · Processing techniques · Mechanical property · Loading · Fibre dimension · Fibre structure · Composite

Preface

This chapter explains about environmental concerns have prompted global efforts to regulate the materials contributing to ecological degradation. In response, researchers have developed natural fibers sourced from plants and animals, which are abundant, eco-friendly, and recyclable. These fibers have found wide applications, from household items to industrial products, due to their low weight, high tensile strength, and sustainability. High-density polyethylene (HDPE) stands out as a natural fiber for its resilience, UV resistance, and structural advantages, finding roles in everything from fuel tanks to cable insulation. Recently, natural fibers like sisal, jute, and rice straw have been incorporated into HDPE composites, enhancing their strength, durability, and eco-friendliness. However, challenges remain in optimizing fiber-matrix compatibility to maximize performance, highlighting a need for continued research on additive selection and fiber treatment.

11.1 Introduction

All over the world, environmental concerns are a prime focus for governments bound to impose strict restrictions on the composites responsible for environmental problems. To mitigate the environmental issues, researchers developed natural fibres that are highly compatible with the environment, easy to recycle, and obtained from (or are produced by) animals and plants (abundance of availability) [1, 2]. The demands of natural fibres are on the zenith due to their vast applications, from home appliances like grocery items and decorative items to industrial needs such as pipes, fuel tanks, cable insulators and many more. There are numerous natural fibres available in the commercial market. Still, high-density polyethene (HDPE) is one of the most exciting and well-known natural fibres due to its unique properties like highly stiff material, leaching not possible, capable of withstanding UV, nonreactive with primary chemical solvents and can withstand up to -148 to 176°F.

The remarkable journey of HDPE started with Low-density polyethene (LDPE) around the 1950s [3], but increasing demand for highly strengthened polymer gave birth to High-density polyethene (HDPE). HDPE's interesting features attract scientists to fabricate denser polyethylene as it has enormous application in industries. Characteristics like higher molecular weight and lesser molecular chain branching prevail in HDPE compared to LDPE. The mechanical properties acquired by HDPE are also much better than other polyethene because carbon and hydrogen are their main constituent—the higher density of HDPE is attributed to its long-linear molecular arrangement. HDPE is found to have high temperature-resistant properties [4]. HDPE acquires unique features with vast applications, including grocery items, pipes, fuel tanks, and cable insulators. HDPE is mainly used as a base material or matrix because of its excellent strength-to-density ratio. It is also equipped with a better specific strength. It is also very cheap and abundantly available in the market.

Furthermore, fibre-reinforced polymer (FRP) constituents are polymer matrix reinforced with strengthened fibre like glass, and carbon forms a composite material. Polymers are divided into two groups: thermosetting plastic and thermoplastics. Thermoplastic materials nowadays play a prominent part in bio-fibres; the most widely used thermoplastics are PE, PVC, and PP, whereas epoxy and polyester resins are the widely utilized thermosetting matrices. Recently, natural fibres as a substitute for additives in polymer composites have gained the attention of several researchers and scholars because of their uniqueness over conventional glass and carbon fibres [5]. The natural fibres are sisal, hemp, kenaf, jute, coir, and banana [6]. Natural fibres have various benefits over synthetic glass and carbon fibres: lower weight-to-density ratio and higher tensile characteristics, low cost, not hazardous to the body, and renewable, recyclable, and biodegradable nature [7, 8]. However, natural fibres have certain limitations, i.e. reinforcement inside a polymer composite creates instability between the hydrophobic thermoplastic matrices and natural hydrophilic fibres. This may lead to the unfavourable properties of the composite developed through reinforcing [9, 10]. Hence, there is a requirement to improve the fibre interface by modifying the chemical composition alongside the selection of a suitable matrix. In this regard, Li et al. [11] developed a composite reinforced with sisal fibre with an HDPE matrix. Researchers observed that sisal fibre provides high strength and durability, and its composites are environmentally friendly. The interfacial shear strength of sisal fibre, including the HDPE matrix, was significantly improved. Yao et al. [12] developed a composite by reinforcing four rice straw components and wood fibre with an HDPE matrix. The lignocellulosic properties governed by rice straw fibres significantly enhanced the properties of composites. The result also showed a larger storage modulus but adversely affected impact strength.

The recycled HDPE and its derivative had much better moduli than pure HDPE. Mohanty and Nayak [13] developed a composite of bamboo fibre-reinforced HDPE. Researchers selected the melt blending technique accompanied by injection moulding. Researchers observed that UTS and flexural strength were improved up to 25.27 MPa and 27.86 MPa, respectively. Furthermore, the composites with 30% fibre loading and 2% MAPE gained maximum strength. Panthapulakkal and Sain [14] prepared a composite with agro-based residue as reinforcement material inside the HDPE matrix. Researchers demonstrated that the wheat straw had more carbon, influencing end properties. The percentage improvement in strength and stiffness of wheat straw is 47% and 36%, respectively.

Hence, the above study reveals ample research to strengthen the HDPE composites by tailoring them with additives. Furthermore, many parameters impact the performance of natural fiber-impregnated composite. Hydrophilic behaviour is not the only property that affects it; the proportion of additives also plays a vital role. Developing excellent fibre content to acquire better mechanical properties of the material is indeed a requirement by industries. Therefore, this work makes every effort to compile valuable works focused on selecting additives and unveiling the physics beneficial for improving properties. This work also addresses the problem that restricts its implications with futuristic approaches. After that, a detailed discussion of a comparative research that illustrates the impact of various natural fibers using HDPE as a polymer matrix follows. Finally, consolidated remarks have illustrated how functionalized HDPE benefits commercial applications such as industrial, biomedical, structural, etc.

11.2 Development of Natural Fiber Polymer Composite and Effect of Additive

There is much variation in the mechanical properties of the reinforcement of different fibre reinforcement. Factors affecting the properties are composition, structure, defects, and behaviour with the base material—additionally, the chemical structure of the matrix and reinforcement significantly cause misleading stress transfers. Hence, there is always an essential chemical treatment for natural fibre to have excellent interfacial properties [15, 16]. While developing natural fibre polymer composites, a weak bond between the hydrophobic polymer and hydrophilic fibre is created. This hydrophilic property in the natural fibre is due to a functional group known as the hydroxyl group, consequences in inferior mechanical properties.

Furthermore, many things have impacted the performance level, such as fibre position, fibre strength, and the physical properties of fibres [17]. The researchers on natural fibre polymer composites showed good mechanical properties compared to pristine matrix, where jute fibres were involved in polylactic acid (PLA). Till now, tremendous growth in composites was seen for jute as reinforced that possesses 121% improvement w.r.t. pristine polyester [18]. Essabir et al. [19] developed composites that included oil palm fibre and clay particles as reinforcement with HDPE. As a reinforced material, Clay reduces thermal interruption alongside improved mechanical properties. The observation was satisfied, with a hike of 11% in mechanical properties. Mazur et al. [20] included NF with an HDPE matrix to consider the effect on the composite.

Researchers showed 40 wt.% of wood flour, flax fibre, and wall nutshell flour were observed as optimum composition. The obtained result exhibits enhanced flexural strength and tensile properties, which increased by 262% and 156%, respectively. Hussain et al. [17] investigated the mechanical properties of green coconut fibre reinforced with HDPE. The UTS was achieved at a maximum of 40% fibre volume. Souza et al. [21] experimented on textile fibre residue reinforced in HDPE. The tensile strength improved by 45.2% compared to the pristine HDPE composite. Babu and SNarayanankutty [22] investigated a coir-reinforced polymer matrix. Raw coir fibre showed a dip in tensile strength, whereas elastic modulus improved with much better thermal stability. Aji et al. [23] prepared a hybrid kenaf/pineapple HDPE composite with different wt.%. Pineapple leaf fibre has better cellulose content and has much tensile strength than Kenaf fibre. The UTS and modulus properties of material developed showed an improvement with the increase in the percentage of Pineapple leaves.

11.3 Effect of Chemical Modification

Chemical modification is required on natural fibres to improve bondamong the base material and natural fibre. It is observed that weak connection among the natural fibre reinforcement and its materials happened owing to the hydrophobic behaviour of matrices and hydrophilic characteristics of the natural fibre. Hereafter, chemical modification improves connection among the natural fibre and its materials. So, researchers carried out ample chemical modifications to obtain superior composite properties [24].

11.3.1 Alkaline Treatment

Under this method, modification is done on the molecular structure using sodium hydroxide to improve the movement of fibre fragmentation and separation [25]. Additionally, hemicellulose and lignin were distant from the fibre by chemical alkali treatment, which also caused the fibre to split into fibrils and reveal more cellulose in the remaining areas of the fibre surface. Using NaOH on the fibres brings advancements in cellulose types and adhesiveness for the discrete fibrils, which follow an enhancement in load capability. As a result of the alkaline inclusion's removal of natural and manufactured contaminants from the fibre surface, the fibrous matrix adheres better [26]. Thus, it brings an enormous increase in mechanical properties. It was also found that alkali treatment enhances the tensile strength of the natural fibre. The alkali inclusion of natural fibre is a beneficial method for improving the surface tension of natural fibre. Alkoxide formation is shown by the below reaction with sodium hydroxide (aqueous) [27].

$$\text{Fiber} - \text{OH} + \text{NaOH} \rightarrow \text{Fiber} - \text{O} - \text{Na} + \text{Water}$$

11.3.2 Effect of Silanes and Natural Fibers

The reactive silanol groups create ASiAOASiA bonds to form H_2 bonds with hydroxyl sites of fibres. Silanols in APS, TAS, MPS contribute to the development of a durable polysiloxane layer on the fiber surface after an initial single layer is established, with further enhancement occurring through a condensation process driven by temperature. The polysiloxane coating on fibre surfaces may prevent silane molecules from diffusing into cell walls, resulting in a higher concentration in the cell walls [28]. To decrease the condensation rate, it is necessary to make adjustments to hydrolysis states, specifically the pH of the solution. This modification allows the reactive silanol to penetrate the cell walls of the fibers and access the hydroxyl sites contained therein. In contrast to MPS, APS, and TAS, the former two demonstrated higher adsorption on cellulose fibers because they can create more robust H_2 connection among amino categories and fiber hydroxyl categories.

During solvent evaporation, the unbound silanol compounds may condense further, developing ASiAOASiA network connections on/in the fibres. The ASiAOASiA bonds produced, unlike the SiAOAC bonds, are highly resistant to hydrolysis [29]. The hydrogen bonds formed at the adsorbent surface between physical adsorption silanols and hydroxyl categories of fibres hardly transformed to inside of the covalent connections of ASiAOACA connections at ambient conditions. So, the adsorbed silanes can be filtered off the fibres using Soxhletremovalthrough ethanol [30]. Heating the fibres (for example, at 110 °C for 2 h) may release water or solvents and initiate the dehydration process at the silanol-fiber hydroxyl adsorbent surface.

11.3.3 Acetylation Treatment

Acetyl groups are fused to the fibres' cellular structure during the acetylation process using acid catalysts [31]. This phenomenon occurs as a result of the interactions between acetic acid and acetic anhydride, which are inseparable. The fibers are immersed in acetic acid and previously subjected to treatment with acetic anhydride for duration of 1–3 h at elevated temperatures to enhance their characteristics. The strength of natural fibers is improved through an esterification reaction that takes place between hydroxyl (OH) and carboxylic/anhydride groups [32]. In the acetylation technique, the contact with acetyl eliminates the wax and cuticle on the fibre's surface, leading to a smoother surface. Hence, Acetylation chemically changes natural lignocellulosic fibres to improve qualities such as dimensional stability, biological attack resistance, and UV degradation. Several laboratories worldwide have examined the technique and considered it for commercialization.

11.3.4 Benzoylation Treatment

Benzoyl chloride is utilised to enhance natural fibres' interfacial bond and thermal steadiness by lowering their hydrophilic character [33]. The first step involves the extraction of lignin, waxes, and surface oils, which uncovers the fiber's surface contains reactive hydroxyl (OH) groups. The next step involves treating the fibers with benzoyl chloride, leading to the substitution of the hydroxyl (OH) groups by benzoyl groups. Finally, the cellulose backbone is bonded with OH groups.

The reaction depicted below illustrates the engagement of the cellulosic hydroxyl group with benzoyl chloride.

$$\text{Fiber} - \text{OH} + \text{NaOH} \rightarrow \text{Fiber} - \text{O} - \text{Na}^{+} + \text{Water}$$

$$\text{Fiber} - \text{O}^{-}\text{Na}^{+} + (\text{C}_6\text{H}_5\text{COCl}) \rightarrow \text{Fiber} - \text{O} - (\text{C}_6\text{H}_5\text{CO}) + \text{sodium chloride.}$$

11.3.5 Maleated Coupling Agents

Carbon-carbon interactions may be formed with the matrix's polymer chains using maleated coupling agents, allowing for improved bond among the fibers' useful parts and the matrix [34]. When maleic anhydride comes into contact with hydroxyl categories (OH) in uncrystalized sections of the cellulose fiber geometry, it generates long-chain polymers that form a coating on the fiber surfaces. This mechanism forms a chemical connection among the fibers' hydroxyl and maleic anhydride categories, effectively lowering hydroxyl group count within the fiber cells and minimizing their hydrophilic inclinations [35]. As a result, a boundary within the chain is created, which enhances the interconnection between the matrix and the fibers. The fibers treated with maleic anhydride exhibit improved bonding with the resin at the interface, leading to tearing failure rather than interfacial failure, ultimately resulting in better mechanical performance of the composite.

11.3.6 Sodium Chlorite Treatment

The bleaching process, especially when sodium chlorite is employed with an acidic solution, is vital for improving the characteristics of natural fibers that are necessary for their function as reinforcements in composite materials. Sodium chlorite, recognized for its strong oxidizing capabilities, is extensively used in the bleaching of natural fibers to eradicate lignin, hemicellulose, and other impurities. The acidic environment aids the oxidation process, leading to the successful breakdown and removal of these non-cellulosic components [36].

A primary benefit of this bleaching method is its effectiveness in lowering the moisture content within the fibers. Natural fibers include hydroxyl groups that attract water, causing expanding, and deterioration of the adhesive nature between the reinforcement and matrix, and a consequent decline in the qualities of the cast material. The application of sodium chlorite in an acidic solution facilitates the removal of these water-attracting components, leading to a significant decrease in the moisture content of the fibers. As a result, the fibers show improved hydrophobic characteristics, indicating a higher resistance to water absorption. This hydrophobic nature is vital for enhancing stability and effectiveness of fiber composites, particularly in damp conditions. Additionally, the bleaching process removes hydrophilic components, which not only enhances moisture resistance but also increases the flexibility of the fibers. The stiff, non-cellulosic materials typically eliminated during bleaching are largely responsible for the rigidity found in untreated fibers.

The reduction in moisture content and increased flexibility allow for more efficient integration of bleached fibres into the composite matrix—this results in improved fibre distribution, enhanced stress transfer, and, ultimately, superior mechanical characteristics. Therefore, the bleaching process using sodium chlorite enhances the hydrophobic properties of natural fibres and maximizes their effectiveness as reinforcements in composite materials.

11.3.7 Acrylation and Acrylonitrile Grafting

The cellulose molecule's free radicals start the acrylation process. In combination with chain scission, cellulose can be associated with particular radiation to create radicals [37]. In their study, Sreekala et al. [38, 39] explored the use of CH_2=CHCOOH to modify surface of natural fibers, particularly oil palm fibers. The treatment involved exposing the fibers to a mixture of acrylic acid at 50 °C for 1 h, using different concentrations. Through this process, acrylic acid can be grafted and polymerized, leading to the desired alteration of glass fibers. The fibres were washed and dried in an aqueous alcohol solution. Oil palm fibre–PE composites' tensile strength is improved. Treatments such as silane and acrylation created strong covalent bonds, which enhanced UTS and YM (young's modulus) of modified fibres. Below is the equation that outlines how AN interacts with the hydroxyl category in fiber [39]:

$$\text{Fiber} - \text{OH} + \text{CH}_2 = \text{CHCN} \rightarrow \text{Fiber} - \text{OCH}_2\text{CH}_2\text{CN}$$

Mishra et al. [40] conducted a study whereby they utilized a mixture of sodium periodate and copper sulfate as an initiator in a water-based solution at temperatures ranging from 50 to 70 °C. Significant factors that influenced the grafting process included the nature of the reaction medium, the length of treatment, the type and concentration of the activator, and the fiber loading. The original fibers underwent modifications in their surface chemistry, which led to a lower affinity for moisture. The finding indicated that synthetically altered fibers grafted with 5%

acrylonitrile had a much higher UTS and YM than those grafted with 10 and 25% AN. This improvement is attributed to the creation of a more organized configuration of polyacrylonitrile units at lower concentrations, which enhances the mechanical properties' fibers.

11.3.8 Isocyanate Treatment

Characterized by the isocyanate functional categories –N=C=O, isocyanates readily interact with the hydroxyl groups in cellulose and lignin found in fibers. For instance, Jabber et al. [41] improved the adhesive properties of composite by applying polymethylene-polyphenyl isocyanate ($C_{15}H_{10}N_2O_2$) to pineapple leaf fiber at 50 °C for 30 min. A comparative evaluation of the mechanical properties of cellulose fiber–polystyrene (PS) cast materials treated with silane and isocyanate showed that isocyanate treatment was more effective. In contrast, TDIC composites treated with oil palm fiber revealed lower UTS and Young's modulus compared to fiber–phenol formaldehyde (PF) composites that were treated with permanganate, peroxide, and alkaline solutions [42]. The reaction illustrated below demonstrates the coupling process between fiber and the isocyanate agent:

$$\mathrm{R-N=C=O+HO-Fiber\rightarrow R-NH-CO-O-Fiber.}$$

11.3.9 Oleoyl Chloride Treatment

The application of fatty acid derivatives, including oleoyl chloride, to natural fibers is a recognized technique for enhancing their wettability and adhesion properties when utilized as reinforcements in composite materials. Oleoyl chloride interacts by way of hydroxyl categories found on surfaces of natural fibers, such as cellulose, which are naturally hydrophilic due to the high concentration of these polar OH groups. This hydrophilicity causes the fibers to absorb moisture, resulting in inadequate compatibility with hydrophobic polymer matrices in composite formulations [43].

The interaction of oleoyl chloride with the fibers causes a chemical reaction between OH groups, leading to in the development of ester bonds. This reaction effectively displaces the hydrophilic OH groups on the fibers' surfaces, replacing them with extended hydrophobic alkyl chains derived from the fatty acid. Consequently, the elimination of OH groups and their substitution with hydrophobic chains greatly reduces the fibers' water affinity, thereby improving their hydrophobic characteristics. This change leads to a distinct enhancement in wettability of the fibers by polymer matrix, a factor that is vital for strengthening composite. Improved wettability enables the polymer to effectively surround and bond with the fibers, resulting in enhanced stress transfer and mechanical performance in the final composite material. In addition, the hydrophobic nature of treated fibers enhances their ability to resist moisture inclusion, which could otherwise result in swelling and negatively affect the reinforcements-matrix interface, ultimately reducing the composite's durability.

In addition, the incorporation of oleoyl chloride significantly augments the hydrophobic characteristics of the reinforcements, enhancing its suitability with hydrophobic polymers. This treatment leads to a more consistent distribution of the reinforcements within the matrix, which is vital for ensuring uniform mechanical properties in the composite. To conclude, utilizing oleoyl chloride to treat fibers is a potent strategy for improving their wettability and adhesion, thereby resulting in enhanced performance in composite applications.

11.3.10 Triazine Treatment

The application of triazine treatment serves as a highly effective approach to strengthen the connection among the fibers and the matrix. in composite materials. This method greatly increases interfacial adhesion by facilitating the formation of covalent connections between the matrix and reinforcements. Triazine molecules possess functional groups capable of reacting within the fibers, one can find hydroxyl groups that are characteristic of cellulose and lignin. The interaction between these functional groups and hydroxyl groups results in the formation of robust covalent bonds, which not only strengthen adhesion properties but also provide moisture resistance to the fibers [44]. Natural fiber composites' performance is heavily reliant on moisture resistance, as they are naturally prone to absorbing water. This absorption may undermine the integrity of the fiber-matrix interaction, resulting in reduced mechanical characteristics for the composite. Triazine treatment addresses this issue by forming covalent bonds that limit the presence of free hydroxyl groups, which are known to attract moisture. Consequently, the fibers exhibit reduced water absorption, thereby enhancing the composite's durability and lifespan in humid conditions. The combination of improved adhesion and moisture resistance positions triazine treatment as a viable technique for developing efficient natural composites with fiber reinforcement.

11.3.11 Fungal Treatment

The process of fungal treatment [45] involves the use of specific enzymes, with white-rot fungus playing a crucial role. This fungus is responsible for generating extracellular oxidase enzymes, which, in conjunction with lignin peroxidase, act as a vital component in eliminating lignin from fibers. The creation of hyphae by the fungus leads to strengthen the relationship between the reinforcements and the matrix, causing tiny pores in the fiber surface [46].

Henceforth, the noteworthy enhancement in the properties alongside adaptability could be achieved by reinforcing natural fiber additives inside HDPE. The incorporation of additives provides imperative results in terms of mechanical properties that are also easy to recycle. Advanced HDPE composites can be treated as future composites with applications in every industry sector.

11.4 Effect of Processing Techniques

The characteristics of natural fibre composites during processing are influenced by several input factors, such as fibre behavior, fibres volume fraction, and the operating temperature. The moistening effect between the fibre and matrix can be reduced before processing for better amalgamation. Other factors affecting composite performance are length, aspect ratio, and chemical proportion. The temperature that must be maintained while processing should also be addressed. Temperatures beyond 200 °C should be avoided if the product is not to disintegrate prematurely. The primarily used methods available in the commercial market are compression, injection, and extrusion moulding [47, 48].

11.4.1 Effect of Compression Molding

Since the 1990s, compression moulding has been the most common method of processing thermoplastic materials. With the passage of time, the appetite for materials that are both lightweight and offer superior performance expanded significantly. In this method, materials are heated to a high temperature before being poured into the mound hole. Afterwards, the moults are squeezed and warped by the mould's middle portion while the cavity is subjected to high pressure [49]. The cavity should be subjected to significant pressure until the mould is released and the composite formed solidifies. For the mechanism to perform well, the heating time, the pressure exerted, and the cooling period must all be regulated. Sheet moulding, a kind of compression moulding, is among the essential processing procedures for producing polymer composites [50]. It is necessary to inject a fluid mass of a specific resin into the plastic film before it can be used to cut the fibres.

11.4.2 Injection Molding

For the large-scale production of polymer composites, injection moulding is considered one of the most efficient manufacturing techniques. The process initiates with feeding the granules of polymers in the hopper. After that, the melting process is

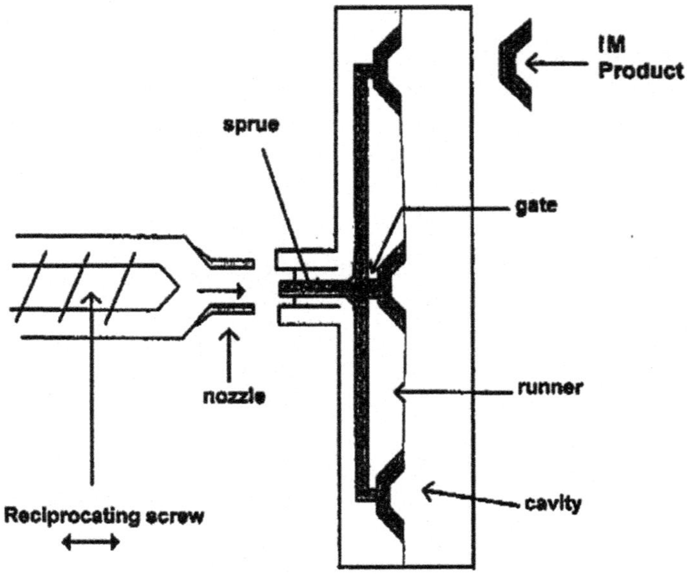

Fig. 11.1 Injection molding process [50]

completed by a heating system inside the hopper, capable of rising at high temperatures. The liquid components are then pumped into an empty cavity. The molten shape solidifies in the mould and is extracted using a split die mould [51–53]. Figure 11.1 elaborates on the detail of the injection moulding press, whose major parts include a sprue, gate, runner, and nozzle. The figure describes the moulded products obtained by injecting plastic materials molten by the heating methods and then letting them cool and get a solidified product.

11.4.3 Extrusion Molding

Extrusion moulding is a well-suited process adopted by the manufacturing industry to produce or develop natural fibre composites. Its suitability is the high stiffness and strength governed by polymer composite. This process starts by keeping the polymer material in the shape of pellets [52], followed by heating to convert it into a molten state. After compaction, the cast components are made to cool down. Table 11.1 depicts a brief literature survey of prior state-of-art work by eminent scholars to exemplify the various processing techniques and its effect on characteristics of finally developed natural fibre composites.

A simple explanation of the life cycle of natural fiber is shown in Fig. 11.2. Initially, the raw fibre gets processed into intermediate products like fibre mats, fleece, etc., and then into the final product. The final product, after usage, when it becomes of

Table 11.1 Natural fiber reinforcement with thermoplastic polymer

Observation	Polymer composite (thermoplastic)	Method	References
Improvement in tensile properties	Coir and oil palm fiber/ polyester composite	Compression molding	[54]
Enhanced UTS	Kenaf fiber/polylactic acid composite	Compression molding	[55]
Improvement in Young's modulus	Jute fiber/polypropylene composite	Compression molding	[57]
Increased tensile strength	Rice hulk/polypropylene composite	Injection molding	[58]

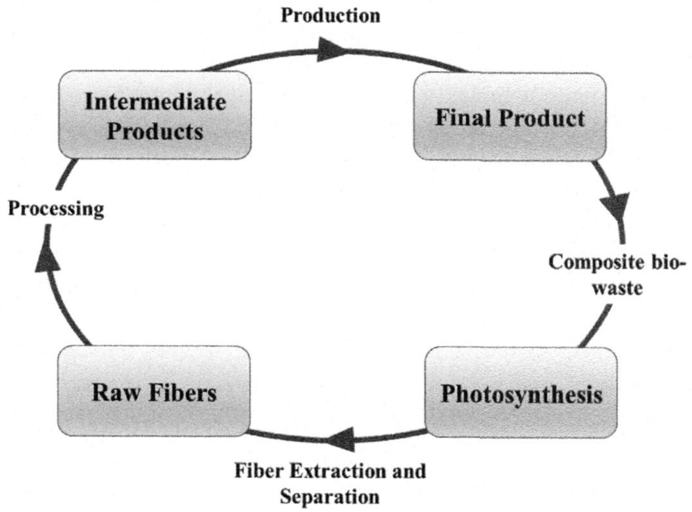

Fig. 11.2 Life cycle of natural fiber

no use and is referred to as waste, gets decomposed into fibre crops after several chemical processes. Hence, fibre extraction is finally done. This process continues cyclically.

11.5 Mechanical Properties

In general, tensile, elongation and hardness of natural fiber composite is less than modified fiber composite. Conversely, their characteristics may be improved through suitable alteration to both composites, as discussed in earlier sections. By implementing targeted treatment processes and refining the composite formulation, one can significantly enhance the strength, durability, and overall performance of natural fiber composites, thereby making them more competitive with synthetic materials.

11.5.1 Tensile Properties

The adhesive connection among fibers and resin has a profound influenced on the UTS and elongation of fiber composites. By implementing physical and chemical modifications to both the fibers and the resin, it is possible to enhance tensile strength. Furthermore, the fiber content within the material is a critical reason in shaping tensile characteristics of these composite. Some studies have shown inconsistent results regarding tensile strength as fiber content changes; however, it is widely observed that decreasing the fiber volume fraction below an optimal level facilitates a better load distribution across the fibers. This enables the matrix to support the load after fiber failure, potentially enhancing tensile strength. In contrast, when the fiber proportion of volume surpasses the ideal limit, brittle fractures in the fibers can occur, leading to the matrix's inability to handle the extra load, this eventually causes the composite to fail. The presence of inconsistent tensile property trends can stem from a range of factors, for example, improper manufacturing methods, fiber deterioration, and inconsistencies among the fiber and matrix. Improving the mechanical characteristics of composites made of natural fibers requires addressing these issues.

A quick summary of the various resins' tensile strengths and elastic moduli is provided in Table 11.2.

11.5.2 Flexural Properties

Flexural rigidity serves as a key mechanical property that reflects a material's resistance to bending. The young modulus and inertia of material are two critical elements

Table 11.2 Mechanical properties (UTS and EM)

Resin	Tensile strength (MPa)	Elastic modulus (GPa)
Low density polyethene	38–76	0.053–0.35
High density polyethene	14.5–38	0.4–1.3
Polystyrene	25–68	4–5
Polyester	41–89	2.07–4.41
Vinylester	69–83	3.1–3.8
Epoxy	55–130	3–6
Polypropylene	26–41.4	0.95–1.77

Table 11.3 Flexural strength and flexural modulus

Resin	Flexural strength (MPa)	Flexural modulus (GPa)
Low density polyethene	8	0.2
High density polyethene	33	1.3
Polypropylene	40	1.4
Polystyrene	70	2.5
Polyester	68–105	2–4
Vinylester	129–139	3
Epoxy	105–145	3–5

that influence flexural properties. With a rise in fiber content, the flexural modulus generally increases, indicating a stronger resistance to bending. However, too much fiber can compromise flexural strength, often resulting from inadequate fiber wetting, which can create stress concentrations within the composite structure. Typically, the flexural modulus of natural fiber composites increases with a higher fiber content, which boosts their capacity to endure bending stresses. Nonetheless, the composite's overall performance is influenced by various additional factors, such as the intrinsic characteristics of the matrix, the fiber volume percentage, and the strength of the interfacial bonds. Together, these factors play a crucial role in defining the composite's fracture toughness, which indicates its ability to resist crack propagation. The flexural performance of natural fiber composites can be improved by effectively balancing fiber content, matrix qualities, and bonding efficiency, thus making these materials advantageous for numerous applications that prioritize both strength and durability.

The Table 11.3 represented below provides the flexural strength as well as flexural modulus of different resin utilized in the composite development.

11.5.3 Compressive Properties

The design of the reinforcing structure and the fiber volume fraction in natural fiber composites significantly affect their compressive performance. Generally, increasing the fiber volume fraction up to an optimal level improves the composite's compressive strength by reducing the number of voids in the material [59–65]. This enhancement is further supported by the uniform distribution and strong bonding among the fibers and the matrix. Nevertheless, when proportion of fiber surpasses a certain threshold—typically about 1.5%—the compressive strength tends to decline. Findings reveal that a 0.5% rise in fiber volume past the optimal point can lead to an 8.5% decrease in compressive strength. While initial increases in fiber volume can improve strength, excessive amounts can adversely affect performance. For example, research by Van et al. [56] highlighted the significance of fiber orientation as a key design factor in hemp fiber/polyester composite tubes created through filament winding. Their results demonstrated that the highest compressive strength and modulus were found at an angle of 10° for the fiber with respect to the loading orientation. The testing process identified multiple fracture modes, as folding buckling, micro-buckling, progressive crushing, and diamond-shaped buckling. These findings highlight the importance of fiber proportion and orientation in determining the compressive characteristics of natural fiber composites. By optimizing these factors, it is possible to improve the performance and broaden the application of these materials in various structural engineering contexts.

11.5.4 Impact Strength

For composites made of natural materials to work as intended, the degree of adherence between the matrix and the fibers is essential, particularly throughout their operational life. Enhancing these bonding properties can lead to improved impact resistance. Several factors, such as bumps, collisions, falling objects, and debris, contribute to the impact loading experienced by these materials [66–72]. Key factors influencing the impact strength of these composites include adhesion levels, efficiently absorbing energy, pulling out fibers, and connecting.

11.6 Applications and Futuristic Aspect of Natural Fiber

Natural fiber composites are said to be mostly used in the automotive, shipbuilding, electrical equipment, aviation, and medical industries because of their excellent mechanical capabilities, lightweight nature, and eco-friendliness, among other advantages. Due to a few drawbacks, such as a lack of endurance in wet environments, piping and boats may eventually be replaced with glass fibre-reinforced composite (GFRC) [73, 74]. In 1941, Henry Ford was the first to use flax and hemp fibres in the automobile industry [75–83]. In construction, natural fibres produce load-carrying elements such as beams, roofs, walls, and bridges. For safety purposes, jute fibre composites are used in indoor parts housing and outdoor housing. Several investigations have been done on using natural fibre in car parts. Interior car components including boot linens, door trim panels, front and rear door linens, and seat backs also use it. Apart from interior parts in the automobile industry, it could also be very effective in using exterior parts.

Some companies have effectively harnessed the cellulose fibre composites in several automobile parts, namely BMW, Audi, Ford, Volkswagen, etc. It was found that Mercedes-Benz used coconut fibres and rubber latex to produce seats in the class model [84–88]. It was also found that model E of Mercedes Benz door panel was made with composite-containing reinforcement, such as a fax sisal fibre mat and a matrix of epoxy. Similarly, a few other companies, such as Ford uses kenaf Fiber, use reinforcement to manufacture door panels in PP composites [85]. Opel Vectras has used a Kenaf and flax mixture to develop trays and door panels. Volkswagen also used cellulose fibre to make Seatback door panels.

There are many other application such as construction, aerospace, ceilings, boats, sports and office product. It is unable to bear heavy load which makes its application limited to non-load bearing components. Few other properties such as lightweight and biodegradability make them better composite for the building industry. Van de Weyenberg et al. [89, 90] studied that sisal fiber reinforced composite having greater strength in tension and compression makes its application in buildings, tanks, roofing elements and pipes. Due to its lightweight it is best suitable for roofing, ceiling and for low -cost house construction.

Because of government rules requiring such goods which do not affect our environment in industry, the market for natural fiber composites was valued at $16.5 billion in 2016 and is projected to reach $50 billion by 2025, according to Marketsand Markets Inc and Grand View Research Inc. Transportation, construction along with consumer goods and other applications each had 35, 9 and 16% of the worldwide natural fibre composite market share in 2016.

11.7 Conclusion

The most striking features of natural fibre are its cost-effectiveness and environment-friendliness, making it a superior option to replace conventional fibres like glass and carbon as an additive inside the polymer matrix. HDPE has proven to be a better matrix than other polymer matrices such as LDPE, polypropylene, and polyester due to its high strength and hardness. The potential applications of HDPE are vast, including pipes, biomedical implantation, packaging, sporting equipment, etc. One of the major concerns encountered in today's world is solid waste management. HDPE-based composite reinforced with natural fibre is most effective due to its eco-friendly and easily decomposable behaviour. Natural fibres such as sisal, wheat, rice, etc., can be added to develop new and advanced composites. Many processing techniques are available for the development of the composite, under which some are discussed as per the requirement. The discussions illustrated that natural fibres reinforced with HDPE have greater mechanical properties of composites with numerous applications. The strength of the HDPE-based composite depends on the fibre loading, which shows improvement up to a certain level when increasing the weight.%.

References

1. Ramesh, M., Palanikumar, K., Reddy, K.H.: Plant fibre based bio-composites: sustainable and renewable green materials. Renew. Sustain. Energy Rev. **79**, 558–584 (2017). https://doi.org/10.1016/j.rser.2017.05.094
2. Elfaleh, I., et al.: A comprehensive review of natural fibers and their composites: an eco-friendly alternative to conventional materials. Results Eng. 101271 (2023). https://doi.org/10.1016/j.rineng.2023.101271
3. Diaz-Silvarrey, L.S., Zhang, K., Phan, A.N.: Monomer recovery through advanced pyrolysis of waste high density polyethylene (HDPE). Green Chem. **20**(8), 1813–1823 (2018). https://doi.org/10.1039/c7gc03662k
4. Sathishkumar, G.K., et al.: Synthesis and mechanical properties of natural fiber reinforced epoxy/polyester/polypropylene composites: a review. J. Nat. Fibers 1–24 (2020). https://doi.org/10.1080/15440478.2020.1848723
5. Yashas Gowda, T.G., et al.: Polymer matrix-natural fiber composites: an overview. Cogent Eng. **5**(1), 1446667 (2018). https://doi.org/10.1080/23311916.2018.1446667
6. Makhlouf, A., et al.: Elaboration and characterization of flax fiber reinforced high density polyethylene biocomposite: effect of the heating rate on thermo-mechanical properties. J. Nat. Fibers 1–14 (2020). https://doi.org/10.1080/15440478.2020.1848737
7. Wan Mohamed, W.Z., et al.: Effects of fiber size and fiber content on mechanical and physical properties of Mengkuang reinforced thermoplastic natural rubber composites. BioResources **13**(2) (2018). https://doi.org/10.15376/biores.13.2.2945-2959
8. Omrani, E., Menezes, P.L., Rohatgi, P.K.: State of the art on tribological behavior of polymer matrix composites reinforced with natural fibers in the green materials world. Eng. Sci. Technol. Int. J. **19**(2), 717–736 (2016). https://doi.org/10.1016/j.jestch.2015.10.007
9. Balaji, R., Sasikumar, M., Jeyanthi, S.: Characterisation of hollow glass fibre reinforced vinyl-ester composites. Indian J. Sci. Technol. **9**(47) (2016). https://doi.org/10.17485/ijst/2016/v9i47/107921
10. Li, Y.-Y., et al.: Review of recent development on preparation, properties, and applications of cellulose-based functional materials. Int. J. Polym. Sci. **2018**, 1–18 (2018). https://doi.org/10.1155/2018/8973643
11. Gautam, S.K., et al.: Wear and mechanical properties of in situ A356/5%TiB2 composite synthesis by cooling slope technique. Int. J. Met. (2022). https://doi.org/10.1007/s40962-022-00931-5
12. Li, Y., Hu, C., Yu, Y.: Interfacial studies of sisal fiber reinforced high density polyethylene (HDPE) composites. Composites A **39**(4), 570–578 (2008). https://doi.org/10.1016/j.compositesa.2007.07.005
13. Yao, F., et al.: Rice straw fiber-reinforced high-density polyethylene composite: effect of fiber type and loading. Ind. Crop. Prod. **28**(1), 63–72 (2008). https://doi.org/10.1016/j.indcrop.2008.01.007
14. Gautam, S.K., et al.: Effect of processing routes on structure-property correlationship of ADC 12 Al alloy. Mater. Res. Express **6**(2), 026518 (2018). https://doi.org/10.1088/2053-1591/aaec2c
15. Panthapulakkal, S., Sain, M.: Agro-residue reinforced high-density polyethylene composites: fiber characterization and analysis of composite properties. Composites A **38**(6), 1445–1454 (2007). https://doi.org/10.1016/j.compositesa.2007.01.015
16. Rajeshkumar, G., et al.: A comprehensive review on natural fiber/nano-clay reinforced hybrid polymeric composites: materials and technologies. Polym. Compos. **42**(8), 3687–3701 (2021). https://doi.org/10.1002/pc.26110
17. Mazur, K., et al.: Green high density polyethylene (HDPE) reinforced with basalt fiber and agricultural fillers for technical applications. Composites B **202**, 108399 (2020). https://doi.org/10.1016/j.compositesb.2020.108399

18. Savini, G., Oréfice, R.L.: Comparative study of HDPE composites reinforced with microtalc and nanotalcs: high performance filler for improving ductility at low concentration levels. J. Mater. Res. Technol. **9**, 16387–16398 (2020). https://doi.org/10.1016/j.jmrt.2020.11.090
19. Aliotta, L., Lazzeri, A.: A proposal to modify the Kelly-Tyson equation to calculate the interfacial shear strength (IFSS) of composites with low aspect ratio fibers. Compos. Sci. Technol. **186**, 107920 (2020). https://doi.org/10.1016/j.compscitech.2019.107920
20. Essabir, H., et al.: Mechanical and thermal properties of hybrid composites: oil-palm fiber/clay reinforced high density polyethylene. Mech. Mater. **98**, 36–43 (2016). https://doi.org/10.1016/j.mechmat.2016.04.008
21. Hussain, S.A., Pandurangadu, V., Kumar, K.P.: Optimization of mechanical properties of green coconut fiber/HDPE composites. Int. J. Adv. Sci. Technol. **92**, 1–8 (2016). https://doi.org/10.14257/ijast.2016.92.01
22. Souza, P.S., et al.: Mechanical properties of HDPE/textile fibers composites. Procedia Eng. **10**, 2040–2045 (2011). https://doi.org/10.1016/j.proeng.2011.04.338
23. Babu, S.A., Narayanankutty, S.K.: Preparation of chemically functionalized and self compatibilized short coir fiber—High density polyethylene composites. J. Appl. Polym. Sci. **138**(42), 51254 (2021). https://doi.org/10.1002/app.51254
24. Khandelwal, H., et al.: Surface modified reinforcements on the structure properties of A356/SiC Stir cast composite. Silicon 1–8 (2024). https://doi.org/10.1007/s12633-024-03160-z
25. Orue, A., et al.: The effect of surface modifications on sisal fiber properties and sisal/poly (lactic acid) interface adhesion. Composites B **73**, 132–138 (2015). https://doi.org/10.1016/j.compositesb.2014.12.022
26. Pantano, A., et al.: Analysis of the parameters affecting the stiffness of short sisal fiber biocomposites manufactured by compression-molding. Polymers. **14**(1), 154 (2021). https://doi.org/10.3390/polym14010154
27. Lucia, A., et al.: A direct silanization protocol for dialdehyde cellulose. Molecules **25**(10), 2458 (2020). https://doi.org/10.3390/molecules25102458
28. Hou, Z., et al.: Effect of different silane coupling agents in-situ modified sepiolite on the structure and properties of natural rubber composites prepared by latex compounding method. Polymers **15**(7), 1620 (2023). https://doi.org/10.3390/polym15071620
29. Vijayasekaran, G., et al.: Mechanical properties evaluation of carbon nanotube/sisal fiber/marble dust reinforced polymer based composites. Mater. Today (2023). https://doi.org/10.1016/j.matpr.2023.06.311
30. Wang, B., et al.: Pre-treatment of flax fibers for use in rotationally molded biocomposites. J. Reinf. Plast. Compos. **26**(5), 447–463 (2007). https://doi.org/10.1177/0731684406072526
31. Kitano, T.: Fiber reinforced thermoplastic composites. Zairyo-to-Kankyo **43**(10), 588–596 (1994). https://doi.org/10.3323/jcorr1991.43.588
32. Sergi, C., et al.: Influence of reprocessing cycles on the morphological, thermal and mechanical properties of flax/basalt hybrid polypropylene composites. Sustain. Mater. Technol. e00648 (2023). https://doi.org/10.1016/j.susmat.2023.e00648
33. Ishak, A., et al.: Silazanes, a novel flax fibers functionalization: effect on silicone-based composites. Composites A **166**, 107382 (2023). https://doi.org/10.1016/j.compositesa.2022.107382
34. Techawinyutham, L., Frick, A., Siengchin, S.: Polypropylene/maleic anhydride grafted polypropylene (MAgPP)/coconut fiber composites. Adv. Mech. Eng. **8**(5), 168781401664544 (2016). https://doi.org/10.1177/1687814016645446
35. Ramamoorthy, S.K., Skrifvars, M., Persson, A.: A review of natural fibers used in biocomposites: plant, animal and regenerated cellulose fibers. Polym. Rev. **55**(1), 107–162 (2015). https://doi.org/10.1080/15583724.2014.971124
36. Wach, R.A., et al.: Dual stimuli-responsive polysaccharide hydrogels manufactured by radiation technique. Appl. Sci. **12**(22), 11764 (2022). https://doi.org/10.3390/app122211764
37. Asim, M., et al.: Effect of surface modified date palm fibre loading on mechanical, thermal properties of date palm reinforced phenolic composites. Compos. Struct. **267**, 113913 (2021). https://doi.org/10.1016/j.compstruct.2021.113913

38. Lila, M.K., Komal, U.K., Singh, I.: Thermal post-processing of bagasse fiber reinforced polypropylene composites. Compos. Commun. 100546 (2020). https://doi.org/10.1016/j.coco.2020.100546
39. Egala, R., Gangi Setti, S.: Influence of surface modifications of Castor oil fibre on mechanical properties of fibre reinforced Epoxy composites. Adv. Mater. Process. Technol. 1–39 (2021). https://doi.org/10.1080/2374068x.2021.1959107
40. Sreekala, M.S., et al.: Environmental effects in oil palm fiber reinforced phenol formaldehyde composites: studies on thermal, biological, moisture and high energy radiation effects. Adv. Compos. Mater. **13**(3–4), 171–197 (2004). https://doi.org/10.1163/1568551042580154
41. Singh, B.K., et al.: Parametric study to investigate mechanical properties of welded dissimilar Al6063 and Al 7073 alloys through FSW process. Eng. Res. Express. (2024). https://doi.org/10.1088/2631-8695/ad5f79
42. Singh, B., Upadhyaya, R.: Fabrication and mechanical testing of hybrid fiber polymer composite materials. Int. Res. J. Eng. Technol. (IRJET) 1290–1296 (2021). https://www.irjet.net/archives/V8/i4/IRJET-V8I4245.pdf
43. Gupta, T.K., Upadhyaya, R.: Fabrication and characterization of silicon carbide reinforced aluminium matrix composites. Int. J. Res. (IJSR) **8**(5), 1324–1325 (2019). https://www.ijsr.net/getabstract.php?paperid=ART20198062
44. Salem, T.F., et al.: Enhancement of mechanical, thermal and water uptake performance of TPU/jute fiber green composites via chemical treatments on fiber surface. e-Polymers **20**(1), 133–143 (2020). https://doi.org/10.1515/epoly-2020-0015
45. Mohit, H., Arul Mozhi Selvan, V.: A comprehensive review on surface modification, structure interface and bonding mechanism of plant cellulose fiber reinforced polymer based composites. Compos. Interfaces. **25**(5–7), 629–667 (2018). https://doi.org/10.1080/09276440.2018.1444832
46. Tanasă, F., et al.: Modified hemp fibers intended for fiber-reinforced polymer composites used in structural applications—A review I. Methods of modification. Polym. Compos. **41**(1), 5–31 (2019). https://doi.org/10.1002/pc.25354
47. Upadhyaya, R., et al.: Investigation of the quality of flywheel sg iron sand casting using the optimized riser dimensions: numerical simulation and experimental validation. Int. J. Met. (2024). https://doi.org/10.1007/s40962-024-01407-4
48. Choubey, O.N., et al.: Cryogenic treatment on M2 HSS tool and its performance. Int. J. Mater. Eng. Innov. **14**(4), 365–379 (2023). https://doi.org/10.1504/ijmatei.2023.136118
49. Ang, H.Y., et al.: Bioresorbable stents: current and upcoming bioresorbable technologies. Int. J. Cardiol. **228**, 931–939 (2017). https://doi.org/10.1016/j.ijcard.2016.11.258
50. Basak, R., Choudhury, P.L., Pandey, K.M.: Effect of temperature variation on surface treatment of short jute fiber-reinforced epoxy composites. Mater. Today **5**(1), 1271–1277 (2018). https://doi.org/10.1016/j.matpr.2017.11.211
51. Gautam, S.K., Singh, B.K.: Investigation on the effects of isothermal holding temperature and time on the coarsening mechanism and rheological properties of ADC12 Al semi-solid slurry. Mater. Chem. Phys. 128813 (2023). https://doi.org/10.1016/j.matchemphys.2023.128813
52. Gautam, S.K., et al.: Rheological properties of ADC12 Al alloy in the semi-solid state. Int. J. Cast Met. Res. 1–9 (2023). https://doi.org/10.1080/13640461.2023.2197808
53. Gautam, S.K., et al.: Determine the rheological parameters for characterization of semi solid slurry of ADC12 Al alloy. J. Inst. Eng. (India). (2023). https://doi.org/10.1007/s40032-023-00961-w
54. Nurazzi, N.M., et al.: Mechanical performance evaluation of bamboo fibre reinforced polymer composites and its applications: a review. Funct. Compos. Struct. **4**(1), 015009 (2022). https://doi.org/10.1088/2631-6331/ac5b1a
55. Anand, A., et al.: Optimization of process variables and characterization of mechanically agitated ceramic-microsphere reinforced aluminium composite. Mater. Und Werkst. **55**(2), 125–140 (2024). https://doi.org/10.1002/mawe.202300179
56. Van Vuure, A.W., et al.: Compressive properties of natural fibre composites. Mater. Lett. **149**, 138–140 (2015). https://doi.org/10.1016/j.matlet.2015.01.158

57. Afroughsabet, V., Ozbakkaloglu, T.: Mechanical and durability properties of high-strength concrete containing steel and polypropylene fibers. Constr. Build. Mater. **94**, 73–82 (2015). https://doi.org/10.1016/j.conbuildmat.2015.06.051
58. Feldmann, M., Heim, H.P., Zarges, J.C.: Influence of the process parameters on the mechanical properties of engineering biocomposites using a twin-screw extruder. Composites A **83**, 113–119 (2016). https://doi.org/10.1016/j.compositesa.2015.03.028
59. Khandelwal, H., et al.: To investigate the effect of varying weight% Cu on the microstructure and mechanical properties of gray cast iron. AIP Conf. Proc. **322**, 1 (2024). https://doi.org/10.1063/5.0235860
60. Gautam, S.K., et al.: Microstructure characterization and mechanical properties of semi solid ADC 12 Al alloy. Int. J. Modern Manuf. Technol. **11**, 36–42 (2019)
61. Gautam, S.K., et al.: Optimization of processing parameters of cooling slope process for semi-solid casting of ADC 12 Al alloy. J. Brazilian Soc. Mech. Sci. Eng. **40**, 1–15 (2018). https://doi.org/10.1007/s40430-018-1213-6
62. Ramesh, S., et al.: Structure property correlation of gravity die-cast and rheocast Al–Mg–Sc–Zr in situ nano-TiB2 composite. Int. J. Met. **18**, 3095–3103 (2024). https://doi.org/10.1007/s40962-023-01223-2
63. Yadav, S., et al.: Die design optimization for connecting rod using simulation modeling and taguchi methodology. Eng. Res. Express **6**, 035424 (2024). https://doi.org/10.1088/2631-8695/ad6d2b
64. Gautam, S.K., et al.: Effect of processing parameters of rheocast on mechanical properties of Al–10.5 Si–1.7 Cu alloy. Int. J. Met. **18**(1), 390–402 (2024)
65. Khandelwal et al.: Numerical simulation and experimental validation of fluidity of AlSi12CuNiMg alloy using multi spiral channel with varying thickness. Int. J. Met. 1–10 (2024). https://doi.org/10.1007/s40962-024-01383-9
66. Mohanty, S., et al.: Taguchi grey relational analysis for optimal process parameter of low carbon steel welded by GTAW by using ER309L filler material. Mater. Today: Proc. **44**, 2556–2561 (2021). https://doi.org/10.1016/j.matpr.2020.12.631
67. Patro, S., et al.: Herpes zoster infection involving the maxillary branch of the right trigeminal nerve-a rare case report. Antiseptic **110**(1), 36–38 (2013). https://www.researchgate.net/profile/Shubhransu-Patro/publication/258343847_
68. Panda, S.K., et al.: Determining the flow correlation for an orifice with a non-dimensional number. Flow Meas. Instrum. **90**, 102338 (2023). https://doi.org/10.1016/j.flowmeasinst.2023.102338
69. Jaiswal, S.K., et al.: E-waste control and its recycling to build sustainable society in the global context. In: Handbook of Research on Applications of AI, Digital Twin, and Internet of Things for Sustainable Development, pp. 200–222. IGI Global (2023). https://doi.org/10.4018/978-1-6684-6821-0.ch012
70. Dwivedi, K., et al.: Free flexural vibration of cracked composite laminated plate using higher-order XFEM. Eng. Fracture Mech. **289**, 109420 (2023). https://doi.org/10.1016/j.engfracmech.2023.109420
71. Das, S.R., et al.: Estimating the effect of machining parameters on surface roughness during machining of hardened EN24 steel using coated carbide inserts. Ind. Eng. **4**(2), 10 (2014). https://d1wqtxts1xzle7.cloudfront.net/77219781/viewcontent-libre.pdf?1640313647
72. Panda, S.K., et al.: The new age of manufacturing from concept to creation: innovative world of 3D printing. In: Evolutionary Manufacturing, Design and Operational Practices for Resource and Environmental Sustainability, pp. 223–244 (2024). https://doi.org/10.1002/9781394198221.ch20
73. Sridevi, G., et al.: Optimization of EDM process parameters with fuzzy logic technique for SS-316 Steel and investigation of microstructural characteristics of EDM machining surface. In: Evolutionary Manufacturing, Design and Operational Practices for Resource and Environmental Sustainability, pp. 245–256 (2024). https://doi.org/10.1002/9781394198221.ch21

74. Mishra, S., et al.: Collaboration of design and manufacturing web service. Int. J. Sci. Res. **8**, 708–712 (2019). https://www.ijsr.net/articlerating.php?paperid=ART20196723
75. Nurul Fazita, M.R., et al.: Green composites made of bamboo fabric and poly (Lactic) acid for packaging applications—A review. Materials **9**(6), 435 (2016). https://doi.org/10.3390/ma9060435
76. Ranakoti, L., et al.: Critical review on polylactic acid: properties, structure, processing, biocomposites, and nanocomposites. Materials **15**(12), 4312 (2022). https://doi.org/10.3390/ma15124312
77. Khang, A., et al.: Machine Vision and Industrial Robotics in Manufacturing: approaches, Technologies, and Applications, p. 252, (2024). https://www.taylorfrancis.com/chapters/edit/10.1201/9781003438137-5
78. Rath, K.C., et al.: Machine Vision and Industrial Robotics in Manufacturing: Approaches, Technologies, and Applications, p. 101 (2024)
79. Harshavardhan, T., et al.: Optimizing smart manufacturing system through advanced robotic kinematics. In: International Conference on Emerging Trends in Mathematical Sciences & Computing, pp. 271–286. Springer Nature Switzerland, Cham (2024). https://doi.org/10.1007/978-3-031-71125-1_23
80. Upadhyaya, R., et al.: Effect of one step in-mould inoculation method on the characterization of thin wall ductile iron. Int. J. Metall. Met. Phys. **3**, 2–8 (2018). https://www.academia.edu/download/107762074/ijmmp-3-024.pdf
81. Muduli, K., et al.: An original assessment of the influence of soft dimensions on implementation of sustainability practices: implications for the thermal energy sector in fast growing economies. Oper. Manag. Res. **14**, 337–358 (2021). https://doi.org/10.1007/s12063-021-00215-x
82. Swain, S., et al.: Blockchain technology for limiting the impact of pandemic: challenges and prospects. In: Computational Modeling and Data Analysis in COVID-19 Research, pp. 165–186 (2021). https://www.taylorfrancis.com/chapters/edit/10.1201/9781003137481-9
83. Muduli, K., et al.: Blockchain technology for enhancing supply chain performance and reducing the threats arising from the COVID-19 pandemic. Sustainability **14**(6), 3290 (2022). https://doi.org/10.3390/su14063290
84. Peter, O., et al.: IoT in combating COVID-19 pandemics: lessons for developing countries. In: Assessing COVID-19 and Other Pandemics and Epidemics Using Computational Modelling and Data Analysis, pp. 113–131 (2022). https://doi.org/10.1007/978-3-030-79753-9_7
85. Swain, S., et al.: Innovations in internet of medical things, artificial intelligence, and readiness of the healthcare sector towards health 4.0 adoption. Int. J. Soc. Ecol. Sustain. Dev. (IJSESD) **13**(1), 1–14 (2022). https://doi.org/10.4018/IJSESD.292078
86. Swain, S., et al.: Intelligent technologies for excellency in sustainable operational performance in the healthcare sector. Int. J. Soc. Ecol. Sustain. Dev. (IJSESD) **13**(5), 1–16 (2022). https://doi.org/10.4018/IJSESD.313963
87. Gautam, S.K., et al.: Studies on mold filling behavior of Al–10.5Si–1.7 Cu Al alloy during rheo pressure die casting system. Int. J. Met. (2023). https://doi.org/10.1007/s40962-023-00958-2
88. Gholampour, A., Ozbakkaloglu, T.: A review of natural fiber composites: properties, modification and processing techniques, characterization, applications. J. Mater. Sci. **55**(3), 829–892 (2019). https://doi.org/10.1007/s10853-019-03990-y
89. Murugu Nachippan, N., et al.: Numerical analysis of natural fiber reinforced composite bumper. Mater. Today **46**, 3817–3823 (2021). https://doi.org/10.1016/j.matpr.2021.02.045
90. Perremans, D., et al.: Investigation of the tensile behavior of treated flax fibre bio-composites at ambient humidity. Compos. Sci. Technol. **159**, 119–126 (2018). https://doi.org/10.1016/j.compscitech.2018.02.038

Rakesh Upadhyay Mr. Rakesh Upadhyay is currently serving as a research scholar at the CSIR-IIP, Dehradun India. Upadhyay has contributed more than 6 research papers to various national and international journals. His areas of expertise include the Metal Matrix Composite and Energy Conservation.

Sujeet Kumar Gautam Dr. Gautam is currently serving as an Research associate at the National Institute of Advanced Manufacturing Technology (NIAMT), formerly known as NIFFT, in Hatia, Ranchi-834003, India. His areas of expertise include the Semi Solid Metal Processing, Numerical Simulation, Composite Fabrication and Characterization.

Rajat Upadhyaya is currently serving as an Assistant Professor at the National Institute of Advanced Manufacturing Technology (NIAMT), formerly known as NIFFT, in Hatia, Ranchi-834003, India. His areas of expertise include the Structure-Property Correlation of Metals and Alloys, as well as Casting Design and Simulation, Composite.

Bipin Kumar Singh is currently serving as an Assistant Professor at the Dept of Mechanical Engineering, SECE, Coimbatore, Tamil Nadu, India. His areas of expertise include the Nano Ceramics, Cutting Inserts, and Evaluation of Mechanical Properties.

Siddharth Yadav is currently serving as an Assistant Professor at the CAS, AKTU, Lucknow-226031, India. His areas of expertise include the Al-Si Alloy Casting, Mechanical Vibration and Grain Refinement.

Kamlesh Kumar Singh is currently serving as a Professor and HOD at the National Institute of Advanced Manufacturing Technology (NIAMT), formerly known as NIFFT, in Hatia, Ranchi-834003, India. His areas of expertise include: Main research interest includes Solidification Behaviour of Metals, Production of S. G. Iron and Austempered Ductile Iron (ADI), Structure-Properties Correlation for Various Metals, Casting Design and Simulation.

Sunil Kumar Sharma is currently serving as a Chair of Rail Vehicle Technology Technische Universitat Dresden, 01069, Germany. His areas of expertise include: Main research interest Metal Casting and Design.

Chapter 12
The Role of Additive Manufacturing in Thermal Applications

Shashikant Das, Benarji Koyilada, Santosh Patro, G. Sridevi, Sujit Mishra, Archana Senapati, and Ankur Srivastava

Abstract This chapter explores the transformative potential of Additive Manufacturing (AM) in thermal management across various industries, including aerospace, automobiles, and electronics. By delving into the fundamentals of heat transfer, material selection, and design strategies, this comprehensive review evaluates the benefits and challenges of various AM techniques, such as Selective Laser Melting and Electron Beam Melting. The applications of advanced materials, including metals, ceramics, polymers, and composites, are investigated, along with their high-temperature properties. Real-world case studies demonstrate the advantages of AM in achieving low-cost, high-performance thermal solutions, while also discussing economic, environmental, and regulatory considerations. Finally, the chapter identifies future research directions, including multi-material printing and functionally

S. Das (✉) · S. Patro · G. Sridevi · S. Mishra
Department of Mechanical Engineering, Centurion University of Technology and Management, Paralakhemundi 761211, Odisha, India
e-mail: shashi350me@gmail.com

S. Patro
e-mail: santosh.patro@cutm.ac.in

G. Sridevi
e-mail: sridevi@cutm.ac.in

S. Mishra
e-mail: sujit.mishra@cutm.ac.in

B. Koyilada
Shri Vishnu Engineering College for Women, Kovvada, India
e-mail: drkbenarjime@svecw.edu.in

A. Senapati
Department of Mathematics, School of Applied Sciences, Kalinga Institute of Industrial Technology (KIIT) Deemed to be University, Bhubaneswar 751024, Odisha, India
e-mail: liti.archu@gmail.com

A. Srivastava
Center of Industrial Electronics, University of Southern Denmark, Alsion 2, 6400 Sonderborg, Denmark
e-mail: srivastava@sdu.dk

© The Author(s), under exclusive license to Springer Nature Switzerland AG 2025

K. C. Rath et al. (eds.), *Engineering, Technology and Management*, Information Systems Engineering and Management 54, https://doi.org/10.1007/978-3-031-93654-8_12

graded structures, highlighting AM's potential to revolutionize thermal management and solve future problems in thermal engineering.

Keywords Additive manufacturing · Thermal applications · Selective laser melting · Electron beam melting · Fused deposition modeling · CFD · Binder jetting process

Preface
The fast evolution of producing technology has paved the manner for brand spanking new processes in addressing complicated engineering challenges. Among these, Additive Manufacturing (AM) stands proud as a groundbreaking generation this is reshaping the panorama of thermal applications. This book chapter is devoted to exploring the multifaceted function of AM in thermal engineering, where green warmth control and progressive thermal answers are essential throughout industries consisting of aerospace, automotive, electronics, and energy.

The chapter delves into the foundational standards of AM and its technological evolution, putting the level for expertise its particular contributions to thermal control. It highlights how AM's capability for layout complexity, customization, and cloth optimization permits the improvement of high-overall performance thermal additives that conventional production strategies battle to achieve. With insights into appropriate materials, AM techniques, and layout considerations, the chapter equips readers with a complete expertise of the way to harness AM for thermal applications.

Case research and real-international examples illustrate the transformative effect of AM, at the same time as discussions on economic, environmental, and regulatory factors offer a holistic angle on its adoption and destiny trajectory. This chapter goals to encourage engineers, researchers, and enterprise specialists to discover the capacity of AM as a effective device for innovation, efficiency, and sustainability in thermal control.

12.1 Introduction

For numerous thermal applications, the additive manufacturing (AM) can also benefit to produce complex structures and better heat transfer properties. Previous work has been developed building heat sinks with high thermal conductivity of AM process such as Vat Photopolymerization (VPP) have also shown that the dissipation rate can be increased by adding graphene based fillers [1]. Moreover, AM makes it possible to produce cold plates with heat pipe integrated for passive thermal management, improving thermal conductivity and lowering weight in electronic cooling systems [2]. To further ensure compatibility with lens materials and avoid misalignment problems, AM techniques can be used to customize the coefficient of thermal expansion in optical components [3]. In summary, AM provides creative ways to thermal

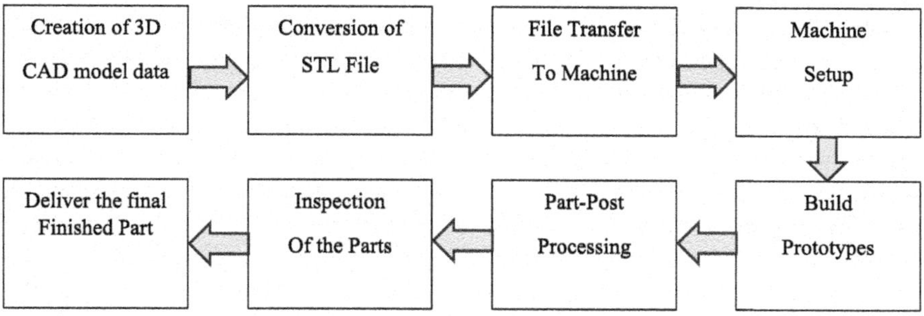

Fig. 12.1 General procedures for 3D printing [9]

management systems, offering effective and tailored methods for various thermal engineering applications.

Broughton et al. [4] have conducted research on additive manufactured foam structures for electric vehicle power electronics, highlighting the efficacy of metal foams with high thermal conductivity for heat transfer. Yagci et al. [5] have also investigated additive manufacturing foam structures for electric vehicle power electronics, demonstrating improved heat transfer capabilities through metal foams with high thermal conductivity. Lin et al. [6] have showcased how additive manufacturing enhances heat dissipation in intricately shaped components, particularly noting the enhanced thermal conductivity and heat dissipation of AlN components. Favero et al. [7] have discovered the strategy flexibility offered using additive manufacturing, underscoring the impact of surface roughness on performance and the necessity of calibrating numerical models to enhance their reliability as design tools in additive manufacturing. Xu et al. [8] have illustrated how additive manufacturing reduces assembly time and streamlines complex fabrication processes, enabling optimized designs that improve thermal performance.

Additive Manufacturing (AM), is usually referred 3D printing, has arisen as a revolutionary technology reshaping various industries, includes thermal applications. In contrast to traditional subtractive manufacturing methods that involve cutting or shaping materials from a solid block, AM constructs objects layer by layer based on digital designs. This approach offers unprecedented design freedom, allowing for the creation of intricate geometries and customized components with unparalleled accuracy. The steps involved in 3D printing are depicted in Fig. 12.1.

12.1.1 Overview of Additive Manufacturing Technologies

AM includes a wide range of technologies, each with special benefits and functionalities appropriate for various applications. Among the most widely utilized AM technologies in thermal applications are the following.

Selective Laser Melting (SLM): SLM creates fully dense metal objects with superior mechanical qualities by layer-by-layer melting and fusing metallic powders with a high-power laser. The production of heat exchangers, turbine parts, and other thermal management systems for the energy, automotive, and aerospace industries is a particularly good fit for it. Electron Beam Melting (EBM): EBM melts metal powders using an electron beam rather than a laser, working on a similar principle as SLM. This process is well-known for its quick build times and is frequently used to produce large, intricate parts for thermal applications in the power and healthcare sectors. Fused Deposition Modeling (FDM) is an additive manufacturing (AM) technique that uses thermoplastic extrusion to create objects layer by layer. The material filament is heated and forced through a nozzle. FDM is mainly used for tooling and prototyping, but it can also be utilized in thermal applications to make molds and plastic parts that can withstand heat. Binder jetting: In order to make 3D objects, binders are selectively deposited layer by layer onto a powder bed.

12.1.2 Importance of AM in Modern Manufacturing Processes

Due to its unrivalled flexibility, efficiency, and customizability, additive manufacturing (AM) has absolutely modified how matters are made today. Using additive manufacturing (AM), thermal packages can produce light-weight substances with complex inner channels, complicated shapes, and thermal overall performance tailor-made structures. Because of the elevated freedom to innovate and decorate thermal control systems, engineers are capable of acquire huge upgrades in product overall performance, sustainability, and strength efficiency. Additionally, AM makes it less complicated to fabricate additives on demand, slicing down on lead instances and facilitating short iterations and prototypes for heat-touchy additives. This flexibility is specially beneficial in sectors like electronics, automotive, and aerospace with short improvement cycles, in which competitiveness relies upon on well timed innovation.

12.1.3 Objective

The goal of the item is to offer a complete knowledge of the function of Additive Manufacturing (AM) in improving thermal packages throughout diverse industries. It targets to discover the essential principles, substances, and AM strategies vital for optimizing thermal overall performance. By imparting exact case research and real-global examples, the item demonstrates the sensible blessings and improvements enabled with the aid of using AM in thermal control. It additionally addresses monetary and environmental implications, emphasizing sustainability and cost-efficiency.

Additionally, the item discusses regulatory compliance and destiny trends, highlighting possibilities for persisted studies and improvement in AM-enabled thermal solutions.

12.1.4 Scope

The scope of this text consists of a complete evaluation of Additive Manufacturing (AM) and its effect on thermal packages throughout diverse industries. It covers essential principles, cloth selection, and layout strategies, whilst additionally addressing the benefits and boundaries of various AM technologies. The article examines monetary, environmental, and regulatory aspects, supported with the aid of using real-global case research.

12.1.5 Novelty

The originality of this paper is located in its all-encompassing and multidisciplinary technique of inspecting Additive Manufacturing (AM) in thermal contexts. It merges technical, financial, environmental, and legislative standpoints, providing an intensive evaluation. Elaborate case research and realistic illustrations make contributions to applicability, even as the emphasis on destiny traits and studies opportunities highlights the forward-questioning issue of the studies. Through tackling present-day subjects like sturdiness and digital amalgamation, the paper promises exclusive views and promotes ingenuity in AM-facilitated thermal resolutions.

12.2 Fundamentals of Thermal Applications

Additive manufacturing enhances thermal applications by enabling the design of complex geometries for improved heat transfer and thermal management. It supports lightweight, efficient, and customized components for industries like aerospace, automotive, and energy systems. Following sections describe the thermal applications.

12.2.1 Basic Principles of Heat Transfer

Heat transfer occurs through conduction (molecular collisions in materials with high thermal conductivity like metals), convection (fluid motion, either natural or forced, as in heat exchangers), and radiation (thermal energy transfer via electromagnetic waves, crucial in high-temperature applications like electronics cooling and solar energy systems) [10].

12.2.2 Importance of Thermal Management in Various Industries

For numerous industries to assure the reliable operation, effectiveness, and sturdiness of structures and equipment, thermal control is crucial. The following are a few vital sectors wherein warmth manage is crucial.

Effective thermal control is vital in electronics, vehicles, aerospace, and the energy industry to prevent overheating, enhance performance, and ensure safety. In electronics, solutions like heat sinks and fans manage heat from components like CPUs and GPUs. In vehicles, thermal systems control engine temperature and optimize HVAC efficiency. Aerospace applications use insulation and cooling to withstand extreme temperature changes. In the energy sector, thermal control is essential for efficient power generation, storage, and distribution.

12.2.3 Challenges in Traditional Manufacturing Methods for Thermal Applications

Traditional manufacturing methods like machining, casting, and forging face challenges in producing complex geometries and optimized internal systems for heat transfer. These methods generate significant material waste, increasing costs and environmental impact. Additionally, they result in longer lead times, limited design flexibility, and high tooling costs, making them less ideal for low-volume production or prototyping of advanced thermal control solutions.

12.2.4 Computational Fluid Dynamics (CFD) Analysis

A necessary technique for resolving conduction and convection troubles is computational fluid dynamics (CFD) analysis, that is specially beneficial for comprehending and forecasting warmth switch and fluid float conduct in engineering applications. CFD affords in-intensity insights into difficult thermal and fluid dynamics that might in any other case be hard to assess experimentally through the use of numerical strategies and algorithms.

Key Concepts

As a end result of a temperature gradient and Fourier's Law, conduction is the time period used to explain the switch of warmth via a stable or desk bound fluid medium. On the opposite hand, convection—that is ruled through Newton's Law of Cooling—entails the motion of warmth both in the fluid or among a flowing fluid and a stable surface.

Governing Equations

These are the primary equations that CFD solves.

Fluid float Navier-Stokes equations:

- Equation of continuity: ensures mass conservation.
- Momentum equation: Controls the fluid's forces, which include viscous, external, and strain forces.

The temperature distribution withinside the fluid and stable domain names is defined through the strength equation.

CFD Analysis Steps

- Problem Definition: Clearly country the geometry, starting conditions, boundary conditions, and cloth attributes of the bodily problem. To placed up an correct model, this step is essential.
- Mesh technology and geometry: Build a geometrical illustration of the area as a way to be examined. After that, the area is split into discrete manage volumes or factors the use of a mesh, which, relying at the geometry's complexity, can be established or unstructured.
- Physics Setup: Enter the cloth's bodily characteristics (viscosity, thermal conductivity, etc.). Assign constant temperatures, warmth flows, and convective warmth switch coefficients as preliminary and boundary conditions. Specify the sort of analysis (temporary or steady-country).
- Solver Settings: Choose the proper solvers for the thermal fields and fluid float (strain, pace). Establish convergence necessities to make certain the answer settles right into a specific depiction of the bodily phenomenon.
- Simulation: Execute the simulation whilst verifying numerical convergence and stability. The governing equations are solved iteratively on this system till the residuals drop underneath predetermined levels.
- Post-processing: Examine temperature distributions, pace fields, strain fields, and warmth float through studying information with graphical tools. Design selections and optimizations may be stimulated through those findings.

Example of Use: Computational Fluid Dynamics (CFD) is a valuable tool for simulating heat transfer processes like convection, as seen in a heated plate cooled by air. By modeling complex interactions between heat transfer and fluid flow, CFD helps identify hot spots and inefficiencies in heat dissipation, providing detailed temperature distribution. Engineers use CFD to optimize thermal systems in applications such as HVAC and electronic cooling, following a structured approach involving

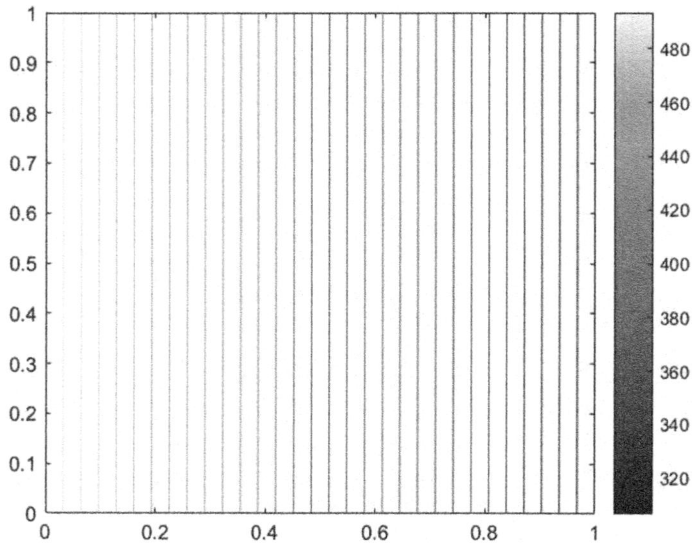

Fig. 12.2 Conduction heat transfer with adiabatic condition at top and bottom surface

problem formulation, mesh generation, physics setup, solver settings, simulation, and post-processing. This approach not only enhances the design of efficient thermal systems but also improves understanding and innovation.

CFD analysis is investigated using a rectangular cross-section with different boundary conditions. This analysis is done using MATLAB code. Figure 12.2 shows a contour plot when conduction heat transfer takes place with adiabatic conditions at the top and bottom surfaces. The left-side temperature is fixed at 300 °C and the right-side temperature at 500 °C. At these given conditions, the constant temperature lines show straight line behaviour. Temperatures at all the 10 nodes of the rectangular cross-section is shown in Table 12.1.

Figure 12.3 shows a contour plot when conduction heat transfer takes place with convection heat transfer from the top and bottom surfaces. The left-side temperature is fixed at 250 °C and the right-side temperature at 500 °C. At these given conditions, the contour plot show maximum temperature at left side and decrease in temperature from convection heat transfer. Temperatures at all the 10 nodes of the rectangular cross-section is shown in Table 12.2.

Table 12.1 Temperatures at different nodes of rectangular cross-section

Node 1	Node 2	Node 3	Node 4	Node 5	Node 6	Node 7	Node 8	Node 9	Node 10
500	477.8	455.6	433.3	411.1	388.9	366.7	344.4	322.2	300
500	477.8	455.6	433.3	411.1	388.9	366.7	344.4	322.2	300
500	477.8	455.6	433.3	411.1	388.9	366.7	344.4	322.2	300
500	477.8	455.6	433.3	411.1	388.9	366.7	344.4	322.2	300
500	477.8	455.6	433.3	411.1	388.9	366.7	344.4	322.2	300
500	477.8	455.6	433.3	411.1	388.9	366.7	344.4	322.2	300
500	477.8	455.6	433.3	411.1	388.9	366.7	344.4	322.2	300
500	477.8	455.6	433.3	411.1	388.9	366.7	344.4	322.2	300
500	477.8	455.6	433.3	411.1	388.9	366.7	344.4	322.2	300
500	477.8	455.6	433.3	411.1	388.9	366.7	344.4	322.2	300

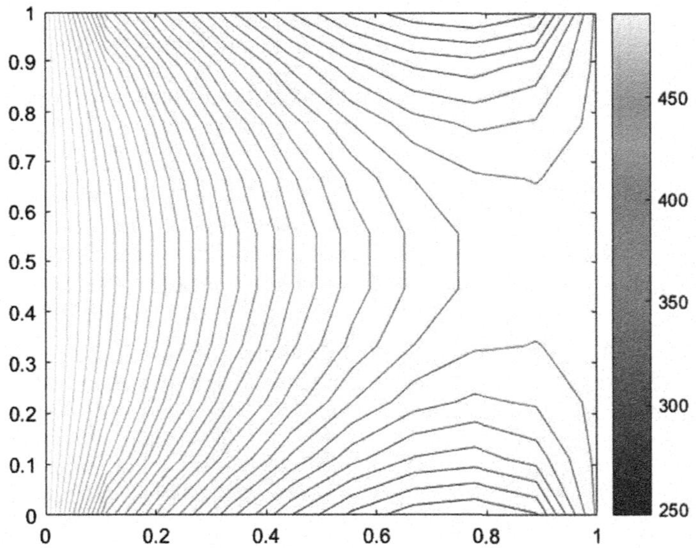

Fig. 12.3 Conduction heat transfer with convection heat transfer from top and bottom surface

12.3 Additive Manufacturing Techniques for Thermal Applications

The time period Additive Manufacturing (AM) refers to a collection of processes, every with unique blessings and downsides for developing thermal components. Here's a precis of some additive production strategies that paintings properly for heated applications, along side a few statistics approximately every one [11].

Table 12.2 Temperatures at different nodes of rectangular cross-section

Node 1	Node 2	Node 3	Node 4	Node 5	Node 6	Node 7	Node 8	Node 9	Node 10
500	387.9	337.8	300.3	272.5	253.1	241.6	238.6	245.8	300
500	426.1	373.2	333.7	304.3	283.8	271.7	268.5	276.1	300
500	443.2	395.3	356.9	327.3	306.1	292.9	287.6	290.3	300
500	451.6	407.9	371.2	342.0	320.4	306.1	298.7	297.3	300
500	455.1	413.7	378.0	349.1	327.3	312.4	303.8	300.4	300
500	455.1	413.7	378.0	349.1	327.3	312.4	303.8	300.4	300
500	451.6	407.9	371.2	342.0	320.4	306.1	298.7	297.3	300
500	443.2	395.3	356.9	327.3	306.1	292.9	287.6	290.3	300
500	426.1	373.2	333.7	304.3	283.8	271.7	268.5	276.1	300
500	387.9	337.8	300.3	272.5	253.1	241.6	238.6	245.8	300

12.3.1 Selective Laser Melting (SLM)

Layer through layer, SLM selectively melts and fuses steel powders the usage of a excessive-energy laser to create absolutely dense steel elements with advanced mechanical features.

Ability to lessen lead instances and manufacturing prices through generating complex, near-net-form elements with little post-processing.

Limitations: Slower construct speeds and a smaller construct extent than different AM techniques may make massive-scale manufacturing unfeasible. excessive device and working prices; implementation requires unique centers and knowledge. vulnerability to residual pressure and distortion because of the melting method's speedy heating and cooling, which requires meticulous method optimization and the development of assist structures.

12.3.2 Electron Beam Melting (EBM) [12]

EBM produces huge, complex objects at fast construct quotes through selectively melting steel debris in a excessive-vacuum surroundings the usage of an electron beam.

Advantages: EBM is desirable for generating massive-scale thermal components, consisting of tooling and aerospace components, due to the fact to its excessive construct volumes and speeds higher mechanical features and excessive fabric purity because of the vacuum surroundings's reduced threat of infection and oxidation. Outstanding fabric utilization performance seeing that there may be little fabric waste due to the fact the entire powder mattress is concurrently heated and melted.

Limitations: Restricted floor end and determination in evaluation to different AM techniques; for a few applications, more machining or completing steps can be

Fig. 12.4 Fused deposition modelling

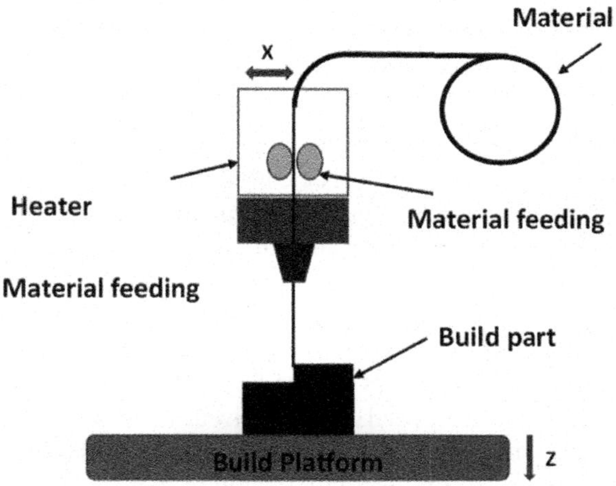

needed restricted fabric picks in evaluation to SLM, with an emphasis on titanium and positive alloys, which limits the selection of substances for precise temperature needs. Increased electricity and device prices associated with electron beam advent and vacuum situations maintenance, requiring a massive capital investment.

12.3.3 Fused Deposition Modeling (FDM) [13]

FDM is a thermoplastic extrusion-primarily based totally additive production era that builds gadgets layer through layer through heating a filament of fabric and extruding it thru a nozzle. Figure 12.4 shows the FDM format diagram.

Advantages: Low-cost, easily operated technology appropriate for low-volume manufacturing of plastic thermal components, tooling, and prototyping. A large number of thermoplastic materials is available, providing flexibility in material selection for thermal applications. These materials include heat-resistant polymers including ABS, polycarbonate, and polyetherimide (PEI). Easy to installation and run; calls for much less help elimination and post-processing than steel AM methods.

Limitations: The use of thermoplastic substances in high-temperature or high-pressure thermal packages is confined due to the fact to their restrained thermal conductivity and mechanical features whilst in comparison to metals and ceramics. Reduced floor polish and determination in assessment to steel AM methods, which in a few conditions can also additionally have an effect on the capability and look of thermal additives. Thermal growth and contraction for the duration of printing can reason warping and dimensional errors, necessitating meticulous procedure manipulate and component layout optimization.

12.3.4 Selection Criteria for Choosing Appropriate AM Technique

When selecting an additive manufacturing (AM) process for thermal applications, several key factors must be considered. These include material requirements, where the AM technology must offer materials with the necessary mechanical and thermal properties such as strength, stiffness, heat resistance, and thermal conductivity. The complexity of the design, including internal channels and surface finish requirements, should be assessed to ensure the chosen AM method can accurately produce the desired geometry with the appropriate resolution. Additionally, production volume and scalability needs should be evaluated to select an AM process that balances build size, speed, and cost-effectiveness for the intended production quantities. Post-processing requirements, such as surface finishing, machining, and heat treatment, must also be considered for their impact on quality, cost, and lead time. Furthermore, the availability of facilities, equipment, energy sources, environmental controls, and the necessary expertise for the chosen AM technology should be reviewed. Finally, application-specific factors, including industry standards, performance objectives, and regulatory compliance, must be taken into account. By carefully evaluating these variables, manufacturers can choose the most suitable AM process for creating advanced thermal components that deliver enhanced performance, efficiency, and capability.

12.3.5 Binder Jetting Procedure

Binder jetting (BJ) became mounted via way of means of the Massachusetts Institute of Technology (MIT) in 1993, and Z Corporation acquired a procedure license from MIT years later. The BJ approach is used first for ceramics and thereafter for polymers. But beginning withinside the early 2000s, steel substances have been processed the usage of the BJ procedure [14]. The BJ approach operates on a comparable idea as PBF, besides in place of a laser, binder droplets (~eighty μm) are used. Figure 12.5 shows the BJ procedure schematic diagram.

According to design data, binder droplets from inkjet nozzles are deposited over the powder, acting as a binding agent between the powder particles and depositing a single layer over the bed [15]. After lowering the powder bed by one layer's thickness, the second layer is deposited. The procedure keeps developing the last green portion in this way. It works quality with substances which can be composites, metals, and polymers. Several post-processing methods, inclusive of curing, de-powdering, sintering, infiltration, annealing, and finishing, are carried out to the very last inexperienced portion. When bronze is applied as an infiltrate, the density of stainless-steel will increase from 60 to ninety five percent. Similarly, cobalt and aluminum are used as infiltrates for tungsten carbide and boron carbide, respectively. Compared to the PBF method, the BJ method can create hole cooling tunnels on a extra scale and gives excessive characteristic resolution. It lets in layout freedom that isn't practicable with conventional methods like casting, machining, and forging. Big construct volumes, speedy print speeds, numerous substances according to construct, self-sufficiency, and absence of guide structures—not like powder mattress fusion—are the advantages of the BJ method. Figure 12.6 presentations the BJ method's method parameters.

12.3.6 Laminated Item Manufacturing (LOM) Method

The uncooked cloth applied withinside the LOM method is a sheet. When the sheet is positioned over the substrate for the primary layer, the laser cuts it in keeping with the vital form from 3-dimensional data. Heat-touchy adhesive paper is activated and the layer is compressed with a warm curler to create a robust bond among the 2 next layers [16]. Figure 12.7 presentations the schematic diagram for the LOM method. The bonding mechanism among sheet layers—adhesive or gluing, clamping, thermal bonding, and ultrasonic welding—is the idea for classifying the LOM method. In contrast to different AM methods, the LOM method eliminates

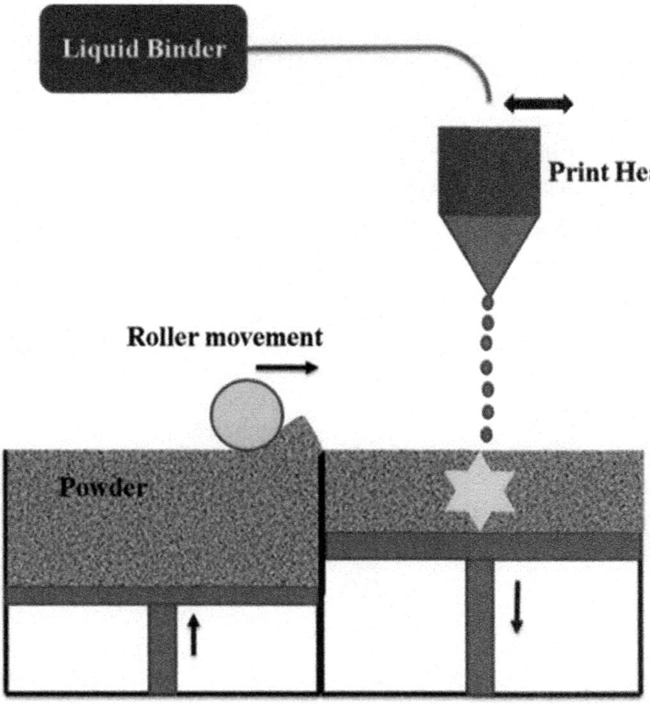

Fig. 12.5 Schematic of binder jetting process [14]

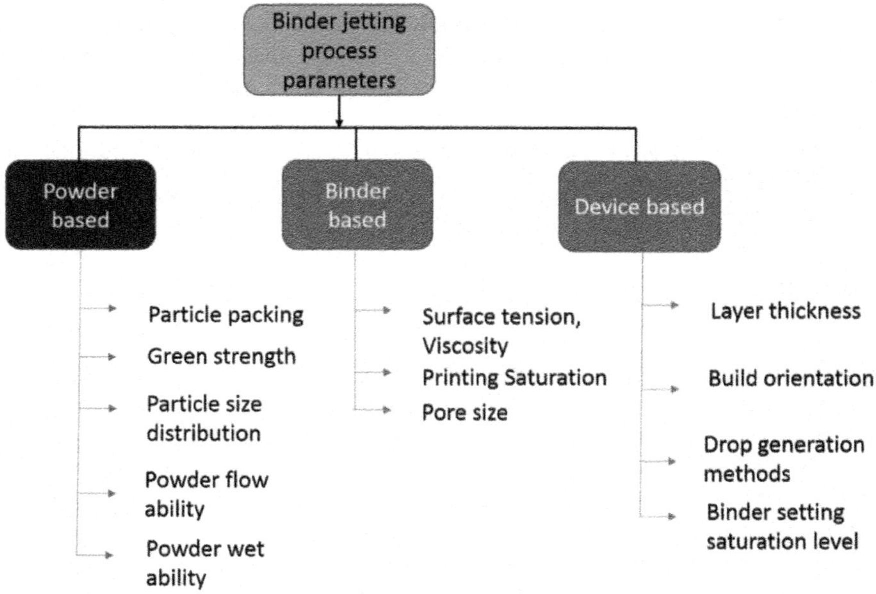

Fig. 12.6 Process parameters of BJ process [14]

distortion, section transformation, and solidification flaws, however it additionally limits element complexity [17]. Depending at the series wherein they're carried out, bonding and glueing may be completed in awesome ways: (a) bond-then-form (first gluing the sheet to the substrate, then forming the vital cross-section), (b) form-then-bond (first reducing the sheet in the correct cross-section, then sticking to the substrate).

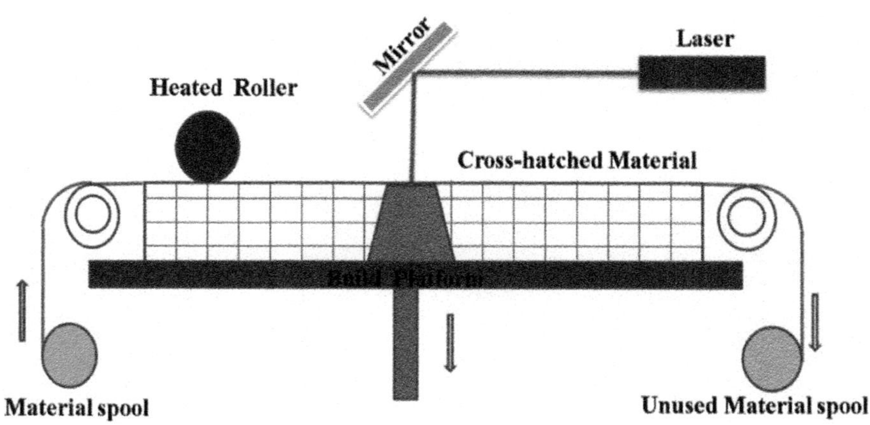

Fig. 12.7 Schematic of laminated object manufacturing process [17]

Instead of using warmness and adhesive bonding to create steel laminates, clamping or bolting is used on this method. One technique that has been correctly used for sheet lamination of steel elements and useful tooling is thermal bonding.

During the ultrasonic additive manufacturing (UAM) method, the bottom plate or formerly deposited layer is included with sheet foil. The sonotrode then oscillates and movements at a frequency of 20 kHz in tandem with the sheet joint. According to [18, 19], the sonotrode oscillations are everyday to the route of motion. Here, pc numerical controller (CNC) milling is used to achieve the vital geometry for every layer. Each sheet of foil is subjected to this hybrid method till the completed product is manufactured. It works nicely on copper, aluminum, and titanium sheet foils. The oscillation amplitude, everyday force, sonotrade tour speed, preheat temperature, and foil thickness are the method parameters hired withinside the UAM method. The UAM method is efficiently used to fabricate aluminum alloy (6061) [20].

12.4 Materials and Design Considerations for Thermal Applications in Additive Manufacturing

12.4.1 Materials

The fabrication of thermal additives has gone through a revolution way to additive production (AM), which presents a extensive variety of substances with specialised traits suitable for plenty of programs. Metals, ceramics, polymers, and composites are a number of those substances; every has positive blessings and problems with reference to warmness control [21].

Metals: Metals such as aluminum, titanium, stainless steel, and nickel-based alloys are commonly used in additive manufacturing (AM) for thermal applications due to their excellent thermal conductivity, mechanical strength, and heat resistance. Key material properties, including thermal conductivity, specific heat capacity, and coefficient of thermal expansion (CTE), influence thermal performance, with higher conductivity enabling efficient heat transfer and lower CTE reducing thermal stresses. Advances in metal AM have led to the development of new alloy compositions for thermal applications like heat exchangers, cooling systems, and turbine components, incorporating elements like copper, tungsten, and chromium to enhance thermal properties and corrosion resistance.

Ceramics: Ceramics such as alumina, silicon carbide, and zirconia offer excellent thermal resistance, chemical stability, and wear resistance, making them ideal for high-temperature applications. While they generally have lower thermal conductivity than metals, ceramics excel in thermal shock resistance and corrosion resistance. Advances in ceramic additive manufacturing, like binder jetting and stereolithography, allow the production of complex components with customized geometries and microstructures, tailored for thermal control applications in industries such as aerospace, electronics, and energy.

Polymers: Polymers like PEEK, PA, and PPS are favored in additive manufacturing for thermal applications due to their lightweight, corrosion resistance, and design flexibility. While they have lower thermal conductivity than metals and ceramics, they offer cost-effectiveness and compatibility with complex designs. Advances in polymer AM have enabled reinforced composites, such as fiber-reinforced nylon, improving strength, stiffness, and thermal conductivity for demanding applications in electronics, automotive, and aerospace.

Composites: Composite materials combine matrices and fibers to achieve enhanced properties for thermal applications. The composition, orientation, and manufacturing methods determine their thermal performance. Reinforcement fibers like carbon, glass, and aramid improve thermal conductivity, strength, and stiffness, while the matrix provides thermal stability and AM compatibility. Advances like continuous fiber reinforcement and hybrid additive/subtractive manufacturing enable the production of lightweight, high-performance composites with complex geometries and tailored thermal properties.

Remarkably, a wide variety of materials, each with specific qualities and performance characteristics, are available for use in thermal applications through additive printing. These materials include ceramics, metals, polymers, and composites. The potential for creating creative thermal management solutions that drive efficiency, sustainability, and performance optimization across a range of industries is continually expanding thanks to developments in material discovery and additive manufacturing technology.

12.4.2 Design Considerations for Thermal Applications in Additive Manufacturing

Using Additive Manufacturing (AM) to optimize thermal overall performance calls for considering quite a few layout principles, standards, and present day strategies which can be acceptable to positive applications. When growing thermal additives with AM, preserve the subsequent factors in mind [22].

12.4.2.1 Design Principles for Optimizing Thermal Performance the Use of AM

Additive manufacturing (AM) allows the creation of complex internal geometries and systems optimized for heat transfer and thermal control, using features like intricate surface textures, conformal cooling channels, and lattice structures. By integrating multiple components into a single design, AM reduces assembly complexity, minimizes thermal resistance, and enhances overall system performance. Selecting materials with specific thermal properties, such as heat capacity, corrosion resistance, and conductivity, is crucial for the intended application. Incorporating thermal insulation materials can further improve energy efficiency and thermal stability in high-temperature environments, reducing heat loss and temperature fluctuations.

12.4.2.2 Design Guidelines for Heat Exchangers, Cooling Systems, and Other Thermal Components

Optimize the design of heat exchanger components like fins, baffles, and tubes to reduce fluid flow resistance and pressure drop while enhancing heat transfer. When designing complex geometries for AM, consider build orientation and support requirements. Cooling systems, including radiators, heat sinks, and liquid channels, should be designed to efficiently disperse heat. AM allows for the creation of intricate cooling channels and maximized surface areas for better heat transfer. Integrating thermal control systems like heat pipes, fans, and pumps ensures temperature regulation and optimal operating conditions, with careful consideration of the placement and orientation of thermal components to improve airflow and coolant circulation.

12.4.2.3 Topology Optimization and Generative Design for Thermal Applications

Iteratively optimize the material distribution and shape of thermal components using topology optimization to meet performance goals while reducing weight, material usage, and manufacturing complexity. Employ generative design to automatically generate optimal solutions for specific thermal needs, considering thermal performance, manufacturability, and structural integrity. Factor in AM-specific elements like build orientation and support structures. Maximize thermal performance and integration while minimizing build time, material consumption, and post-processing by tailoring designs for additive manufacturing.

12.5 Case Studies, Future Guidelines and Economic Implications

Case studies showcase AM's success in optimizing thermal systems, particularly in aerospace and electronics. Future guidelines emphasize advancements in materials and design strategies to enhance AM's potential. Economically, AM offers cost savings through reduced material waste and simpler assembly, though initial investment may be high, making it a transformative tool for thermal system design and production.

12.5.1 Case Studies

Following case study is described in details as follow.

12.5.1.1 Heat Exchanger Optimization in Aerospace

An industry-main aerospace company optimized warmness exchanger designs for plane environmental manipulate structures thru the usage of additive manufacturing. Complex indoors geometries had been advanced to boom warmness switch performance and store weight via way of means of making use of AM's layout freedom. The warmness exchanger's form turned into iteratively optimized the use of generative layout algorithms to attain higher thermal overall performance. Improved aircraft gasoline economic system and operational price financial savings resulted from the first rate improvements in warmness switch performance, weight reduction, and universal gadget overall performance exhibited via way of means of the AM-optimized warmness exchangers [23].

12.5.1.2 Automotive Cooling Systems Innovation

A automobile corporation used additive printing to create present day hybrid and electric powered car (EV) cooling structures. With problematic indoors capabilities like coolant channels and warmth sinks which might be designed for powerful warmness dissipation and thermal control, AM turned into applied to create light-weight, compact cooling additives. By effectively dispersing the warmth produced via way of means of electric powered drivetrains and strength electronics, the AM-enabled cooling structures expanded thermal overall performance, extended battery life, and expanded car range, thereby selling the uptake and improvement of sustainable mobility solutions.

12.5.1.3 AM-Enabled Innovations in Aerospace, Automotive, Electronics, and Energy Sectors

Aerospace: To create light-weight warmness exchanger panels with complicated inner channels for powerful thermal control in plane environmental manipulate structures, aerospace industries have grew to become to additive manufacturing. By combining numerous additives right into a single, monolithic structure, additive manufacturing (AM) lowers meeting complexity and weight whilst improving warmness switch performance.

Automobiles: Engine Cooling Parts: Automobile producers use additive manufacturing (AM) to create engine cooling elements with higher sturdiness and thermal overall performance, like radiators and coolant manifolds. AM makes it feasible to layout problematic inner passageways and capabilities that decrease element weight and enhance warmness switch performance, which increases the performance and dependability of vehicles.

Electronics: Customized Thermal Management Solutions, which include warmness sinks, thermal interface substances, and liquid cooling structures, are advanced via way of means of electronics corporations the use of additive manufacturing (AM) to fulfill the particular wishes of digital gadgets and structures. Through the usage of additive manufacturing (AM), problematic geometries and reliable, progressed designs that enhance warmness dissipation can be produced, giving electric additives extra overall performance and an extended lifespan.

Energy: Additive Manufacturing (AM): AM is utilized by power corporations to provide warmness exchangers and different thermal elements for strength distribution and technology structures. High-overall performance warmness exchangers with most reliable shapes and substances for powerful warmness switch may be produced way to additive manufacturing (AM), which lowers power intake and the environmental impact of strength technology operations.

12.5.2 Challenges and Future Directions

There are nonetheless regulations on cloth characteristics inclusive of thermal conductivity, warmness resistance, and compatibility with AM procedures, despite the fact that additive production offers a massive sort of substances suitable for thermal programs. It remains hard to create new substances and decorate those who exist already to meet the stern specs of thermal control systems. Particularly in excessive-temperature programs, floor roughness and porosity in elements made via way of means of additive production can effect thermal overall performance and durability. To offer perfect warmness transmission and gadget efficiency, AM operations need to reap constant floor end and best control. Advanced layout and simulation strategies are had to broaden complicated geometries and inner additives which can be optimized for thermal overall performance making use of additive production. Engineers and architects need to triumph over layout constraints such thermal stresses, construct orientation, and aid systems a good way to maximize warmness switch efficiency. It's nonetheless hard to assure consistency and repeatability in AM strategies, specifically while generating thermal additives on a massive scale [24].

12.5.2.1 Future Trends and Advancements in AM Technology for Thermal Management

The capacity to manufacture multi-cloth, functionally graded systems with custom designed thermal houses is made viable via way of means of improvements in additive production technology. Thermal conductivity, thermal expansion, and different thermal houses may be exactly managed via way of means of combining diverse substances and gradients internal a unmarried part, growing new possibilities for thermal control gadget optimization. Active thermal control and real-time tracking are made viable via way of means of the direct integration of sensors and embedded cooling channels into additively constructed elements. With AM, complicated indoors systems with included sensors and fluid pathways may be created, enhancing gadget overall performance, dependability, and potential for predictive maintenance.

The aim of studies and improvement is to create excessive-temperature substances and additive production strategies which can face up to extraordinarily excessive temperatures. Heat-resistant alloys and ceramic composites suitable for automotive, aerospace, and strength programs may also now be produced way to tendencies in metallic additive production (AM) technology, which include directed strength deposition and powder mattress fusion. The use of virtual twins and simulation-primarily based totally layout gear is turning into an increasing number of important for AM-primarily based totally thermal control gadget optimization. By comparing diverse layout configurations, cloth opportunities, and running situations previous to bodily production, engineers can reduce down on time-to-marketplace and improvement charges thru the usage of digital prototyping, predictive modeling, and simulation-pushed layout.

12.5.2.2 Opportunities for Research and Development in AM-Enabled Thermal Solutions

To growth the opportunities of AM-enabled thermal solutions, studies into new substances with specialised thermal houses is important. Examples of those houses encompass excessive thermal conductivity, low thermal expansion, and corrosion resistance. Innovation in cloth improvement for unique thermal programs may be fueled via way of means of cooperation among cloth scientists, engineers, and AM researchers. Process Control and Optimization: To get beyond the existing limitations and bring elements with constant best, studies endeavors aimed toward improving system optimization, control, and reliability in AM technology are essential. Improvements in closed-loop comments systems, in-situ sensing, and system tracking can enhance the reproducibility and balance of procedures in AM-enabled thermal production.

To absolutely make use of additive production, studies into layout concepts and suggestions for AM-enabled thermal packages is necessary. The improvement of DfAM methodologies, which include conformal cooling channels, lattice structures, and topology optimization, that contain thermal issues into the layout method can useful resource withinside the improvement of novel thermal answers with maximized overall performance and performance. Sustainability and Circular Economy: There is a developing frame of studies being accomplished to higher recognize the environmental impact and sustainability of AM-enabled thermal production.

12.5.3 Economic and Environmental Implications

12.5.3.1 Cost Analysis of Additive Manufacturing Compared to Traditional Manufacturing Methods in Thermal Applications

Compared to conventional production strategies, additive production regularly necessitates a good sized in advance funding in software, hardware, and skills. However, as generation develops and turns into greater extensively available, the barrier of access for AM has been reducing over the years. The fee of using additive production to supply thermal additives can range primarily based totally on numerous parameters, inclusive of fabric costs, construct extent, layout complexity, and post-processing requirements. Because of economies of scale, conventional production strategies might also additionally offer less expensive unit expenses for high-extent manufacturing, despite the fact that additive production (AM) may be greater cheap for low-extent manufacturing or complicated geometries. In a few situations, additive production can offer a aggressive area whilst taking into consideration the overall fee of ownership, which incorporates the authentic funding, manufacturing costs, preservation expenses, and lifespan costs. Compared to conventional production tactics, additive production (AM) reduces lead times, stock costs, and deliver chain dangers with the aid of using allowing speedy prototyping, layout iteration, and on-call for manufacturing [25].

12.5.3.2 Environmental Impact Assessment of AM Processes for Thermal Components

Compared to standard manufacturing strategies, additive production reduces fabric waste with the aid of using most effective the amount of fabric required to finish the part. However, waste from AM strategies also can consist of leftover powder, guide structures, and post-processing leftovers that want to be recycled or disposed of. Depending on variables inclusive of system type, construct duration, and fabric qualities, additive production approaches use exceptional quantities of power. The environmental effect of AM tactics may be lessened over the years with the aid of using upgrades in method performance and power-saving strategies, despite the fact that AM may be power-intensive, specially for high-extent manufacturing.

Particularly throughout fabric processing and post-processing activities, additive production strategies have the capacity to supply emissions and pollutants, which include unstable natural compounds (VOCs), particulate matter, and risky with the aid of using-products. In order to lessen environmental pollutants and fitness worries associated with AM operations, it's miles vital to put into effect suitable ventilation, filtration, and waste control strategies [26].

12.5.3.3 Sustainability Considerations and Circular Economy Principles in AM for Thermal Applications

Reducing waste and useful resource intake via the software of recycling and reuse answers for AM materials, which include steel powders and thermoplastics, can guide the improvement of a round economy. For use in upcoming AM builds, leftover powder and post-processing residues may be recycled way to closed-loop fabric structures and powder restoration generation. Designing thermal additives with sustainability in thoughts for additive production includes optimizing designs for end-of-life, power conservation, and fabric performance. Techniques for generative layout, lightweighting, and topology optimization can restrict cloth use, keep electricity, and decorate product overall performance and recyclability.

Lifecycle assessments (LCAs) of AM-enabled thermal additives resource in assessing the environmental effect of those additives throughout the complete lifecycle of the product, from extraction and production of uncooked substances to apply and disposal. LCAs manual decision-making closer to extra sustainable production methods through considering variables which includes electricity use, greenhouse fueloline emissions, useful resource depletion, and waste generation. In order to increase sustainability applications and create enterprise-huge requirements and pleasant practices for additive production in thermal packages, collaboration amongst stakeholders—manufacturers, academics, legislators, and enterprise organizations—is crucial.

12.6 Conclusion, Discussion and Future Scope

This paper gives a complete exploration of the giant affect of Additive Manufacturing (AM) on thermal packages inside numerous sectors. It delves into how AM allows the improvement of intricate, optimized thermal additives which can be difficult to supply the usage of traditional production techniques. The exam encompasses center ideas of warmth switch, cloth choices, in addition to the blessings and constraints related to exclusive AM methods. Practical times from real-global situations exhibit the utility and improvements in industries which includes aerospace, automotive, electronics, and electricity, underscoring the tangible blessings of AM in enhancing thermal law systems. Furthermore, the paper delves into the economic, environmental, and regulatory dimensions, imparting a well-rounded attitude on the prevailing and potential roles of AM in thermal engineering.

Additive Manufacturing (AM) has been a game-converting era that has advanced thermal packages in quite a few industries. Among AM's maximum essential discoveries and contributions to thermal engineering are. Complex geometries, inner structures, and specialised thermal answers that meet specific utility desires can all be created with AM. The layout flexibility that additive production (AM) gives enables advanced warmth switch overall performance, improved efficiency, and innovation in thermal control systems. AM makes it less difficult to create state-of-the-art substances, which includes metals, ceramics, polymers, and composites, with specialised thermal properties. The improvement of latest substances makes it feasible to supply high-overall performance, light-weight thermal additives that could live on harsh operating environments and temperatures. When it involves cloth efficiency, electricity savings, and waste reduction, additive production (AM) has blessings over conventional production techniques.

The improvement of AM substances, procedures, and technology particularly desirable for thermal packages may be fueled through studies and improvement activities. The packages of additive production (AM) in thermal engineering will develop with the creation of multi-cloth printing, functionally graded structures, and high-temperature substances. More powerful and optimized thermal answers may be made feasible through the convergence of additive production with virtual layout tools, simulation software, and information analytics. Predictive modeling, simulation-pushed layout, and virtual twins will enhance AM-enabled thermal thing layout, optimization, and overall performance forecasting.

Additive Manufacturing has the capacity to absolutely remodel warmness control in a number of industries even as providing formerly unheard-of possibilities for sustainability, efficiency, and creativity. With AM, incredibly custom designed thermal answers that meet unique software needs, preferences, and overall performance requirements can be produced. AM permits designers and engineers to supply thermal answers which are tailor-made for particular requirements, starting from custom designed scientific gadget to custom automobile cooling systems. Because AM permits for short prototyping, iterative layout iterations, and on-the-fly modifications, it quickens the product improvement time.

Future studies withinside the area must be directed closer to the development of substances and approaches mainly tailor-made for Additive Manufacturing (AM) packages in thermal settings, with a specific recognition at the improvement of high-temperature substances and systems with functionally graded properties. The integration of virtual technologies, which include virtual twins and predictive modeling, is predicted to complement the system of designing and optimizing thermal additives synthetic via AM. Additionally, there's a want to prioritize sustainability aspects, encompassing fabric recycling, energy-green procedures, and complete life-cycle evaluations. The status quo of universally normal requirements withinside the enterprise and the cultivation of collaborative relationships amongst manufacturers, researchers, and regulatory our bodies are critical to bolstering the reputation and modern capacity of AM in thermal control practices. The ongoing exploration of those thematic regions is poised to propel the evolution of AM-enabled thermal answers, thereby fostering efficiency, sustainability, and technological progress.

References

1. Schleifer, S.L., Regev, O.: Additive manufacturing of anisotropic graphene-based composites for thermal management applications. **70**, 103567 (2023). https://doi.org/10.1016/j.addma.2023.103567
2. Chen, J., Pierce, J., Williams, G., Simpson, T.W., Meisel, N., Prabha Narra, S., McComb, C.: Accelerating thermal simulations in additive manufacturing by training physics-informed neural networks with randomly-synthesized data. **1**, 1–4 (2024). https://doi.org/10.1115/1.4062852
3. Rani, P., Deshmukh, K., Kesarla, M.K., Kar, T., Pasha, S.K.: Applications of additive manufacturing techniques in the fabrication of thermoelectric materials and devices. **2**, 527–560 (2023). https://doi.org/10.1002/9783527835478.ch18
4. Broughton, J., Joshi, Y.: A numerical investigation of additive manufactured foam structures for single phase hotspot thermal management. **59322**, V001T06A023 (2019). https://doi.org/10.1115/IPACK2019-6519
5. Yagci, V., Parlak, M., Cadirci, S.: Investigation of grooved heat pipe embedded cold plate design with additive manufacturing, pp. 1–5 (2022). https://doi.org/10.1109/THERMINIC57263.2022.9950694
6. Lin, L., Wu, H., Ni, P., Chen, Y., Huang, Z., Li, Y., Lin, K., Sheng, P., Wu, S.: Additive manufacturing of complex-shaped and high-performance aluminum nitride-based components for thermal management. **52**, 102671 (2022). https://doi.org/10.1016/j.addma.2022.102671
7. Favero, G., Bonesso, M., Rebesan, P., Dima, R., Pepato, A., Mancin, S.: Additive manufacturing for thermal management applications: from experimental results to numerical modeling. **10**, 100091 (2021). https://doi.org/10.1016/j.ijft.2021.100091
8. Xu, J., Tarau, C., Ellis, M., Alvarez-Hernandez, A.: Design of an additively manufactured vapor chamber heat exchanger for space applications. 1091–1097 (2021). https://doi.org/10.1109/ITherm51669.2021.9503233
9. Gross, B.C., Erkal, J.L., Lockwood, S.Y., Chen, C., Spence, D.M.: Evaluation of 3D printing and its potential impact on biotechnology and the chemical sciences. 3240–3253 (2014). https://doi.org/10.1021/ac403397r
10. Das, S., Subudhi, S.: Studying the physiology of the human body using a mathematical model of transient energy balance. **48**, 245 (2023). https://doi.org/10.1007/s12046-023-02295-w
11. Das, S., Kumar, D., Subudhi, S.: A mathematical steady-state energy balance model for studying the physiology of the human body. 701–712 (2020). https://doi.org/10.1007/978-981-99-1288-9_73
12. Murr, L.E.: Metallurgy of additive manufacturing: examples from electron beam melting. **5**, 40–53 (2015). https://doi.org/10.1016/j.addma.2014.12.002
13. Ning, F., Cong, W., Qiu, J., Wei, J., Wang, S.: Additive manufacturing of carbon fiber reinforced thermoplastic composites using fused deposition modeling. **80**, 369–378 (2015). https://doi.org/10.1016/j.compositesb.2015.06.013
14. Ziaee, M., Crane, N.B.: Binder jetting: a review of process, materials, and methods. **28**, 781–801 (2019). https://doi.org/10.1016/j.addma.2019.05.031
15. Li, C., Liu, Z.Y., Fang, X.Y., Guo, Y.B.: Residual stress in metal additive manufacturing. **71**, 348–353 (2018). https://doi.org/10.1016/j.procir.2018.05.039
16. Zhang, X., Wu, X., Shi, J.: Additive manufacturing of zirconia ceramics: a state-of-the-art review. **9**, 9029–9048 (2020). https://doi.org/10.1016/j.jmrt.2020.05.131
17. Diegel, O., Nordin, A., Motte, D.: Additive Manufacturing Technologies, pp. 19–39 (2019). https://doi.org/10.1007/978-981-13-8281-9_2
18. Shahrubudin, N., Lee, T.C., Ramlan, R.: An overview on 3D printing technology: technological, materials, and applications. **35**, 1286–1296 (2019). https://doi.org/10.1016/j.promfg.2019.06.089
19. Robinson, C., Stucker, B., Branch, K.C., Palmer, J., Strassner, B., Allen, S., Bugos, R., Navarrete, M., Lopes, A., Macdonald, E., Medina, F., Wicker, R.B.: Fabrication of a mini-SAR

antenna array using ultrasonic consolidation and direct-write. 30 (2007). https://www.osti.gov/servlets/purl/1267093
20. Gussev, M.N., Sridharan, N., Norfolk, M., Terrani, K.A., Babu, S.S.: Effect of post weld heat treatment on the 6061 aluminum alloy produced by ultrasonic additive manufacturing. **684**, 606–616 (2017). https://doi.org/10.1016/j.msea.2016.12.083.
21. Bhatia, A., Sehgal, A.K.: Additive manufacturing materials, methods and applications: a review. **81**, 1060–1067 (2023). https://doi.org/10.1016/j.matpr.2021.04.379
22. Wrobel, R., Hussein, A.: Design considerations of heat guides fabricated using additive manufacturing for enhanced heat transfer in electrical machines, pp. 6506–6513 (2018). https://doi.org/10.1109/ECCE.2018.8557559
23. Mekki, B.S., Langer, J., Lynch, S.: Genetic algorithm based topology optimization of heat exchanger fins used in aerospace applications. **170**, 121002 (2021). https://doi.org/10.1016/j.ijheatmasstransfer.2021.121002
24. Bacciaglia, A., Ceruti, A., Liverani, A.: Additive manufacturing challenges and future developments in the next ten years, pp. 891–902 (2019). https://doi.org/10.1007/978-3-030-31154-4_76
25. Thomas, D.S., Gilbert, S.W.: Costs and cost effectiveness of additive manufacturing. **1176**, 12 (2014). https://doi.org/10.6028/NIST.SP.1176
26. Khalid, M., Peng, Q.: Sustainability and environmental impact of additive manufacturing: a literature review. **18**, 1210–1232 (2021). https://doi.org/10.14733/cadaps.2021.1210-1232

Dr. Shashikant Das obtained his B.Tech from NIT Calicut in 2011 and his M.Tech from IIT Kharagpur in 2015. Further, he obtained his Ph.D. from the Department of Mechanical and Industrial Engineering, IIT Roorkee, in 2022. His area of research is thermal comfort, where he focused on the impact of a hot climate on the human body in terms of analytical and experimental investigation. In his analytical work, he has gone through steady-state, transient state, and exergy analyses of the human body. In experimental work, he has gone through the effect of a hot environment on occupants' performance, sensation, and physiology. He has academic experience in teaching subjects like heat transfer, thermodynamics, air conditioning, I-C engines, power plants, and fluid mechanics at the UG level.

Dr. Benarji Koyilada is an accomplished Assistant Professor in the Mechanical Engineering Department at Shri Vishnu Engineering College for Women's, Andhra Pradesh. I received my doctoral degree from the Department of Mechanical Engineering, NIT Warangal, India, in July 2021, The thesis title is "Processing and Characterization of Laser directed energy deposition Based Additive Manufactured SS 316 and its Composites". M.Tech in Production Engineering from NIT Rourkela. His research primarily focuses on laser additive manufacturing, metal cutting, and sustainable manufacturing processes, with notable expertise in laser surface treatment and material characterization. Dr. Benarji has published several articles in high-impact international journals which consists of 5 SCI journals, 2 Scopus, 1 Book chapter and 3 international conferences and also received awards for his research contributions, including a Best Paper award at an international conference.

Dr. Santosh Patro is currently working as an Assistant Professor in the Department of Mechanical Engineering at Centurion University of Technology and Management, Odisha. He completed his B.Tech in Mechanical Engineering from the Biju Patnaik University of Technology, Odisha in 2010 and his M.Tech in Mechanical Engineering (Specialization—Production Engineering) from Veer Surendra Sai University of Technology, Odisha in 2014. He completed his Ph.D. from Centurion University of Technology and Management, Odisha in 2021. He has more than 10 years of teaching experience. He has taught various subjects like Manufacturing Technology, Non-Traditional Machining, Production Technology, Engineering Metrology, and Optimization Techniques at the Diploma and Undergraduate level. His area of research is Multi Response Optimization of Wire Electrical Discharge Machining using Hybrid Optimization techniques. He has published more than 12 papers in different International and National Journals and 2 book chapters. He is a Member of the Indian Society of Technical Education.

G. Sridevi is currently working as an Assistant Professor in the Department of Mechanical Engineering at Centurion University of Technology and Management, Odisha. She has completed her B.Tech in Mechanical Engineering from the Biju Patnaik University of Technology, Odisha in 2003 and her M.Tech in Mechanical Engineering (Specialization—Design and Manufacturing Engineering) from Centurion University of Technology and management, Odisha in 2016. She has 9 years of industrial experience in EPC bulk material handling turnkey projects and more than 10 years of experience in teaching and research. She is presently pursuing her Ph.D. in Centurion University of technology and Management. Her area of research is material characterisation, Additive manufacturing and Advanced machining .She has published more than 10 papers in International SCI/ Scopus journals, 2 book chapters and 1 book.

Dr. Sujit Mishra is presently an Assistant professor and Head of the Mechanical Engineering Department at Centurion University of Technology and Management, Paralakhemundi campus. He obtained his Ph.D. from Centurion University of Technology and Management, M.Tech and B.Tech from Biju Patnaik University of Technology, Odisha. He has ten years of teaching experience with subject interests such as Machine Design, Fluid mechanics, Computational Fluid Dynamics, Computer-Aided Drafting, Computer-Aided Manufacturing, and Computer-Aided Engineering at both UG and PG levels. His Ph.D. research area includes the influence of dust particles in predicting aerodynamic parameters for Simplified Cars using the Computational Fluid Dynamics approach. He is also keen on providing simulation solutions for industrial problems such as heat transfer enhancement in various electronic devices, and heat exchangers using nanofluids/hybrid nanofluids. He has published more than

11 international SCI/ Scopus journals, three book chapters, and one book. He is also an active life member of the Indian Society of Technical Education (ISTE) and Indian Mathematical Society (IMS).

Dr. Archana Senapati earned her Ph.D. in Mathematics from KIIT Deemed to be University under the supervision of Dr. Saumya Ranjan Jena. She completed her M.Sc. from Utkal University and her B.Sc. from Ramadevi Women's University, both located in Bhubaneswar, Odisha, India. She is currently serving as an Assistant Professor in the Department of Mathematics at KIIT Deemed to be University, Bhubaneswar, Odisha. Before joining KIIT, she held the position of Assistant Professor in the Department of Mathematics at Centurion University of Technology and Management, Paralakhemundi, Odisha. Her research interests focus on Numerical Approximation. To date, she has published 10 research papers in international journals and conferences and has also contributed to two book chapters.

Dr. Ankur Srivastava received the Ph.D. degree in electrical engineering from the Department of Electrical Engineering, Indian Institute of Technology Roorkee (IIT Roorkee), Roorkee, India, in 2023. He is now Postdoc researcher in center for industrial electronics, South Denmark University (SDU), Sonderborg, Denmark. His area of interest is in design of grid connected converters for distributed renewable sources, switched capacitor based multi-level inverter, WBG (Wide Band Gap) devices, EMI (Electromagnetic Interference) issues with ac drives and three phase pfc rectifier.

Chapter 13
Aiot-Driven Zero-Waste Manufacturing in Biotechnology and Pharmaceuticals: Challenges, Solutions, and Future Perspectives

Polaki Suman, B. Rabi Prasad, G. Pushpalatha, Ch. Mohan Babu, and Islauddin Khan

Abstract This chapter examines the synergistic interplay between Artificial Intelligence (AI) and the Internet of Things (IoT), highlighting their combined potential to revolutionize technological ecosystems as catalysts for advancing zero-waste manufacturing within the biotechnology and pharmaceutical sectors. It begins by delineating the concept of zero-waste manufacturing and its critical role in reducing environmental impacts. The discussion then addresses these industries' specific challenges, including hazardous waste management, regulatory compliance, and the economic and technological barriers to adoption. Critical strategies for sustainable process design are examined, focusing on applying green chemistry principles, process integration, and Design for Environment (DfE) approaches. The chapter also delves into waste minimization techniques, such as source reduction, material substitution, process optimization, and recycling, while highlighting the contributions of AI and IoT in enhancing these strategies through advanced process monitoring, optimization, and data-driven decision-making. Additionally, the chapter reviews

P. Suman (✉) · G. Pushpalatha
Department of Biotechnology, M.S. Swaminathan School of Agriculture, Centurion University of Technology and Management, Parlakhemundi, Odisha, India
e-mail: polaki.suman@cutm.ac.in

G. Pushpalatha
e-mail: pushpalathag@cutm.ac.in

B. R. Prasad
Department of Biotechnology, School of Engineering and Technology, GIET University, Gunupur, Odisha, India
e-mail: brabiprasad@giet.edu

Ch. M. Babu
School of Management, Centurion University, Vizianagaram, Andhra Pradesh, India
e-mail: mohanababu@cutmap.ac.in

I. Khan
Postdoc Research Associate, UIC, Chicago, USA
e-mail: khan99@uic.edu

© The Author(s), under exclusive license to Springer Nature Switzerland AG 2025
K. C. Rath et al. (eds.), *Engineering, Technology and Management*, Information Systems Engineering and Management 54, https://doi.org/10.1007/978-3-031-93654-8_13

innovative technologies, including biocatalysis, single-use systems, and advanced purification methods, supported by industry case studies. The regulatory and policy frameworks underpinning zero-waste practices are evaluated, emphasizing waste reduction's economic and environmental benefits. Finally, the chapter considers future trends and calls for industry-wide collaboration to realize the full potential of zero-waste manufacturing.

Keywords Zero-waste manufacturing · Biotechnology · Pharmaceuticals · Sustainability · Waste reduction · Environmental impact

Preface

This chapter provides the integration of Artificial Intelligence (AI) and the Internet of Things (IoT) as powerful drivers of zero-waste manufacturing in the biotechnology and pharmaceutical industries. It begins by defining zero-waste manufacturing and its vital role in mitigating environmental impact. The chapter navigates through the unique challenges these industries face, including hazardous waste management, stringent regulatory requirements, and economic and technological hurdles. It explores sustainable strategies rooted in green chemistry principles, process integration, and Design for Environment (DfE) approaches, offering practical solutions for waste minimization through techniques such as source reduction, material substitution, and recycling. The transformative role of AI and IoT in enhancing these efforts through advanced monitoring, optimization, and data-driven decision-making is thoroughly analyzed. Further, the chapter highlights cutting-edge technologies like biocatalysis, single-use systems, and advanced purification methods, illustrated with real-world case studies. By evaluating regulatory frameworks and emphasizing the economic and environmental advantages of waste reduction, the chapter advocates for industry-wide collaboration to fully realize the vision of zero-waste manufacturing while setting a course for future innovations.

13.1 Introduction

Zero-waste manufacturing is a forward-thinking strategy developed to reduce waste generation while optimizing resource utilization throughout production workflows. This approach is intrinsically aligned with the principles of sustainable development, emphasizing the elimination of waste through redesigning systems and processes to boost efficiency and environmental friendliness. Zero-waste manufacturing aims to establish a self-sustaining system in which all materials are efficiently repurposed, recycled, or biodegraded, drastically minimizing the environmental impact of manufacturing processes [1]. Its core, zero-waste manufacturing, is a comprehensive strategy that seeks to eradicate waste at every stage of the manufacturing process. It encompasses several guiding principles: Design for Sustainability, which focuses on creating products and processes that significantly reduce waste and lower environmental impact by selecting sustainable materials and planning for product dismantling and reuse, thereby reducing energy consumption throughout the product's lifecycle [2]. It also includes Resource Efficiency, optimizing materials and energy to ensure that resources are utilized as efficiently as possible, thereby minimizing waste.

A vital component of this approach is embracing a Circular Economy, which shifts away from the traditional linear production shifts from a linear take-make-dispose framework to a circular approach where materials are persistently reused, recycled, and reintegrated into manufacturing cycles. [3]. It adheres to the Waste Hierarchy and emphasizes a structured approach to managing waste, prioritizing prevention, followed by reuse, recycling, and recovery, with disposal regarded as a last-resort option. This principle guides decision-making processes to reduce waste generation at the source [4]. Furthermore, Lifecycle Thinking is employed, considering the complete lifecycle of a product, from raw material extraction to final disposal, is analyzed to uncover opportunities for minimizing waste and conserving resources [5]. Pollution Prevention is central to implementing practices and technologies that prevent pollution and minimize the discharge of harmful substances into the environment [1].

The importance of zero-waste manufacturing in sustainable production is immense. As global populations and industrial activities expand, so do the demands on natural resources and waste generation. Zero-waste manufacturing addresses these challenges by promoting environmental sustainability, reducing the environmental impact of production activities, helping to conserve natural resources, and protecting ecosystems [2]. It also enhances economic efficiency, as waste reduction translates into cost savings through improved resource efficiency, lower disposal costs, and enhanced productivity. Companies can achieve substantial financial benefits by optimizing their use of materials and energy. Moreover, adopting zero-waste practices helps companies comply with increasingly stringent environmental regulations, avoiding fines and penalties. Embracing zero-waste manufacturing demonstrates the organization's dedication to upholding corporate social responsibility (CSR), enhancing brand reputation, attracting environmentally conscious consumers, and improving stakeholder relations [6]. It drives innovation by promoting the creation of innovative, sustainable technologies and processes, offering businesses that implement these practices a competitive edge in the market [7].

In the biotechnology and pharmaceutical industries, zero-waste manufacturing presents unique opportunities and challenges due to the complex processes, stringent regulatory requirements, and high product quality and safety standards. However, the principles of zero-waste manufacturing are highly relevant and beneficial in these contexts, helping streamline processes, reduce waste, and improve efficiency [8]. These sectors can realize significant cost reductions by optimizing resource utilization and reducing waste, improving adherence to regulations, and fostering the innovation of sustainable technologies and products that support overarching sustainability objectives. Integrating AI and IoT into zero-waste manufacturing processes enhances effectiveness by facilitating real-time monitoring, proactive maintenance, and efficient resource optimization, as shown in Fig. 13.1. This transformative

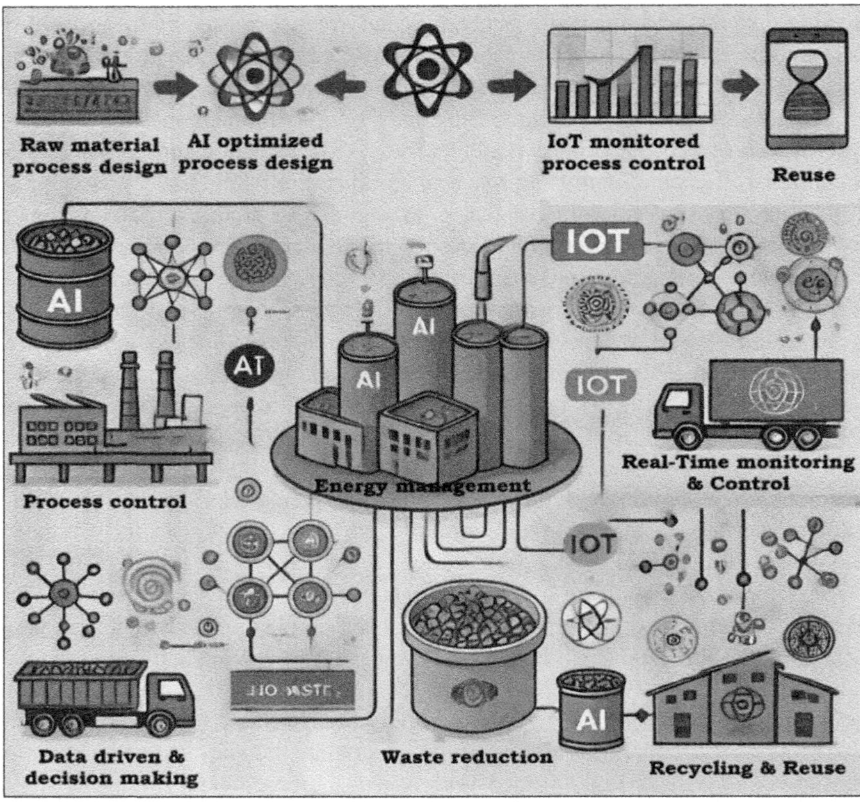

Fig. 13.1 Integration of AI and IoT in zero-waste manufacturing processes

approach to sustainable manufacturing leads the biotechnology and pharmaceutical industries to achieve significant environmental, economic, and social benefits, promoting more sustainable and resilient production systems.

13.2 Challenges in Zero-Waste Manufacturing

While the benefits of zero-waste manufacturing are significant, achieving this goal is fraught with challenges, particularly in complex and highly regulated industries like biotechnology and pharmaceuticals [9]. These challenges encompass a range of issues, from waste generation to regulatory compliance and economic and technological barriers. Addressing these obstacles is crucial to advancing zero-waste initiatives within these sectors. Biopharmaceutical manufacturing, by its very nature, involves complex, multi-step processes that inherently produce various types of waste. Understanding these sources is essential for devising effective waste-reduction strategies.

One significant source of waste is raw material waste. The production of biopharmaceuticals often requires high-purity raw materials, which can lead to significant waste if these materials do not meet stringent quality requirements. Any deviation from specified standards can result in the disposal of entire batches of raw materials, contributing to considerable waste [10]. Additionally, process-related waste is a significant concern. For example, fermentation and cell culture processes generate substantial biological waste, including spent media, biomass, and by-products. The purification and filtration stages, critical for ensuring product safety and efficacy, also produce waste in the form of used filters, chromatography columns, and other disposable components [11]. Furthermore, using solvents in extraction, purification, and cleaning processes contributes to hazardous waste streams, posing additional challenges for waste management.

(a) ***Packaging waste*** is another significant issue in biopharmaceutical manufacturing. High product safety and sterility standards necessitate the extensive use of disposable plastics and packaging materials, contributing to a significant waste burden. Moreover, expired products and inventory waste present considerable challenges. Products not meeting quality specifications or expiring before use represent a significant waste source, including finished products and intermediary components.
(b) ***Regulatory and compliance challenges*** further complicate the implementation of zero-waste manufacturing in the biotechnology and pharmaceutical industries. These industries operate under stringent regulatory frameworks to ensure product safety, efficacy, and quality. While necessary, these regulations often present unique challenges for zero-waste initiatives. For instance, stringent quality standards mandated by regulatory bodies require high-quality control, which can lead to the disposal of materials and products that do not meet exacting standards. This focus on quality can sometimes conflict with waste minimization goals. Additionally, implementing new waste reduction practices often requires extensive validation to prove they do not compromise product quality or safety. The time and resources needed for this validation can be significant, making it a barrier to adopting zero-waste practices. Moreover, many materials used in biopharmaceutical manufacturing are classified as hazardous, and regulations governing the disposal and treatment of hazardous waste are stringent, adding complexity to waste reduction efforts. Compliance with Good Manufacturing Practices (GMP) standards necessitates using specific materials and processes that may generate waste. Any changes to these practices to reduce waste must still comply with GMP guidelines, which can be a significant hurdle.
(c) ***Economic and technological barriers*** also contribute significantly to addressing the challenges associated with zero-waste manufacturing. Transitioning to zero-waste manufacturing often requires substantial upfront investment in new technologies, equipment, and training. This financial burden can be a significant barrier for many companies, especially smaller firms. Furthermore, technological limitations present additional challenges. Advanced technologies for

zero-waste manufacturing, such as waste-to-energy systems, recycling technologies, and sustainable materials, may not be fully developed or commercially viable, impeding progress. The cost of sustainable materials and alternatives to single-use components can also be more expensive than their conventional counterparts, impacting the overall cost structure of manufacturing processes. Moreover, the economic benefits of zero-waste manufacturing, such as cost savings from reduced waste disposal and increased resource efficiency, may not be immediately realized. The long-term return on investment may deter companies from adopting these practices. Integrating zero-waste practices into existing manufacturing processes can also be technically challenging. Legacy systems and infrastructure may not be compatible with new waste reduction technologies, requiring significant modifications or replacements.

(d) ***Integration of AI and IoT:*** Despite these challenges, integrating AI and IoT offers promising solutions for overcoming the obstacles associated with zero-waste manufacturing. AI and IoT technologies can offer real-time monitoring capabilities of manufacturing operations, enabling the ability to the immediate detection and correction of deviations that could lead to waste. This real-time monitoring ensures that production processes remain within optimal parameters, minimizing waste generation. Predictive maintenance, enabled by AI, can further minimize downtime and prolong the lifespan of equipment, thereby minimizing waste from premature equipment failure. IoT sensors are essential in tracking the usage and condition of raw materials and components, enabling the improvement of resource allocation efficiency and the minimization of waste. By providing detailed insights into the production process, these sensors help optimize resource management and ensure materials are used as efficiently as possible. Additionally, AI algorithms are capable of processing large volumes of data to detect patterns and streamline processes, leading to reductions in waste and enhancements in efficiency

While zero-waste manufacturing presents significant challenges, particularly in the biotechnology and pharmaceutical industries, integrating AI and IoT technologies offers innovative solutions to overcome these obstacles. By addressing the issues of waste generation, regulatory compliance, and economic and technological barriers, these industries can move closer to achieving zero-waste manufacturing, ultimately fostering a more sustainable and efficient production system.

13.3 Sustainable Process Design

Achieving zero-waste manufacturing in the biotechnology and pharmaceutical industries necessitates the integration of sustainable process design principles. These principles are the foundation for transforming manufacturing processes to minimize environmental impact and enhance overall sustainability. Sustainable process design focuses on adopting green chemistry and engineering principles, process integration and intensification, and Design for Environment (DfE) strategies, all critical components in this endeavor [12].

(a) ***Green chemistry and engineering principles*** provide a comprehensive framework for designing manufacturing processes that prioritize waste reduction, resource conservation, and use safer materials and methods. One of the fundamental tenets of green chemistry is the prevention of waste. Manufacturers can effectively manage and reduce waste from the outset by designing processes that minimize waste generation at every production stage. This is further complemented by the concept of atom economy, which emphasizes maximizing the utilization of all materials in the production process to create the final product, thereby minimizing waste through the efficient use of raw materials. Safer chemical syntheses and designing safer chemicals are crucial, ensuring that synthetic methods and chemical products are developed to minimize toxicity and environmental impact [13].

Additionally, using safer solvents and reaction conditions, such as water or other benign solvents, helps minimize risks to human health and the environment, avoiding using volatile organic compounds (VOCs). Energy efficiency is another vital aspect of green chemistry, where processes are designed to require less energy, often achieved by conducting reactions at ambient temperature and pressure. Furthermore, using renewable feedstocks, instead of depleting nonrenewable resources, is encouraged, along with reducing derivatives, which involves minimizing auxiliary substances like solvents and separation agents. Catalysis, which employs catalytic reagents that are typically more efficient than stoichiometric amounts of reagents, is employed to reduce waste and improve selectivity. Additionally, designing chemical products for degradation ensures that they decompose into harmless by-products at the end of their use, avoiding environmental persistence. Real-time analysis for pollution prevention is another critical strategy involving the creation of analytical methods for real-time monitoring and control during synthesis to prevent the formation of hazardous substances. Finally, inherently safer chemistry for accident prevention involves designing chemicals and their physical states to mitigate the risk of chemical accidents, such as explosions, fires, and environmental releases [12].

(b) ***Process integration and intensification*** are other critical strategies for making manufacturing more efficient and sustainable. Process integration involves optimizing the use of resources by utilizing waste heat from one part of the process to meet the thermal energy requirements of another, thereby reducing overall energy consumption. On the other hand, mass integration focuses on integrating material flows to maximize the reuse and recycling of by-products and intermediates, further reducing waste. Process intensification, meanwhile, involves the development of more efficient equipment that requires less space and energy, such as microreactors and high-shear mixers. Enhanced reaction conditions enhance reaction rates and yields using advanced techniques like high-pressure and ultrasonic reactors. Multifunctional reactors, which combine multiple process steps into a single piece of equipment, reduce the number of unit operations and streamline the overall process [12].

(c) ***Design for Environment (DfE)*** strategies are integral to creating products and processes that minimize environmental impact across their entire lifecycle, from raw material extraction to end-of-life disposal. Lifecycle Assessment (LCA) plays a crucial role in this regard, involving comprehensive assessments of the environmental impacts of all stages of a product's lifecycle, as shown in Fig. 13.2. LCA helps identify areas for improvement and supports decision-making for more sustainable practices. Eco-design, which integrates environmental factors into product design, emphasizes using recyclable materials, minimizing material consumption, and designing products for easy disassembly. Material selection, another critical aspect of DfE, involves choosing renewable, recyclable, or biodegradable materials to reduce the environmental impact of raw material extraction and disposal.

(d) ***Energy and water efficiency*** are also critical components of DfE strategies. Manufacturers can significantly reduce their environmental footprint by designing processes that minimize energy and water use, such as implementing

Fig. 13.2 Lifecycle assessment (LCA) integration in zero-waste manufacturing

closed-loop water systems and using energy-efficient equipment. Waste minimization, which involves designing processes to generate less waste and finding ways to repurpose or recycle waste materials, is also crucial. This includes implementing waste-to-energy technologies and using by-products as raw materials for other processes. Sustainable packaging, which involves developing solutions that reduce waste and environmental impact, such as using biodegradable materials, reducing packaging size, and improving packaging efficiency, is another critical consideration.

(e) ***Integrating AI and IoT technologies*** into sustainable process design further enhances the effectiveness of zero-waste manufacturing. Real-time monitoring and control enabled by AI and IoT ensure that manufacturing processes are maintained under optimal conditions, minimizing waste generation. Predictive AI-driven analytics can forecast equipment maintenance needs and process adjustments, preventing waste from equipment failures or inefficiencies. Intelligent resource management, facilitated by IoT sensors, tracks resource usage and identifies opportunities for conservation, reducing material and energy waste. Data-driven optimization, powered by AI algorithms, analyzes vast datasets to identify inefficiencies and optimize processes for better resource utilization and waste reduction [8].

Sustainable process design is fundamental to zero-waste manufacturing in the biotechnology and pharmaceutical industries. By integrating green chemistry and engineering principles, process integration and intensification, and Design for Environment strategies, along with the advanced capabilities of AI and IoT, these industries can significantly reduce their environmental impact, enhance efficiency, and move closer to the goal of zero-waste production.

13.4 Waste Minimization Techniques

Waste minimization is a cornerstone of zero-waste manufacturing, playing a critical role in reducing the environmental impact of industrial processes while enhancing efficiency and sustainability. In the biotechnology and pharmaceutical sectors, implementing effective waste minimization techniques is essential for achieving the ambitious goals of zero-waste manufacturing. These techniques include minimizing source waste, substituting materials, optimizing processes, efficiency improvements, and the recycling and reuse of by-products, each contributing to a more sustainable manufacturing process by reducing waste generation at its source, improving process efficiency, and creating value from waste materials [14].

(a) ***Source reduction and material substitution*** are fundamental strategies in waste minimization. Source reduction involves designing products and processes to generate less waste. For example, product redesign can be employed to require fewer materials or to simplify manufacturing processes, thereby reducing the complexity of product components and enabling modular designs. Precision manufacturing techniques, such as advanced machining and 3D printing technologies, further minimize excess material usage and reduce scrap rates, ensuring that materials are used more efficiently. Batch size optimization is another critical aspect of source reduction, where adjusting batch sizes to align with demand helps minimize overproduction and the waste associated with expired or unsold products.

Material substitution replaces hazardous or non-renewable materials with safer, more sustainable alternatives. Adopting green chemistry principles is central to this approach, allowing manufacturers to replace hazardous chemicals with safer, more sustainable alternatives, such as bio-based solvents and renewable raw materials. Using biodegradable materials is another critical strategy, where non-biodegradable materials are substituted with alternatives that break down more efficiently, thereby reducing environmental impact. Additionally, incorporating recycled content into manufacturing streamlining processes reduces reliance on raw materials and decreases the overall environmental footprint of production by minimizing resource extraction and waste generation.

(b) ***Process optimization and efficiency improvements*** are also crucial in waste minimization, focusing on refining manufacturing processes to operate more sustainably. Lean manufacturing principles, for instance, are employed to identify and eliminate waste in production processes. Techniques such as value stream mapping, Kaizen (ongoing improvement), and 5S (Sort, Set in Order, Shine, Standardize, Sustain) focus on enhancing efficiency and organization in production systems to streamline operations and reduce waste, ensuring that resources are utilized efficiently [15]. Energy and water efficiency are further enhanced by utilizing the use of energy-efficient technologies, like high-efficiency motors and LED lighting, alongside the implementation of energy management systems, which help monitor and reduce energy usage. Similarly, water-saving innovations, such as closed-loop cooling systems and water recycling, contribute to waste reduction and enhance the sustainability of processes [16, 17].

(c) ***Process control and automation*** are essential for enhancing precision and reducing variability in manufacturing processes. Advanced process control and automation technologies facilitate real-time tracking and modification of process parameters, ensuring that manufacturing operations remain within optimal conditions and minimize waste. Yield improvement is another significant focus of process optimization, where optimizing reaction conditions in chemical synthesis, enhancing fermentation efficiency in bioprocessing, and refining purification steps help reduce product losses and maximize output [18].

(d) ***Recycling and reusing by-products*** are critical components of waste minimization, transforming waste into valuable resources and minimizing the demand for raw materials. By-product utilization can be achieved through industrial symbiosis, where partnerships with other industries allow by-products from one manufacturing process to be used as feedstock for another. Internal recycling systems within the same manufacturing facility further reduce waste by reprocessing scrap materials and reclaiming solvents for reuse. Waste-to-energy technologies, including anaerobic digestion, incineration, and gasification, transform waste materials into usable energy, generating electricity, heat, or biofuels from waste and reducing reliance on fossil fuels.

(e) ***Composting and biodegradation processes*** are also employed to convert organic waste into valuable compost or bio-based products, particularly relevant for biodegradable materials and biological waste [19]. Circular economy practices are integral to this approach, involving the development of products and processes that enable the continuous reuse and recycling of materials. This involves designing products for straightforward disassembly and recycling implementing take-back programs to reclaim materials at the end of their lifecycle.

(f) ***Integrating AI and IoT technologies*** plays a transformative role in waste minimization efforts. Predictive maintenance, driven by AI can predict equipment failures and arrange maintenance proactively, preventing breakdowns and minimizing downtime. and waste from defective products. Intelligent inventory management is another critical application, where IoT sensors and AI algorithms optimize inventory levels to prevent overproduction and minimize waste from expired products. Real-time process optimization, enabled by AI systems, continuously analyzes data from IoT sensors to optimize process parameters, ensuring efficient resource utilization and waste reduction. Additionally, AI and IoT technologies provide detailed tracking and analysis of waste streams, identifying opportunities for recycling and reuse, further enhancing waste minimization efforts.

(g) ***Waste minimization techniques*** are vital for achieving zero-waste manufacturing in the biotechnology and pharmaceutical industries. By implementing strategies such as source reduction and material substitution, process optimization and efficiency improvements, and the recycling and reuse of by-products, these industries can significantly reduce their environmental impact, improve process efficiency, and move closer to zero-waste production. Integrating AI and IoT technologies further enhances these efforts, providing real-time monitoring, optimization, and decision-making tools, ultimately contributing to a more sustainable and efficient manufacturing process.

13.5 Innovative Technologies for Waste Reduction

Incorporating innovative technologies that minimize waste and enhance efficiency is essential to achieve zero-waste manufacturing in the biotechnology and pharmaceutical industries. These industries, known for their complex processes and stringent regulatory requirements, benefit greatly from advancements in biocatalysis and enzymatic processes, single-use technologies, and advanced separation and purification technologies. These innovations reduce waste generation and improve process sustainability and manufacturing efficiency, making them critical components of a zero-waste strategy [20].

Biocatalysis and enzymatic processes are pivotal in waste reduction because natural catalysts, such as enzymes, are used to conduct specific and efficient chemical reactions. These processes offer several advantages, starting with high specificity and efficiency. Enzymes are highly selective, catalyzing only the desired reactions without producing unwanted by-products, significantly reducing waste generation and improving the desired product's yield. Furthermore, enzymatic processes typically function under gentle conditions, such as room temperature and neutral pH, which reduces the need for harsh chemicals and minimizes energy consumption. This not only leads to less waste but also lowers the environmental impact of manufacturing processes. Renewable catalysts, such as biodegradable enzymes produced from renewable resources, further enhance sustainability by reducing the waste associated with catalyst disposal. Biocatalysis is widely used in the production of pharmaceuticals, including the production of active pharmaceutical ingredients (APIs) and chiral intermediates, and bioconversion processes, all of which help reduce waste and improve process sustainability [21].

Single-use technologies (SUTs) have gained significant traction in the biotechnology and pharmaceutical industries due to their potential to reduce waste, improve process efficiency, and enhance flexibility. One of the primary benefits of single-use systems is the reduction in cleaning and sterilization requirements between production batches. By eliminating the need for extensive cleaning and sterilization, single-use systems significantly reduce water and chemical usage, leading to less waste generation and a lower environmental impact. Moreover, using disposable components for each batch reduces the likelihood of cross-contamination, thereby improving product safety and quality while reducing waste from contaminated batches and improving overall process efficiency [22]. Single-use technologies also provide greater flexibility and scalability, allowing manufacturers to quickly adapt to changing production needs without extensive reconfiguration. This adaptability reduces waste associated with equipment changeovers and downtime, making single-use systems especially beneficial in sectors where production demands fluctuate significantly.

Examples of single-use technologies include disposable bioreactors used for cell culture and fermentation processes, disposable filters for purification and separation steps, and disposable mixers for formulation and blending operations. While single-use technologies offer numerous benefits, they generate solid waste as disposable components. However, advances in materials science and recycling technologies are helping to mitigate this impact by developing recyclable and biodegradable single-use components, further enhancing the sustainability of these systems [23].

Advanced separation and purification technologies are critical in biopharmaceutical manufacturing, where these steps often involve multiple stages that can generate significant waste, as shown in Fig. 13.3. Innovations in this area, such as membrane filtration and chromatography, have substantially reduced waste and improved process efficiency. Membrane filtration technologies, including ultrafiltration and microfiltration, concentrate and purify biomolecules with high selectivity and efficiency, minimizing extensive solvent usage and reducing waste. Nanofiltration, another membrane-based technology, removes minor impurities and concentrates high-value products with minimal energy input. Chromatography techniques, such as continuous chromatography, enhance the efficiency of separation processes by continuously feeding and eluting products, thereby reducing solvent consumption and waste generation. Affinity chromatography, which uses specific interactions between biomolecules and ligands, achieves high-purity separations with minimal by-product formation, further reducing waste.

Advanced adsorption technologies also contribute significantly to reducing waste during purification processes. Developing novel adsorbent materials with high capacity and selectivity for target molecules reduces the need for extensive downstream processing, thereby minimizing waste. Integrated adsorption processes, combining adsorption with other unit operations, streamline purification and reduce waste. Process intensification techniques, such as continuous processing, equipment miniaturization, and integrating multiple process steps into a single unit, have also

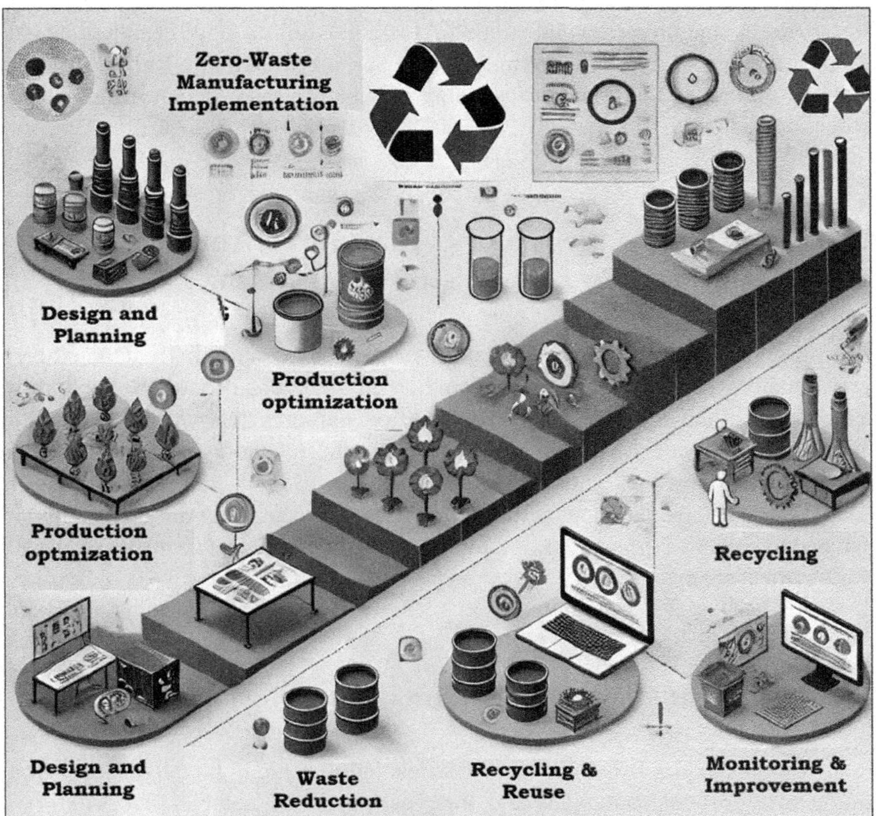

Fig. 13.3 Stages of Zero-Waste Manufacturing Implementation

effectively reduced waste and improved the efficiency of separation and purification processes [20].

Integrating AI and IoT technologies further enhances waste reduction efforts in these innovative processes. AI-driven predictive maintenance, for instance, can predict equipment failures and plan maintenance in advance, preventing breakdowns and minimizing downtime and waste from defective products. Intelligent inventory management, enabled by IoT sensors and AI algorithms, optimizes inventory levels to prevent overproduction and minimize waste from expired products. Real-time process optimization, powered by AI systems, continuously analyzes data from IoT sensors to optimize process parameters, ensuring efficient resource utilization and waste reduction. Additionally, AI and IoT technologies provide detailed tracking and analysis of waste streams, identifying opportunities for recycling and reuse, further contributing to waste minimization in manufacturing processes.

Innovative technologies such as biocatalysis and enzymatic processes, single-use technologies, and advanced separation and purification technologies are essential for achieving zero-waste manufacturing in the biotechnology and pharmaceutical industries. These technologies minimize waste generation and enhance process sustainability and efficiency. Integrating AI and IoT further amplifies these benefits by facilitating real-time monitoring, process optimization, and informed decision-making, ultimately contributing to a more sustainable and efficient manufacturing process.

13.6 Case Studies, Regulatory Frameworks, And Future Perspectives In Zero-Waste Manufacturing

The biotechnology and pharmaceutical industries have increasingly adopted zero-waste manufacturing practices to reduce their environmental footprint and improve operational efficiency. This integrated section explores real-world case studies demonstrating successful zero-waste strategies, the regulatory and policy frameworks that support these initiatives, and future trends that will further drive the adoption of sustainable manufacturing practices. By examining these elements, we can understand the current state and future direction of zero-waste manufacturing in these critical sectors.

13.6.1 Case Studies in Zero-Waste Biomanufacturing

Real-world examples from the pharmaceutical industry highlight the practical application of zero-waste strategies and their benefits. One notable success story is Amgen's implementation of single-use technologies (SUTs), which led to a 60% reduction in water usage and a 40% decrease in chemical use. This transition enhanced operational flexibility, resulted in quicker changeovers, and reduced downtime, illustrating the significant environmental and operational benefits of SUTs [1]. Another example is Genzyme's application of green chemistry principles to enzyme production, which achieved a 30% reduction in hazardous waste and a 25% increase in yield. These improvements reduced the environmental impact of their operations and enhanced process efficiency. Pfizer's Kalamazoo facility offers another compelling case: adopting a waste-to-energy program converted organic waste into biogas, leading to a 20% reduction in fossil fuel consumption and a 15% decrease in greenhouse gas emissions. These case studies demonstrate that with the right strategies and technologies, significant environmental and economic benefits can be realized in biomanufacturing.

13.6.2 Lessons Learned and Best Practices

Several key best practices can be derived from these case studies as critical for successful zero-waste initiatives. Cross-functional collaboration requires early engagement from all stakeholders to address challenges and integrate solutions effectively. Continuous improvement is also vital, as regular process reviews and optimizations lead to incremental waste reduction and sustainability gains. Training employees in waste reduction techniques ensure their buy-in and active participation, which is essential for recognizing and applying waste reduction strategies and opportunities. Additionally, maintaining regulatory compliance involves thorough documentation, process validation, and clear communication with regulatory bodies. Investing in advanced technologies such as biocatalysis and single-use systems offers significant long-term benefits, enhancing waste reduction efforts across the board.

13.6.3 Quantitative Analysis of Waste Reduction Achievements

The quantitative impact of these initiatives is significant. For instance, Amgen's implementation of SUTs resulted in a 60% reduction in water use and a 40% decrease in the need for cleaning chemicals. Genzyme's green chemistry approach led to a 30% reduction in hazardous waste and a 25% increase in yield. Pfizer's waste-to-energy initiative reduced fossil fuel consumption by 20% and decreased greenhouse gas emissions by 15%. These metrics underscore the effectiveness of targeted waste reduction strategies in achieving environmental and economic benefits. Furthermore, Pfizer's biogas production from organic waste significantly reduced landfill dependency while providing renewable energy, illustrating the broader impact of innovative waste management solutions.

13.6.4 The Role of AI and IoT in Waste Reduction

Advanced technologies such as AI and IoT are pivotal in enhancing waste reduction efforts. AI-driven predictive maintenance minimizes downtime by Predicting equipment failures and arranging maintenance before breakdowns happen. This not only reduces waste from defective products but also extends equipment lifespan. IoT devices are critical for real-time monitoring, ensuring optimal environmental conditions, and minimizing waste throughout the manufacturing process. Additionally, AI and IoT are instrumental in optimizing resource management, reducing overproduction, and minimizing waste from expired products. These technologies provide detailed tracking and analysis of waste streams, identifying opportunities for recycling and reuse and contributing significantly to manufacturing processes' overall efficiency and sustainability.

13.6.5 Regulatory and Policy Frameworks

A robust regulatory and policy framework is essential for guiding and supporting zero-waste manufacturing initiatives in the biotechnology and pharmaceutical industries. Various regulations and guidelines have been established by governmental and international bodies to ensure environmental protection, waste management, and sustainable manufacturing practices. For example, the Environmental Protection Agency (EPA) administers the Resource Conservation and Recovery Act (RCRA), which regulates the management of both hazardous and non-hazardous solid waste, emphasizing waste minimization and proper disposal to protect human health and the environment [6]. The Clean Air Act and Clean Water Act also set standards to limit pollutants from industrial processes. In Europe, the European Medicines Agency (EMA) provides Good Manufacturing Practice (GMP) guidelines that include waste management practices to ensure safety and environmental compliance, along with Environmental Risk Assessment (ERA) guidelines for pharmaceutical companies to assess and mitigate their environmental impact. The International Conference on Harmonisation (ICH) also offers guidance through ICH Q7 and ICH Q9, focusing on GMP for active pharmaceutical ingredients and quality risk management, respectively, which includes waste reduction measures.

13.6.6 Policies Promoting Zero-Waste Manufacturing

Governments and international organizations have introduced policies and initiatives to encourage zero-waste manufacturing and support the shift towards a circular economy. The European Green Deal is a notable policy initiative aiming to make

Europe climate-neutral by 2050, including measures to promote sustainable manufacturing, reduce waste, and increase recycling and reuse of materials. The Circular Economy Action Plan, part of the European Green Deal, details strategies for shifting from a linear to a circular economy, emphasizing sustainable product design, waste reduction, and resource efficiency. In the United States, the EPA's Sustainable Materials Management (SMM) program encourages the sustainable use of materials throughout their lifecycle, promoting waste minimization, recycling, and using renewable resources in manufacturing processes. Extended Producer Responsibility (EPR) policies require manufacturers to take responsibility for the full lifecycle of their products, including their management at the end of life, encouraging waste reduction and recycling efforts.

13.6.7 Industry Standards and Certifications

Industry standards and certifications are crucial in promoting and validating zero-waste manufacturing practices, ensuring compliance with best practices and regulatory requirements. ISO 14001 provides a framework for organizations to manage and reduce their environmental impact through systematic waste management, resource conservation, and continuous improvement. ISO 50001 offers guidelines for establishing, implementing, and maintaining an energy management system designed to enhance energy efficiency and lower greenhouse gas emissions. LEED Certification (Leadership in Energy and Environmental Design) is an internationally recognized standard for sustainable buildings and manufacturing facilities, promoting sustainable site development, water and energy conservation, material selection, and waste minimization. Green Chemistry Certifications, such as the Green Chemistry Challenge Awards, recognize innovations in chemical processes that reduce or eliminate hazardous substances, promoting safer and more sustainable manufacturing practices. B Corp Certification recognizes companies that meet stringent social and environmental performance criteria, showcasing their dedication to sustainability and ethical business practices, sustainability, including waste reduction and resource efficiency. The Global Reporting Initiative (GRI) offers a robust framework for sustainability reporting that enables companies to evaluate and convey their environmental, social, and economic performance, including initiatives related to waste management.

13.6.8 Integration of AI and IoT in Regulatory and Policy Frameworks

AI and IoT technologies are increasingly integrated into regulatory and policy frameworks to enhance monitoring, compliance, and efficiency of zero-waste manufacturing. AI algorithms can analyze large datasets to track adherence to environmental regulations. and waste management practices, helping identify deviations and optimize processes for reduced waste generation. IoT Sensors and devices facilitate real-time tracking and monitoring of environmental parameters, waste streams, and resource usage in manufacturing facilities, facilitating proactive decision-making and continuous improvement in waste reduction efforts. AI-driven predictive models forecast waste generation patterns based on historical data and operational variables, enabling proactive adjustments in Manufacturing processes to reduce waste and maximize resource efficiency. Automation and optimization powered by AI and IoT systems streamline operations, optimize energy and resource usage, and reduce human error in waste management practices, enhancing overall efficiency and sustainability in manufacturing.

13.6.9 Economic and Environmental Benefits

Zero-waste manufacturing in the biotechnology and pharmaceutical industries promotes environmental sustainability and offers significant economic advantages. Implementing zero-waste strategies can result in significant cost savings by minimizing waste production, optimizing resource use, and improving process efficiencies. For example, efficient use of raw materials through process optimization and waste minimization reduces the need for additional material purchases, leading to lower production costs and improved profitability [24]. Minimizing waste generation also reduces the costs associated with waste disposal, including transportation, treatment, and landfill fees, while energy-efficient technologies and water-saving measures reduce utility bills. Reducing waste into valuable by-products or energy can create new revenue streams, offsetting production costs. Adhering to environmental regulations and reducing hazardous waste helps companies avoid fines and penalties associated with non-compliance, enhancing a company's reputation and market position.

13.6.10 Environmental Impact Assessment

Assessing the environmental impact of manufacturing processes is crucial for understanding and mitigating their ecological footprint. Zero-waste manufacturing significantly reduces environmental impact in several key areas. Implementing waste reduction and energy-efficient practices lowers greenhouse gas emissions associated with manufacturing processes while reducing hazardous waste generation and adopting green chemistry principles to help prevent contamination of water bodies and soil. These practices also contribute to biodiversity preservation by reducing pollution and conserving natural resources, minimizing the impact on wildlife and habitats [25]. Zero-waste practices focus on recycling and reusing waste materials, diverting them from landfills, which reduces landfill usage, decreases methane emissions from waste decomposition, and extends landfill lifespans.

13.6.11 Long-Term Sustainability Benefits

Zero-waste manufacturing practices offer long-term benefits beyond immediate economic and environmental gains. These include enhanced resilience, improved stakeholder relationships, and contributions to global sustainability goals. Zero-waste practices improve a company's resilience to supply chain disruptions and resource scarcity, allowing better navigation of raw material availability and price fluctuations [26]. Demonstrating a commitment to sustainability strengthens relationships with stakeholders, including customers, investors, and regulatory bodies, leading to enhanced brand loyalty, customer trust, and investor confidence [27]. Zero-waste manufacturing aligned with global sustainability efforts, such as the United Nations Sustainable Development Goals (SDGs), which play a crucial role in addressing climate change, preserving ecosystems, and ensure responsible consumption and production. Although the initial investment in zero-waste technologies and practices may be high, the long-term cost savings from reduced waste disposal, lower raw material usage, and increased process efficiency outweigh the upfront costs, leading to sustained economic benefits and improved financial performance [28].

13.6.12 Future Perspectives and Innovations

The drive towards zero-waste manufacturing in the biotechnology and pharmaceutical industries continues to evolve with emerging trends, digital technologies, and collaborative approaches. A key emerging trend is the adoption of the circular economy model, which focuses on reusing, recycling, and repurposing materials. Companies increasingly integrate lifecycle considerations into product and process design to reduce waste production and enhance resource efficiency. Advances in green chemistry enable the development of safer, more sustainable chemical processes and reduce manufacturing processes' environmental impact [29]. Biotechnology innovations, including synthetic biology, metabolic engineering, and cell-free synthesis, allow more precise and efficient production processes, reducing waste and improving yield. Eco-friendly packaging solutions, including biodegradable and compostable materials, are increasingly gaining popularity, further minimizing waste in the manufacturing process. Emerging energy-efficient technologies, such as advanced bioreactors, improved fermentation methods, and low-energy separation techniques, are reducing the energy consumption of manufacturing processes, contributing to overall sustainability.

13.6.13 Role of Digital Technologies and Collaborative Approaches

Digital technologies, particularly those associated with Industry 4.0, are transforming manufacturing processes. Intelligent sensors and real-time data analytics enable precise monitoring and control, optimizing resource use and minimizing waste. Digital twin technology, which creates virtual replicas of physical manufacturing processes, allows companies to optimize processes, predict and mitigate potential issues, and reduce waste before implementing changes in the real world [30]. Additive manufacturing (3D printing) is being used to produce customized components with minimal material waste, reducing the need for excess inventory and storage. Blockchain technology improves supply chain transparency and traceability, safeguarding the integrity of transactions and data quality of materials and products that are sustainably sourced and managed, verifying compliance with zero-waste practices, and improving overall supply chain efficiency. Advanced process control (APC) systems, which optimize manufacturing processes using real-time data and advanced algorithms, continuously adjust process parameters to minimize waste, improve product quality, and enhance efficiency.

Collaborative approaches, including industry consortia, partnerships, and public–private collaborations, are essential for advancing zero-waste manufacturing. These collaborations facilitate sharing best practices, technological innovations, and research findings, accelerating the adoption of sustainable manufacturing practices. Open innovation platforms encourage exchanging ideas and solutions across the industry, enabling companies to collaborate on developing and implementing zero-waste technologies and processes. Standardization and harmonization of industry-wide standards and regulations can further accelerate the adoption of zero-waste practices by providing clear guidelines and benchmarks for companies to measure and improve their sustainability performance. Knowledge-sharing initiatives, such as workshops, webinars, and training programs, are also crucial for educating industry professionals on the latest zero-waste technologies and practices, fostering a culture of sustainability within organizations.

Integrating case studies, regulatory frameworks, and future perspectives highlights the significant progress toward zero-waste manufacturing in the biotechnology and pharmaceutical industries. By adopting innovative technologies, leveraging digital tools, and fostering collaborative approaches, these industries are well-positioned to continue leading the way in sustainable manufacturing, contributing to economic growth and environmental stewardship.

13.7 Conclusion

Zero-waste manufacturing is essential for the biotechnology and pharmaceutical industries to meet sustainability goals. Companies can significantly reduce waste and minimize environmental impact by optimizing resource use, integrating green chemistry, and adopting advanced technologies. Adopting Industry 4.0 technologies such as IoT, AI, and digital twins further enhances these efforts; blockchain technology improves supply chain transparency and traceability, safeguarding the integrity of transactions and data. Collaborative approaches, including industry and public–private collaborations, are critical for scaling these practices and establishing sustainable industry standards. The future of zero-waste manufacturing is promising, driven by continuous technological innovation and a solid commitment to sustainability. To fully realize its potential, industry stakeholders Must allocate resources towards research and development, continuously refine processes, and provide their workforce with the essential skills. Supportive regulations and public–private partnerships will accelerate the transition to zero-waste manufacturing. By embracing Industry 4.0, fostering collaboration, and committing to sustainability, the biotechnology and pharmaceutical industries Can improve operational efficiency, lower costs, and support a more sustainable future.

References

1. Zaman, A.: A new sustainability paradigm for addressing the global waste problem. In: Björnberg, K.E., Matts-Åke, B., Hansson, S.O., Tingvall, C. (eds.) The Vision Zero Handbook: Theory, Technology, and Management for a Zero Casualty Policy. Springer, Singapore (2022)
2. Demets, R., Roosen, M., Vandermeersch, L., Ragaert, K., Walgraeve, C., De Meester, S.: Development and application of an analytical method to quantify odour removal in plastic waste recycling processes. Resour. Conserv. Recycl.. Conserv. Recycl. **161**(104907), 104907 (2020). https://doi.org/10.1016/j.resconrec.2020.104907
3. Singh, S., Ramakrishna, S., Gupta, M.K.: Towards zero waste manufacturing: a multidisciplinary review. J. Clean. Prod. **168**, 1230–1243 (2017). https://doi.org/10.1016/j.jclepro.2017.09.108
4. Yaashikaa, P.R., Priyanka, B., Senthil Kumar, P., Karishma, S., Jeevanantham, S., Indraganti, S.: A review on recent advancements in recovery of valuable and toxic metals from e-waste using bioleaching approach. Chemosphere **287**(Pt 2), 132230 (2022). https://doi.org/10.1016/j.chemosphere.2021.132230

5. Awogbemi, O., Kallon, D.V.V., Bello, K.A.: Resource recycling with the aim of achieving zero-waste manufacturing. Sustainability **14**(8), 4503 (2022). https://doi.org/10.3390/su14084503
6. Bhatti, R.M.: Assessing the role of policy frameworks in sustainable food systems for zero hunger by 2030 (Unpublished) (2023). https://doi.org/10.13140/RG.2.2.36393.36968
7. Frank, A.G., Dalenogare, L.S., Ayala, N.F.: Industry 4.0 technologies: implementation patterns in manufacturing companies. Int. J. Prod. Econ. **210**, 15–26 (2019). https://doi.org/10.1016/j.ijpe.2019.01.004
8. Arden, N.S., Fisher, A.C., Tyner, K., Yu, L.X., Lee, S.L., Kopcha, M.: Industry 4.0 for pharmaceutical manufacturing: preparing for the smart factories of the future. Int. J. Pharm. **602**(120554), 120554 (2021). https://doi.org/10.1016/j.ijpharm.2021.120554
9. Pietzsch, N., Ribeiro, J.L.D., de Medeiros, J.F.: Benefits, challenges and critical factors of success for zero waste: a systematic literature review. Waste Manage. **67**, 324–353 (2017). https://doi.org/10.1016/j.wasman.2017.05.004
10. Geigert, J.: Starting materials for manufacturing the biopharmaceutical drug Substance. In: The Challenge of CMC Regulatory Compliance for Biopharmaceuticals, pp. 183–229. Springer Nature Switzerland, Cham (2023)
11. Jozala, A.F., Geraldes, D.C., Tundisi, L.L., Feitosa, V.A., Breyer, C.A., Cardoso, S.L., Mazzola, P.G., et al.: Biopharmaceuticals from microorganisms: from production to purification. Braz. J. Microbiol. **47**, 51–63 (2016). https://doi.org/10.1016/j.bjm.2016.10.007
12. Singhal, T.S., Jain, J.K., Ramacharyulu, D.A, Jain, A., Abdul-Zahra, D.A., Manjunatha, M., Srivastava, A.P.: Eco-design of products and processes: a review on principles and tools for sustainable manufacturing. E3S Web of Conferences **505**, 01033 (2024). https://doi.org/10.1051/e3sconf/202450501033
13. Kumar, N., Hasan, S.S., Srivastava, K., Akhtar, R., Yadav, R.K., Choubey, V.K.: Lean manufacturing techniques and its implementation: a review. Mater. Today Proc. **64**, 1188–1192 (2022). https://doi.org/10.1016/j.matpr.2022.03.481
14. Gorina, L., Korneeva, E., Kovaleva, O., Strielkowski, W.: Energy-saving technologies and energy efficiency in the post-covid era. Sustain. Dev. (2024). https://doi.org/10.1002/sd.2978
15. Kandpal, V., Jaswal, A., Santibanez Gonzalez, E.D.R., Agarwal, N.: Circular economy principles: shifting towards sustainable prosperity. In: Circular Economy and Sustainability, pp. 125–165. Springer Nature Switzerland, Cham (2024)
16. Betchem, G., Dabbour, M., Tuly, J.A., Billong, L.F., Ma, H.: Optimization of fermentation conditions to improve the functional and structural characteristics of rapeseed meal with a mutant bacillus subtilis species. Ind. Crops Prod. **205**(117424), 117424 (2023). https://doi.org/10.1016/j.indcrop.2023.117424
17. Hassan, N.Y.I., Abd El Wahed, N.H., Abdelhamid, A.N., Ashraf, M., Abdelfattah, E.A.: Composting: an eco-friendly solution for organic waste management to mitigate the effects of climate change. Innovare J. Soc. Sci., 1–7 (2023). https://doi.org/10.22159/ijss.2023.v11i4.48529
18. Rena, S.Y., Patel, S., Killedar, D.J., Kumar, S., Kumar, R.: Eco-innovations and sustainability in solid waste management: an indian upfront in technological, organizational, start-ups and financial framework. J. Environ. Manage. **302**(113953), 113953 (2022). https://doi.org/10.1016/j.jenvman.2021.113953
19. Zhang, N., Domínguez de María, P., Kara, S.: Biocatalysis for the synthesis of active pharmaceutical ingredients in deep eutectic solvents: state-of-the-art and prospects. Catalysts **14**(1), 84 (2024)
20. Mizuno, M., Yori, K., Takeuchi, T., Yamaguchi, T., Watanabe, K., Tomaru, Y., Shimizu, N., Sekiya, I.: Cross-contamination risk and decontamination during changeover after cell-product processing. Regen. Ther. **22**, 30–38 (2023). https://doi.org/10.1016/j.reth.2022.12.003
21. Mazzei, H.G., Specchia, S.: Latest insights on technologies for the treatment of solid medical waste: a review. J. Environ. Chem. Eng. **11**(2), 109309 (2023). https://doi.org/10.1016/j.jece.2023.109309

22. Kar, K., Chowdhury, S., Chakraborty, P., Saha, A.: Sustainable practices in the pharmaceutical industry: development and adoption. In: Preserving Health, Preserving Earth, pp. 11–33. Springer Nature Switzerland, Cham (2024)
23. Hegab, H., Shaban, I., Jamil, M., Khanna, N.: Toward sustainable future: strategies, indicators, and challenges for implementing sustainable production systems. Sustain. Mater. Technol. **36**(e00617), e00617 (2023). https://doi.org/10.1016/j.susmat.2023.e00617
24. Katsaliaki, K., Galetsi, P., Kumar, S.: Supply chain disruptions and resilience: a major review and future research agenda. Ann. Oper. Res.Oper. Res. **319**(1), 965–1002 (2022). https://doi.org/10.1007/s10479-020-03912-1
25. Nonet, G.A.H., Gössling, T., Van Tulder, R., Bryson, J.M.: Multi-stakeholder engagement for the sustainable development goals: introduction to the special issue. J. Bus. Ethics **180**(4), 945–957 (2022). https://doi.org/10.1007/s10551-022-05192-0
26. Dahmani, N., Benhida, K., Belhadi, A., Kamble, S., Elfezazi, S., Jauhar, S.K.: Smart circular product design strategies towards eco-effective production systems: a lean eco-design industry 4.0 framework. J. Clean. Prod. **320**(128847), 128847 (2021). https://doi.org/10.1016/j.jclepro.2021.128847
27. Pandey, K., Pandey, M., Kumar, V., Aggarwal, U., Singhal, B.: Bioprocessing 4.0 in biomanufacturing: paving the way for sustainable bioeconomy. Syst. Microbiol. Biomanuf. (2023). https://doi.org/10.1007/s43393-023-00206-y
28. Attaran, M., Celik, B.G.: Digital twin: benefits, use cases, challenges, and opportunities. Decis. Anal. J. **6**(100165), 100165 (2023). https://doi.org/10.1016/j.dajour.2023.100165
29. Kharat, V.J., Singh, P., Raju, G.S., Yadav, D.K., Gupta, M.S., Arun, V., Majeed, A.H., Singh, N.: Additive manufacturing (3D printing): a review of materials, methods, applications, and challenges. Mater. Today Proc. (2023). https://doi.org/10.1016/j.matpr.2023.11.033
30. Varma, A., Dixit, N., Ray, S., Kaur, J.: Blockchain technology for sustainable supply chains: a comprehensive review and future prospects. World J. Adv. Res. Rev. **21**(3), 980–994 (2024). https://doi.org/10.30574/wjarr.2024.21.3.0804

Dr. Polaki Suman is an Assistant Professor in the Department of Biotechnology at M.S. Swaminathan School of Agriculture, Centurion University of Technology and Management, Parlakhemundi, Odisha, India. His research interests include biotechnology applications in agriculture, aquaculture, and the pharmaceutical industry. He has actively contributed to academic and research activities. He has published 26 research articles, 6 conference papers, 5 book chapters, and 1 book.

Dr. B. Rabi Prasad serves as an Assistant Professor in the Department of Biotechnology, School of Engineering and Technology, GIET University, Gunupur, Odisha, India. He specializes in biotechnology and its interdisciplinary applications, contributing to various research endeavors.

Dr. G. Pushpalatha is a Professor in the Department of Biotechnology and Dean of Centurion School of Smart Agriculture, Centurion University of Technology and Management, Parlakhemundi, Odisha, India. Coordinator for Tissue Culture and Vegetative Propagation Research Center. She is dedicated to advancing education and research in biotechnology, focusing on sustainable agricultural practices.

Dr. Ch. Mohan Babu, Assistant Professor and Head of the School of Management, Centurion University, Vizianagaram, Andhra Pradesh, India. With a focus on management education and interdisciplinary approaches, he brings a unique perspective to academic research and institutional development.

Dr. Islauddin Khan is a Postdoctoral Research Associate at UIC, Chicago, USA. His expertise lies in advanced research methodologies and innovative biotechnological applications. He is an active contributor to international scientific research communities.

Chapter 14
Sustainable Cooling and Lubrication Techniques Used in Grinding for Improving Machinability

Sunil Kumar Singh, Ashok Kumar Jha, and Amitava Ray

Abstract Grinding is a versatile and popular finishing operation for achieving high surface quality products within a precise tolerance. It is primarily used in modern industries to grind superalloys, ceramics, composites and many difficult-to-cut materials that cannot easily machined by other traditional machining methods. Despite positive grinding outputs, this process suffers serious drawbacks, including excessive heat generation, which is more than ten times that of other traditional machining operations. To degrade temperature, conventional cutting fluid is applied in the compact grinding zone. Unfortunately, conventional cutting fluids are mostly petroleum-based and mainly extracted from crude oil. Furthermore, conventional fluids react at elevated temperatures, leading to environmental pollution and increases workers health concerns. All these factors are responsible for exploring sustainable alternatives to improve cutting fluid management and detract all issues by maintaining maximum efficiency and quality. The sustainable concept reduces environmental impact, minimizes fluid wastage, uses biodegradable cutting oils and enhances economic and social benefits. This chapter highlights the important sustainable concepts used in modern grinding processes, including vegetable oil lubricants, minimum quantity lubrication (MQL), green nanofluids, cryogenic grinding, ultrasonic-assisted grinding, and hybrid grinding systems. Results show that sustainable processes decrease heat generation, improve surface texture, increase abrasive wheel life and reduce grinding forces. Therefore, it can be an alternative to conventional mineral-based cutting fluid for increasing grindability and minimizing environmental problems.

S. K. Singh (✉) · A. K. Jha
Department of Production and Industrial Engineering, National Institute of Technology, Jamshedpur, Jharkhand, India
e-mail: singhsunilkumar555@gmail.com

A. K. Jha
e-mail: ashokjha.prod@nitjsr.ac.in

A. Ray
Department of Mechanical Engineering, Jalpaiguri Government Engineering College, Jalpaiguri, West Bengal, India

© The Author(s), under exclusive license to Springer Nature Switzerland AG 2025

K. C. Rath et al. (eds.), *Engineering, Technology and Management*, Information Systems Engineering and Management 54, https://doi.org/10.1007/978-3-031-93654-8_14

Keywords Grinding · Conventional fluids · Sustainable · Vegetable oil-based lubricants · Minimum quantity lubrication · Cryogenic

Abbreviations

CDG	Conventional dry grinding
UADG	Ultrasonic-assisted dry grinding
CG	Conventional grinding
CCF	Conventional cutting fluid
MMF	Metal working fluid
UAG	Ultrasonic-assisted grinding
SQL	Small quantity lubrication
CFG	Creep feed grinding
MQL	Minimum quantity lubrication
UACFG	Ultrasonic-assisted creep feed grinding
LCO_2	Liquid carbon dioxide
CBN	Cubic boron nitrite
LN_2	Liquid nitrogen
NFs	Nanofluids
NPs	Nanoparticles
CNT	Carbon nanotube
$scCO_2$	Supercritical carbon dioxide
CA	Cryogenic air
CA+NFMQL	Cryogenic air-assisted Nanofluid minimum quantity lubrication
NFMQL	Nanofluids minimum quantity lubrication

Preface

This chapter addresses thermal challenges in the grinding process, particularly when machining hard-to-cut materials. The excessive heat generated during grinding necessitates a decrease in surface textures. Therefore, effective cooling and lubrication are used to enhance grindability and to ensure superior performance. Traditionally, conventional cutting fluids have been employed in large quantities to address these challenges. However, their usage poses significant environmental risks due to chemical reactions at high temperatures and disposal issues. This chapter underscores the importance of sustainable, eco-friendly cutting fluids as a viable alternative. By exploring innovative cooling and lubrication techniques, it emphasizes how adopting sustainable practices can eliminate reliance on conventional fluids, minimize fluid wastage, and pave the way for greener manufacturing processes.

14.1 Introduction to Grinding

Grinding is an important machining process that removes unwanted materials from the workpiece to obtain desired dimensions with a high surface quality. Material is removed as microchips by sharp and hard abrasive grits of multiple point cutting tool rotated at high velocity. These grits are firmly held in a circular wheel by suitable bond materials. Different types of abrasive grits and bond materials are selected by considering the mechanical properties of workpieces. Depending upon the machining and surface required, five types of grinding operations are mostly used in modern industries including, surface grinding, centreless grinding, internal grinding, creep-feed grinding, and cylindrical grinding. In recent years, significant progress in grinding has ever developed due to rapid growth of industrialization worldwide, which accounts for about 30% of the total machining costs.

Material removal in grinding is achieved mainly through shearing action of sharp and hard abrasive grits of grinding wheel, followed by ploughing and rubbing phenomenon. Due to continuous machining, the cutting edge of many grits becomes blunt, which is responsible for increased grinding forces. This force is further used by grits to remove it completely from the wheel and expose a new one. Thus, a grinding wheel possesses self-sharpening characteristics.

The grinding wheels abrasive grits rake angle primally lies between −30 and −75°, which is highly negative. Hence, it is mainly responsible for high specific energy. To minimize its effects, the wheel velocity is kept very high. As the rake becomes less negative during grinding operations, specific energy starts decreasing, which accounts for an increase in MRR. However, specific energy starts increasing after a certain value of MRR, as demonstrated in Fig. 14.1. This may be because of chip accommodation problems, which are responsible for high rubbing action in the grinding zone. The chip formation system in conventional machining is straightforward and predictable due to well-defined tool geometry. However, the geometry of cutting edges is unfavourable and random in grinding, making the chip formation process more complex [1].

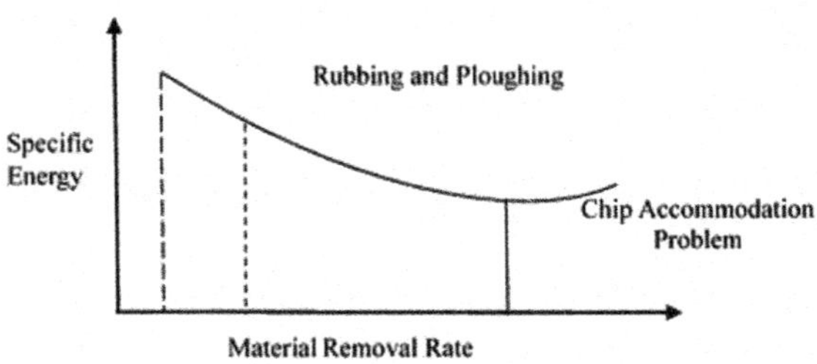

Fig. 14.1 Fluctuation of specific energy concerning material removal rate (MRR)

14.1.1 Thermal Problems in Grinding

Grinding inherently generates high heat levels, more than ten times that of the traditional metal-cutting processes like turning and milling. This occurs mainly due to a lack of proper lubrication, which causes random rubbing action of grits against workpiece surfaces. Furthermore, the rotational velocity of grinding wheel is very high and contains extreme kinetic energy (K.E). During grinding, maximum K.E is converted into heat energy and transferred to the grind materials finished surface, which deteriorates surface integrity and reduces grindability performance. High thermal-related problems are the root cause of most grinding processes, especially for advanced materials like ceramic, titanium, and superalloy. In modern industrialization, the demand for these materials is receiving more attention, and they are extensively used in aviation, biomedical, nuclear, automobile, defence, and other high-production industrial sectors. This is due to their unique characteristics, like high mechanical strength, lightweight, ductility, excellent corrosion resistance, good fatigue strength at low temperatures, and high melting point. Unfortunately, advanced materials typically exhibit high hardness and low thermal conductivity, thus reducing grindability performance.

The thermal effect is mainly generated from the resistance encountered by the work material during the shearing process, along with the frictional resistance. The heat produced is distributed in several ways. Generally, it is said that if the cut arc is less than 1.5 mm, only about 8% to 10% of heat energy is removed by the chip or coolant, and the remaining (more than 90%) is transferred between the wheel and workpiece. However, the absorbability of heat by coolant can be increased if cut arc is more. Depending upon the thermal conductivity of workpiece and wheel, the heat is distributed, but majority is transferred to the finished portion of work material. Consequently, the finished surface may experience high temperatures, which increases grinding forces, surface burn, extensive wheel loading, excessive residual stresses, and phase deformation. Furthermore, elevated temperatures at the grinding zone can lead to dimensional deviation, subsurface microcracks, high thermal residual stress, and rapid oxidation of the workpiece. Therefore, effective cooling and lubrication are vital, and modern research in this area is needed to minimise these issues.

14.1.2 Conventional Cutting Fluids and Its Challenges

Conventional cutting fluids (CCF) or metalworking fluids (MMF) are used to overcome thermal-related problems. For years, substantial amounts of cutting fluids have been used to enhance grindability. Its function is to cool the grinding area, remove chips, and provide sufficient lubrication. Generally, semi-synthetic, synthetic and emulsion oils are mostly used as grinding lubricants. The lubricant also provides a thin layer that protects the freshly ground surface from contamination by harmful gases present in the atmosphere. Moreover, grinding advanced material requires a significantly high amount of cutting fluid, which increases expenses by 20–30% of total manufacturing costs. The majority of grinding lubricants are extracted from mineral oils containing toxic chemicals. The presence of nitrosamines, hydrocarbons, alkanolamines, chlorinated additives, bactericides and heavy metals in cutting fluid leads to various techno-environment issues, including:

(a) At elevated temperatures, cutting fluid reacts, and chemical breakdown takes place, which pollutes the environment by releasing harmful gases like CO and CO_2
(b) Health problems for operators who inhale toxic chemical gases.
(c) Contamination of soil and water during disposal.
(d) Needs extra floor space for pumping, storage, cooling, filtration systems, etc.

On the other hand, conventional cutting fluid is delivered in large quantities with a certain amount of pressure in the grinding zone. However, most cutting fluid cannot penetrate the compact grinding zone and is wasted. This occurs due to the formation of a rigid air layer surrounding the grinding wheel. It is formed due to the centrifugal force of high rotating speed of porous grinding wheel, which deviates grinding fluid away from the machining area.

Additionally, the wheel abrasive grits, which gets detached during the grinding operation, contribute an extra environmental burden. During grinding, only small amounts of abrasive cutting edge particles are used in machining, and the majority of cutting edges produce heat through frictional effects on the freshly ground surface of work materials. Grinding also related to high rotational speed, which accounts for high noise and vibration. These overall difficulties deteriorate surface integrity and increase overall machining costs. To address the limitations of traditional grinding methods, several researchers have focused on sustainable techniques to raise grindability and develop environmentally friendly alternatives [2].

14.1.3 Sustainability in Grinding

Sustainability in the machining process involves maintaining the social, economic, environmental, and technological aspects over an extended duration. It refers to an organization's capability to meet necessary demands while preserving the environment and natural resources. Sustainable fluids are mostly vegetable oil-based liquids derived from plants, water, and other raw ingredients. These biodegradable fluids are used to cool and lubricate the grinding zone.

Commonly used vegetable-based cutting fluids that are extracted from plants are coconut, groundnut, palm, soybean, castor, and sunflower. Sustainable or green grinding involves various methods to maintain ecological factors without hampering grindability performance; popularly used sustainable processes are vegetable oils-based cutting fluids, MQL, green nanofluid, cryogenic cooling, and ultrasonic assisted grinding, as shown in Fig. 14.2.

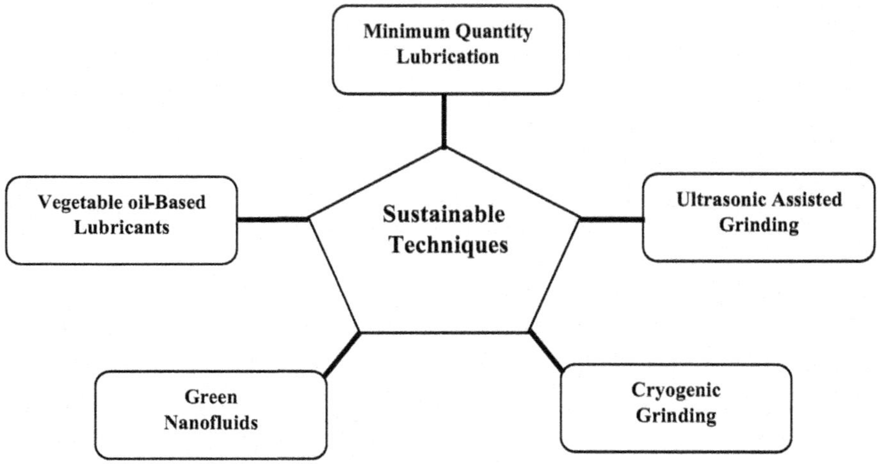

Fig. 14.2 Sustainable techniques used in grinding

14.1.4 Objectives

Sustainable grinding primarily aims to minimize waste and reduce harmful environmental impact while conserving and recycling from natural resources. The key objectives of the sustainable grinding process are:

(a) To replace mineral-based cutting fluid with biodegradable cutting fluid.
(b) Reduce harmful environmental impact.
(c) Increase safety for the operators.
(d) To reduce wastage of cutting fluid using MQL or Nanofluid-based MQL technology.
(e) To maximize sustainable hybrid cooling and lubrication techniques for grinding advanced materials.

14.1.5 Scope of the Work

Sustainable cooling and lubrication in grinding have been the subject of recent research. Modern manufacturing faces several sustainability-related challenges. Adopting this technology to minimize time, costs, and waste is crucial while considering superior surface quality and environmental impact. Many industries still rely on conventional cutting fluids and use them in large quantities. All these shortcomings significantly represent a higher scope to replace them with biodegradable fluids to enhance economic and environmentally friendly machining processes in grinding. Effective techniques like MQL and ultrasonic-assisted wet grinding, which reduces the amount of cutting fluid, must be replaced with flood cooling systems. Hybrid cooling and lubrication systems such as MQL+cryogenic or nanofluids+cryogenic must be more precisely developed and should be implemented in modern industries. This chapter elaborated on different sustainable techniques. It highlights the research findings and significant advancements that can provide a clear insight into the sustainable grinding process.

14.2 Sustainable Lubrication in Grinding

Sustainable lubrication is mainly made from vegetable oils, animal fats and water-based liquids. These are biodegradable and green lubricants that possess a high capacity to reduce friction and transfer heat from the machining area. Different types of vegetable oil and their effects on the material are elaborated. This section also highlights modern techniques to increase the efficiency of sustainable lubricants.

14.2.1 Vegetable Oil-Based Lubricants in Grinding

During second half of nineteenth century, the requirement for lubricating oils increased due to rapid growth of industrialization globally. Many industries use mineral oils as they provide many favourable properties of lubricants at a low cost and are produced in large quantities. Still, many top industries use petroleum-based mineral oils for cooling and lubrication purposes. Due to environmental protection, eco-friendly fluids have been in great demand. This biodegradable fluid is prepared mainly from plants and water, known as vegetable oils. Moreover, cutting fluids based on vegetable oil decreases environmental pollution and enhances sustainability [3]. Some crucial considerations are taken into account when choosing cutting fluid based on vegetable oil, including:

(a) It should be resistant to oxidation, or an additive should be used to increase its oxidation stability.
(b) It should not react with the workpiece or the grinding wheel.
(c) It should have the lowest possible viscosity.
(d) Vegetable oils and additives should be cheaper than conventional cutting fluids.

Generally, vegetable oils can be used in two different ways for cooling and lubrication purposes: firstly, by direct utilization without mixing, and secondly, by mixing a small portion of nanoparticles or commercial lubricant, which increases heat-carrying capacity and overall maintaining a good tribological properties. The primary objective is to make vegetable oil cutting fluid as a replacement for non-biodegradable lubricants. Six main types of vegetable oil that are typically used in grinding include soybean, coconut, castor, rapeseed, palm, and sunflower oil. Among these, castor oil exhibits superior cutting properties with a maximum energy ratio coefficient [4]. Figure 14.3 shows the crucial advantages of vegetable oil lubricants.

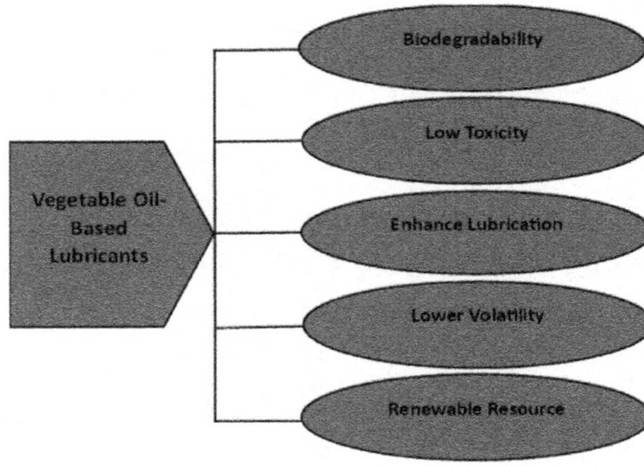

Fig. 14.3 Advantage of vegetable oil cutting fluids

Eco-friendly cutting fluid is a better option for reducing environmental pollution and provides workers safety. Its ability to lubricate decreases the frictional force, consequently lowering the grinding forces and minimizing surface degradation. Therefore, it can be stated that vegetable oil lubrication results in better surface finishes and reduces specific energy consumption. A recent study has revealed that utilizing biodegradable cutting fluid for surface grinding of titanium grade 5 alloys enhances surface quality and significantly outperforms dry grinding methods. This method could lead to more sustainable and efficient grinding [5]. Different types of vegetable oil-based cutting fluids perform different actions on work material. Castor oil is mainly used to increase the grindability of nickel-based alloy (GH4169), which gives a higher performance than other vegetable oils [6]. These lubricants also provide a thin layer of film that protects the nascent ground surface from contamination from harmful gases present in the atmosphere (Table 14.1).

The fluid properties can be enhanced by mixing sulfuric acid with pure castor oil, which is called sulfonate castor oil. This fluid is non-toxic and biodegradable, with no corrosion-inhibiting properties. Table 14.2 shows the properties of different vegetable oil lubricants primarily used in machining.

Table 14.1 Key research on vegetable oil-based grinding

Authors	Workpiece	Grinding wheel/ specifications	Sustainable techniques	Key findings
Alves et al. [4]	Steel	CBN	Sulfonate castor oil	Grinding wheel wear reduces
Li et al. [5]	Nickel-based alloy	CBN	Vegetable oil-based MQL	Castor oil decreases grinding forces and heat generations
Wang et al. [6]	GH4169	White corundum grinding wheel	Vegetable oil-based MQL	Using castor oil as a cutting fluid leads to an improved surface finish

Table 14.2 Tribological properties of vegetable oil lubricants

S/N	Vegetable oils	Viscosity index	Flash point (°F)	Pour point (°F)	Density Kg/m^3
1	Castor oil	85	500	−23.8	970
2	Palm oil	189	579.3	53.8	910
3	Sunflower oil	205	521.2	5.1	890
4	Coconut oil	255	464	68	915
5	Soybean oil	224	489.1	10.5	907
6	Rapeseed oil	208	600.8	−5.8	918

14.2.1.1 Challenges of Vegetable Oil-Based Cutting Fluids

Vegetable oil-based cutting fluids have numerous tribological benefits compared to petroleum-based fluids but still possess inherent properties and challenges. Some of the important drawbacks are:

(a) Poor oxidation properties
(b) Low thermal stability
(c) Poor hydrolytic stability
(d) Low pour point
(e) High cost.

14.2.1.2 Drawback of Rigid Air Layer Formation

Different techniques have been used for many years to control high localized temperatures during grinding. Most industries still rely on flood cooling systems to control temperature, but much lubricant is wasted. This phenomenon mainly occurs because a rigid air layer forms around the rotating wheel, as shown in Fig. 14.4. Consequently, the air layer deviates most of the cutting fluid, and only 10–40% of the applied fluid passes through these layers [7]. To minimize its effect, a painted wheel and scraperboard break the rigid air layer and are positioned just ahead of the fluid nozzle. Maintaining a small gap between the scraper plate and the wheel's periphery is essential and requires frequent adjustment.

Using an aerodynamic deflector is also an effective method to minimize rigid air layers. The high-velocity air jets aimed at the cutting arc can disrupt this rigid layer. Moving at a significantly higher velocity than the grinding wheel, these jets can easily penetrate by breaking air layer boundaries and reaching the cutting area. Enhancing the flow can be achieved by increasing the porosity of the wheel and positioning the fluid delivery nozzle closer to the grind area. Another method is the use of a pneumatic barrier that effectively diminishes the rigid air layer by placing the barrier at a polar angle of $\theta = 30°$ and a swivel angle of $\alpha = 50°$ [8]. Its impact is particularly notable at higher grinding wheel speeds, as the fluid penetrates deep inside the grinding zone, resulting in reduced grinding forces and improved workpiece surface finish. However,

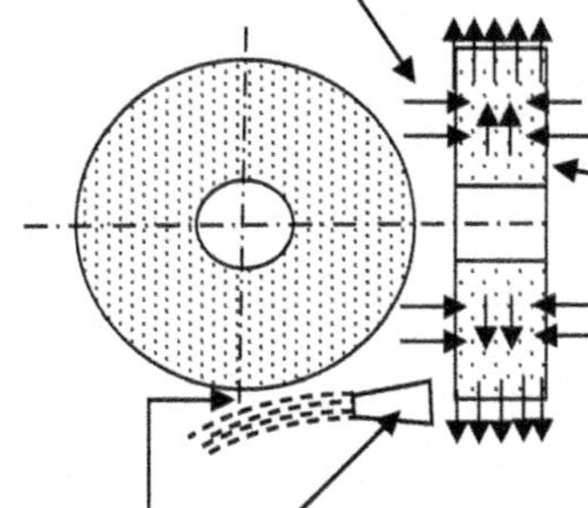

Fig. 14.4 Effect of rigid air layer formation

flood cooling systems are still inefficient and should be replaced by environment-friendly MQL concepts to increase vegetable oil efficiency.

14.2.2 Environment-Friendly Minimum Quantity Lubrication (MQL)

Biodegradable vegetable oil lubricants effectively prevent thermal damage and provide lubrication effects. The efficiency can be increased using suitable techniques to carry atomized lubrication droplets into a compact machining area without any wastage. Therefore, an attractive alternative is the MQL grinding method. This technique is also known as near-dry grinding (NGD). This technique has gained attention for its ability to lower environmental impacts. With increasing regulations to limit health risks and environmental pollution, many modern grinding operations utilize biodegradable cutting fluids. Implementing the MQL technique not only mitigates operational health and safety risks but also cuts down machining costs. This process uses a very low quantity of lubricant and mixed with compressed air in the MQL chamber and delivers this mixture into the machining area as microdroplets through a high-pressure nozzle. This mixture, also called aerosol, is sufficient for deep penetration in compact grinding zones, producing cooling and lubrication effects. MQL grinding provides several advantageous effects. It minimizes the quantity of lubricant, reduces cost and increases safety for the operator and environment. The MQL process depends upon many parameters and factors, including depth of cut, material's mechanical properties and wheel specifications. Therefore, proper selection is crucial to increase the performance of the MQL system.

Nowadays, many conventional cutting fluids have been replaced by vegetable oils due to their extensive benefits, drawing researcher's attention to develop effective eco-friendly MQL systems. The experiment is performed on AISI H13 steel with an alumina grinding wheel using MQL techniques. Groundnut oil is used as a lubricant at a flow rate of 200 ml/h and a pressure of 4 bar. Their findings indicated that groundnut oil significantly reduced grinding energy, forces, temperature, and surface roughness [9]. Vegetable oil demonstrates excellent biodegradability, fulfils lubrication requirements, and has great potential for MQL applications. Among all vegetable oils, Castor oil is renowned for its exceptional lubrication properties. However, its high viscosity can pose challenges for industrial applications, making it less versatile. To overcome these limitations, castor oil is mixed with various eco-friendly oils, resulting in a more effective lubrication solution that maximizes efficiency and environmental benefits. MQL system reduces forces and friction when grinding soft and hard steel compared to dry and flood environmental conditions [10]. Vegetable oil is an outstanding, biodegradable, and renewable energy source with low toxicity, making it a perfect choice for a base lubricating oil in machining processes. Generally, castor oil performance is exceptionally high compared to other vegetable oil-based lubricants [11]. However, it is restricted in many industrial applications due to its high viscosity properties. Therefore, it is mixed with other eco-friendly oils to increase its performance.

14.2.3 Nanofluids (NFs)

In recent studies, vegetable oil-based nanofluid has emerged as an innovative cutting lubricant for machining processes. Nanofluids (NFs) present vast potential in improving the tribological characteristics of lubricants. Incorporating nanomaterials elevates the thermal conductivity of the NFs, thereby significantly enhancing their cooling performance. By blending nanoparticles with base fluids (vegetable oil), nanofluids are created, which makes them a sustainable alternative to conventional cutting fluids. Commonly used nanoparticles are Al_2O_3, MoS_2, WS_2, CNT, TiO_2, ZnO, Cu, Al, Fe, SiO_2, etc. In modern industries, nanofluids are widely used for their exceptional cooling and lubricating properties and their environmental friendliness properties. Nanofluids improve surface texture and reduce friction when grinding hard-to-machine titanium alloy. The nanofluid forces depend on the nozzle orifice and inlet pump pressure. Higher the orifice outlet pressure value, more and better penetration of fluid takes place in machining zone [12].

By adding MoS_2 nanoparticles (NPs) to soybean oil decreased energy consumption and wheel wear [13]. This is because of the increased thermal conductivity of soyabean oil, which is achieved by adding nanoparticles, thus reducing grinding temperature. A similar experiment was conducted by using eight different concentrations of carbon nanotube (CNT) added to palm oil. They find that at 2vol% NFs, minimum temperature was observed while grinding nickel-based alloys [14]. Other researchers [15] investigate grinding performance of different workpiece materials using nanofluids (Al_2O_3 NPs) MQL method. Their result indicates that Al_2O_3 NPs increase grindability for high-strength materials like Inconel, ceramic, titanium, etc.

Generally, vegetable oil + Al_2O_3 NPs achieve lower friction coefficients, grinding forces, and surface roughness [16].

The grindability of C/SiCs alloys under dry, wet, and MQL-assisted carbon nanofluid (CN + MQL) methods was done, and it was observed that specific grinding energy is reduced by using CN + MQL under the same grinding parameters [17]. When nanofluid MQL techniques are used by using Al_2O_3 and diamond nanoparticles, a significant reduction in grinding forces and wheel wear was noted [18]. Similarly, Al_2O_3 nanofluid MQL reduce grinding forces, temperature and surface roughness compared to dry, flood and pure water MQL [19]. The tribological behaviour of various nanoparticles, including Al_2O_3, MoS_2, graphite, and SiO_2, as additives in environmentally friendly vegetable oils, have been studied extensively by researchers and have shown numerous advantages. Therefore, it is critical to analyse the concentration of nanoparticles for different materials, as this can impact the thermophysical properties and the effectiveness of the base fluids (Table 14.3).

Table 14.3 Key research on vegetable oil-based MQL and nanofluid grinding

Authors	Workpiece	Grinding wheel/ specifications	Sustainable techniques	Key findings
Awal et al. [9]	AISI H13 tool steel	AA-54-K-5-V6	Groundnut oil-based MQL	Surface roughness and grinding force decrease
Rabiei et al. [10]	100Cr6	Aluminium oxide	Dry, flood and MQL	MQL reduces grinding force and friction
Jia et al. [11]	Nickel-based alloy GH4169	WA80MV12P	Vegetable oil-based MQL	Castor oil reduces oil viscosity and increases fluid flow
Kalita et al. [13]	EN 24 alloy steel	32A46-HVBEP	Nanofluid MQL	MoS_2 nanoparticle in soyabean oil reduces grinding temperature, wear and friction
Li et al. [14]	Ni-Based alloy	Alumina ceramic bound grinding wheel	Nanofluid MQL	The lowest grinding force was obtained in SiO2 nanoparticles mixed in palm oil
Virdi et al. [16]	Inconel 718	Alumina wheel	Nanofluids Minimum Quantity Lubrication Technique	Experimentally observed, NFMQL lowers friction coefficients, grinding forces, and surface roughness when Al_2O_3 nanoparticles are used
Qu et al. [17]	2.5-C/SiCs	Resin-bonded diamond wheel	Dry, wet and carbon nanofluid MQL	Specific energy is reduced by using carbon nanofluid MQL as compared to dry and wet grinding

14.3 Sustainable Cooling in Grinding

The sustainable cooling process includes the use of extremely cold gases without the use of lubricants. Cryogenic gases like LN_2 are mostly used. In some ductile and low-strength alloys, compressed cold air is used as a cooling medium, which is discussed below in detail.

14.3.1 Cryogenic Grinding

Cryogenic coolants are receiving increasing interest in the manufacturing sector because of their distinct benefits compared to traditional grinding processes. "cryogenic" comes from a Greek word meaning creation or production using cold. Cryogenic technology has emerged as the leading approach to achieve efficient, cost-effective and environmentally friendly outcomes. Generally, LN_2, LCO_2 and $scCO_2$ cryogenic fluids are used in the machining process. Cryogenic grinding is used to grind materials at very low temperatures. This type of cooling has been applied in many modern manufacturing sectors, including aerospace, defence, automotive, and electronic industries. Advanced materials like superalloys and titanium have very low thermal conductivity. Therefore, cryogenic cooling can be one of the best options to enhance cooling capacity. It increases cutting speed, obtains a better surface finish, and improves wheel life.

Cryogenic cooling is also used to grind hardened steel and heat-resistant alloys in the automotive industry. The low temperatures of cryogenic fluid range from −78 °C to −195.8 °C at atmospheric pressure. They are injected directly into the grinding zone. It freezes the abrasive grain and reduces temperature. Thus, it minimises the use of conventional cooling and also benefits sustainability. Cryogenic cooling enables an operator to increase the depth of cut without creating excessive heat at machining zone. This method is typically applied to heat-sensitive materials where high temperatures cause several thermal damage.

Liquid nitrogen (LN_2) impacts the grinding of various steel alloys. It significantly enhances the surface finish compared to other workpieces, which are ground using soluble oil [20]. This improvement can be attributed to nitrogen's extremely cold and eco-friendly nature, which prevents oxidation and other chemical reactions on the freshly ground surface. Furthermore, their findings indicated that cryogenic cooling positively influenced tool longevity, as the grinding wheel's abrasives maintained their sharpness for extended periods. Cryogenic grinding makes materials more brittle, improving machining efficiency, reducing wheel wear, and ultimately making the process eco-friendly. Another experiment was conducted using liquid nitrogen for grinding steel and concluded that there was no surface burn or oxidation of the ground surface, indicating better penetration of LN_2 coolant [21].

Cyogenic cooling decreases the surface roughness of nickel-based superalloys by about 30% to 50%. This cooling technique also minimizes tensile residual stress in the workpiece and enhances work hardening significantly. Additionally, it lowers the temperature by 45% to 50% during higher infeed operations [22]. Another study performed on Ti-6Al-4 V using liquid cooling and LN_2 cryogenic environment cooling conditions. They find a significant decrease in forces when LN_2 is used [23]. Another researcher compared the grindability of stainless steel using an Al_2O_3 grinding wheel under three different environmental conditions. Experimental results stated that surface roughness decreased by 50% using LN_2 compared to dry and wet conditions. Cryogenic grinding lowers grinding forces and enhances surface quality. It offers unique benefits compared to traditional cooling methods. However, using cryogens in a humid environment can lead to ice buildup around the delivery nozzle and piping system, potentially causing blockages in the cryogenic delivery system.

14.3.2 Compressed Cold Air Grinding (CCAG)

Some materials, such as superconducting magnet ceramic, are susceptible and cannot be grounded by grinding fluids. In such conditions, compressed cold air grinding (CCAG) is used. It lowers heat generation and prevents deterioration of work material's physical and chemical properties. The effectiveness of compressed cold air in generating compressed residual stress mostly depends on the types of material being processed and the grinding parameters being employed. Compressed cold air grinding is mainly done on heat-sensitive materials and semiconductors. Generally, at small grinding depths, CCAG shows remarkable cooling efficiency, allowing for optimal temperature control and surface integrity during the grinding process. However, as the grinding depth increases, the cooling performance of CCAG diminishes. This decline can be attributed to several factors, including:

a. CCAG has a lower heat capacity, making it less effective at a higher depth of cut.
b. It struggles with chip-clearing ability.
c. Its capacity to penetrate the grinding zone is weak.

14.4 Hybrid Grinding Process

To increase the effectiveness of cryogenic cooling, a hybrid system known as cryogenic + MQL has been introduced, which can significantly improve heat dissipation efficiency and provide an outstanding cooling and lubrication effect. A hybrid cryogenic + MQL technique has been used to grind difficult-to-cut materials like titanium, Inconel, ceramic, etc. Many researchers only use cryogenic air (CA) as a cooling medium in a grinding process. CA provides effective cooling, strong versatility and low cost. Unfortunately, CA lacks lubrication properties and is limited to a few materials. This shortcoming led to the development of a new hybrid technology called cryogenic air + MQL.

Table 14.4 Key research on cryogenic and hybrid cryogenic grinding

Authors	Workpiece	Grinding wheel/ specifications	Sustainable techniques	Key findings
Paul et al. [21]	Steel	A60K5V8	Cryogenic LN_2	Elimination of surface burn and oxidation on the ground surface
Lee et al. [22]	Nickel alloys	Monolayer CBN	Cryogenic LN_2	Surface roughness is decreased by 30% to 50% in cryogenic conditions
Elanchezhian et al. [23]	Ti-6Al-4 V	Electroplated CBN	Wet, Cryogenic LN_2	Grinding force and surface roughness are reduced using LN_2 conditions
Zhang et al. [24]	Titanium grade 5 alloy	GC80K12 V	Cryogenic air, MQL and cryogenic air nanofluids MQL	Cryogenic air nanofluids MQL resulted in excellent cooling and lubrication performance
Cui et al. [25]	Ti-6Al-4 V	Silicon carbide grinding wheel	Cryogenic air, NFMQL, and CA + NFMQL	CA + NFMQL provides better grindability. The surface roughness is decreased by 25.6% compared with NFMQL

Other researchers developed a hybrid cooling and lubrication system using cryogenic air (CA) and nanofluid MQL. They used three environmental conditions: CA, NFMQL and CA + NFMQL for the surface grinding of titanium grade 5 alloys as the workpiece material. Cryogenic air-assisted nanofluids minimum quantity lubrication (CA + NFMQL) shows excellent performance and obtained the lowest specific grinding energy (51.96 J/mm^3) and friction coefficient (0.60) as compared with CA and NFMQL, respectively [24]. Further experiments were conducted and concluded that cryogenic air cooling and nano lubricant minimum quantity lubrication (NMQL) are ineffective due to their inadequate cooling and lubrication performance. Therefore, compiling both modes (CA + NFMQL) shows the lowest tangential force (46.8 N) and normal force (61.4 N), as compared with CA and NFMQL [25]. Another hybrid grinding concept uses Cryo + MQL and Cryo + MQL with 1% Al_2O_3 NPs. Results show that Cryo + MQL with 1% Al_2O_3 NPs lowers the temperature by providing maximum cooling and lubrication effect (Table 14.4).

14.5 Ultrasonic-Assisted Grinding (UAG)

Ultrasonic-assisted grinding (UAG) is a combination of general grinding process incorporating ultrasonic vibration. This method is generally used to grind brittle and hard materials. In UAG process, a very high frequency, more than 20 kHz, is applied by keeping the value of amplitude as low as possible (2 to 35 μm). Generally, USG is classified into 1-D and 2-D. In one-dimensional (1D) USG, ultrasonic vibrations are provided in any one direction, either vertically to the workpiece or tangential to the wheel axis, and in two-dimensional (2D) USG, vibrations are provided to the workpiece simultaneously in both directions.

Researchers are increasingly studying dry grinding to minimize harmful effects of cutting fluid on environment and lower manufacturing costs. In conventional dry grinding (CDG), no cutting fluid is used, leading to increased thermal issues. More wear of abrasive cutting tool occurs in dry conventional grinding, generally at higher infeed due to lack of lubrication. Dry grinding generates larger cuts, wider grooves, and surface cracks on the workpiece. Therefore, its adoption in industrial environments has been limited. An effective ultrasonic-assisted dry grinding (UADG) technology is implemented to overcome these constraints.

During the grinding of steel bearing (100Cr6) material, by applying an amplitude of 10 μm and ultrasonic frequency of 23 kHz, the grindability increases by decreasing surface roughness and frictional effects compared to dry grinding [26]. Similarly, using UAG on Ti-6Al-4 V alloy reduces surface roughness by 10%, normal force by 13.5%, and tangential force by 14.5% compared with CG [27]. Another experiment signifies that an ultrasonic amplitude of 4 μm produces optimal surface contact performance. The workpiece's surface roughness is at least 24% lower than a traditional ground surface [28].

Compared to other grinding processes, UAG efficiency becomes low when the depth of cut increases. Therefore, to improve its efficiency, UAG is integrated with minimum quantity lubrication (MQL) or nanofluids (NFs), making it a hybrid grinding process. By integrating UAG with MQL and UAG with eco-friendly nanofluids MQL (UAG + NFMQL) in their experimental work, they found that using UAG + NFMQL significantly enhances surface smoothness. The improvement in surface smoothness is attributable to the reduced friction in the horizontal vibration of the grit, which increases the cutting ability. Combining nanofluid-MQL with UAG presents a promising avenue for enhancing grinding efficiency by decreasing the quantity of lubricants required and reducing overall costs [29].

The fresh grind surface is also affected by the UAG process, which can be studied using bearing and retention indexes. As compared to the conventional grinding process, UAG index values are higher. Furthermore, the significant increase in bearing and retention indexes can be increased by increasing the ultrasonic amplitude value [30]. Another experiment was conducted on γ- titanium aluminide alloys using two grinding processes. Firstly, by using creep feed grinding (CFG) and secondly, by using a hybrid grinding mode, i.e. ultrasonic assisted creep feed grinding (UACFG). The experiment results showed that grinding forces significantly decreased in UACFG process using SiC wheel [31].

The grinding wheel is typically affected by ultrasonic vibration in the UAG process. High vibration causes wear on the wheel, improving abrasive grits self-sharpening properties, which improve surface texture. Unfortunately, at higher infeed, larger cuts, wider grooves, and surface crack marks were observed on the workpiece. The surface residual stress of titanium alloy (Ti-6Al-4 V) under high

Table 14.5 Key research on ultrasonic-assisted grinding

Authors	Workpiece	Grinding wheel / specifications	Sustainable Techniques	Key Findings
Tawakoli et al. [26]	100Cr6	Vitrified bond Al_2O_3 grinding wheel	Ultrasonic-assisted dry grinding	The UADG process reduces grinding forces by more than 50%
Nik et al. [27]	Ti-6Al-4 V	89A60K5AV217	Ultrasonic-assisted dry grinding	Surface roughness is reduced by 10%, normal force by 13.5% and tangential force by 14.5%
Wen et al. [28]	12CrNi4A	CBN	Ultrasonic-assisted grinding	Surface roughness value is reduced by 24% using UAG process
Molaie et al. [29]	Alloy steel	5SG46-JVS	UAG with MQL UAG with nanofluids MQL	UAG with nanofluids MQL reduces normal force up to 61% and tangential force up to 58%
Chen et al. [30]	12Cr2Ni4A steel	CBN	Ultrasonic-assisted dry grinding	Surface morphology improved with higher ultrasonic amplitude
Maroju et al. [32]	Ti-6Al-4 V	CBN	Vibration-assisted grinding	Vibration increases compressive residual stress

frequency and low amplitude mode can be studied by using a finite element model. During contact between workpiece and abrasive tool, the residual stress value continuously fluctuates at initial phase and stabilises after 15 passes. Due to vibration, the compressive residual stress value increases, and hence, it positively impacts fatigue life of workpiece. This increase in value is mainly due to indentation effects [32] (Table 14.5).

14.6 Conclusion and Future Scope

Green grinding" or "eco-friendly grinding" generally denotes a grinding process or technique that focuses on environmental sustainability and ecological impact. This chapter focused on sustainable cooling and lubrication processes to improve grindability. Today, critical sustainable processes, including vegetable oil-based lubricants, MQL, nanofluids, cryogenic cooling, ultrasonic grinding, and hybrid processes, have been a subject of great interest. All these processes facilitate the provision of an alternative to synthetic or petroleum-based lubricants. Many researchers have stated that adopting green fluids has numerous benefits. Enhancing vegetable oil with nanoparticles can boost its efficiency. For minimizing lubrication quantities, Minimum Quantity Lubrication (MQL) is an emerging and effective technology while maintaining maximum machining efficiencies. Generally, hybrid methods of cooling and lubrication are adopted for grinding hard-to-cut material. All this process reduces or eliminates the use of traditional and non-biodegradable cutting fluids. From the above-mentioned sustainable technology, it can be concluded that.

- Vegetable oil-cutting fluids are characterized by biodegradability and renewability. It provides safety for the operator and environment. During grinding, vegetable oil-based lubricants significantly decrease cutting forces, improving tool life and effectively minimizing thermal issues. Different vegetable oils vary in properties; therefore, choosing the appropriate one is crucial. Six different types of vegetable oil are used in grinding: soybean, castor, sunflower, palm, coconut, and rapeseed oil. Many researcher's results show that castor oil performance is exceptionally high and possesses maximum lubricity.
- MQL is a popular and preferable technique in most modern manufacturing industries. It provides an economically and environmentally friendly impact by cutting the quantity of lubricating liquids. Vegetable oil-based MQL reduces surface roughness, cutting temperature, wheel wear, and grinding forces. MQL performance varies based on workpiece material, grinding wheel specification, nozzle angle, and input grinding parameters. Therefore, proper selection of MQL fluid is essential for maximum efficiency.
- Nanoparticles are dispersed in vegetable oil lubricants to create eco-friendly nanofluids. It significantly enhances tribological properties and heat-carrying capacity. Integrating nanofluid with MQL significantly reduces friction and enhances thermal conductivity. Generally, aluminium oxide (Al_2O_3), molybdenum disulphide (MoS_2) and carbon nanotubes (CNTs) are the most common nanoparticles used for making nanofluids.
- Cryogenic cooling is effective in grinding difficult-to-cut materials like titanium, ceramics, composites, and other superalloys with low thermal conductivity properties. The extremely cold temperature sufficiently reduces heat generation and grinding forces. This is one of the most impactful sustainable techniques for reducing temperature and improving surface morphology.
- Ultrasonic-assisted grinding is an alternative to dry grinding. It eliminates the use of conventional fluid and reduces harmful effects on the environment. USG generally used low amplitude and very high frequency. Ultrasonic vibrations lead to acoustic softening of materials being ground, which reduces the yield strength of the workpiece, resulting in better and smoother cutting. Many researchers found that USM effectively reduces wheel wear, surface roughness and grinding forces (up to 60%). A hybrid process, including MQL + USG and nanofluid-MQL with UAG, was introduced to boost its efficiency.
- Hybrid cooling and lubrication is the newest and most effective process. It improves grinding efficiency, enhances surface quality, reduces grinding forces and increases tool life. Hybrid processes are generally used to grind advanced materials. Commonly used hybrid systems are MQL + Cryogenic and Nanofluids MQL + Cryogenic.

Many researchers have noticed that vegetable oil lacks lubrication properties and other critical properties compared to petroleum-based cutting oil. Additional components, such as nanoparticles, are integrated to improve its performance. However, it is still inefficient. Therefore, much work must be done in this area to increase its efficiency. Further research can be conducted on biodegradable nanofluids, in which multiple types of nanoparticles can be implemented to maximise lubricants' thermal conductivity. Effective hybrid grinding processes, including nanofluids-assisted grinding, cryogenic-assisted MQL and ultrasonic-assisted nanofluids MQL, should be of great interest and could be investigated more in future to develop effective systems, especially for newly developed and advanced materials.

References

1. Adibi, H., Hatami, O., Rezaei, S.M.: Effects of minimum quantity lubrication (MQL) on grinding processes using eco-friendly nanofluids: a review. Adv. Mater. Proc. Technol. **10**(3), 1666–1707 (2023). https://doi.org/10.1080/2374068X.2023
2. Kumar Singh, S., Kumar Jha, A., Ray, A., Dewangan, S.: Enhancing grinding parameters of TC4 titanium superalloy by using hybrid eco-friendly cooling system. Mater. Today Proc. **62**, 1505–1509 (2022). https://doi.org/10.1016/j.matpr.2022.02.214
3. Vasudevan, B., Nagarajan, L., Nachippan, N.M., Mahadevan, S.: Vegetable oils in minimum quantity lubrication: a comparative analysis of properties and performance. Interactions **245**(1), 253 (2024). https://doi.org/10.1007/s10751-024-02070-1
4. Alves, S.M., De Oliveira, J.F.G.: Development of new cutting fluid for grinding process adjusting mechanical performance and environmental impact. J. Mater. Process. Technol. **179**(1–3), 185–189 (2006). https://doi.org/10.1016/j.jmatprotec.2006.03.090
5. Li, B., Li, C., Zhang, Y., Wang, Y., Jia, D., Yang, M.: Grinding temperature and energy ratio coefficient in MQL grinding of high-temperature nickel-base alloy by using different vegetable oils as base oil. Chin. J. Aeronaut. **29**(4), 1084–1095 (2016). https://doi.org/10.1016/j.cja.2015.10.012
6. Thekkoot Surendran, S.B., Sooraj, V.S.: Into the boundary layer behavior of segmented grinding wheels and its illustration on Ti6Al4V. Mach. Sci. Technol. **28**(3), 267–298 (2024). https://doi.org/10.1080/10910344.2024.2311371
7. Wang, Y., Li, C., Zhang, Y., Y ang, M., Li, B., Jia, D., Hou, Y., Mao, C.: Experimental evaluation of the lubrication properties of the wheel/workpiece interface in minimum quantity lubrication (MQL) grinding using different types of vegetable oils. J. Cleaner Prod. **127**, 487–499 (2016). https://doi.org/10.1016/j.jclepro.2016.03.121
8. Mandal, B., Singh, R., Das, S., Banerjee, S.: Improving grinding performance by controlling air flow around a grinding wheel. Int. J. Mach. Tools ManufManuf **51**(9), 670–676 (2011). https://doi.org/10.1016/j.ijmachtools.2011.06.003
9. Awale, A.S., Vashista, M., Khan Yusufzai, M.Z.: Multi-objective optimization of MQL mist parameters for eco-friendly grinding. J. Manuf. Processes **56**, 75–86 (2020). https://doi.org/10.1016/j.jmapro.2020.04.069
10. Rabiei, F., Rahimi, A.R., Hadad, M.J., Ashrafijou, M.: Performance improvement of minimum quantity lubrication (MQL) technique in surface grinding by modeling and optimization. J. Clean. Prod. **86**, 447–460 (2015). https://doi.org/10.1016/j.jclepro.2014.08.045
11. Jia, D., Li, C., Zhang, Y., Yang, M., Wang, Y., Guo, S., Cao, H.: Specific energy and surface roughness of minimum quantity lubrication grinding Ni-based alloy with mixed vegetable oil-based nanofluids. Precis. Eng. **50**, 248–262 (2017). https://doi.org/10.1016/j.precisioneng.2017.05.012

12. Panda, S.K., Rath, K.C., Choudhury, B.K.: Determining the flow correlation for an orifice with a non-dimensional number. Flow Meas. Instrum.Instrum. **90**, 102338 (2023). https://doi.org/10.1016/j.flowmeasinst.2023.102338
13. Kalita, P., Malshe, A.P., Arun Kumar, S., Yoganath, V.G., Gurumurthy, T.: Study of specific energy and friction coefficient in minimum quantity lubrication grinding using oil-based nanolubricants. J. Manuf. Process. **14**(2), 160–166 (2012). https://doi.org/10.1016/j.jmapro.2012.01.001
14. Li, B., Li, C., Zhang, Y., Wang, Y., Jia, D., Yang, M., Zhang, N., Wu, Q., Han, Z., Sun, K.: Heat transfer performance of MQL grinding with different nanofluids for Ni-based alloys using vegetable oil. J. Clean. Prod. **154**, 1–11 (2017). https://doi.org/10.1016/j.jclepro.2017.03.213
15. Tiwari, A., Singh, D.K., Mishra, S.: A review on minimum quantity lubrication in machining of different alloys and superalloys using nanofluids. J. Braz. Soc. Mech. Sci. Eng. **46**, 112 (2024). https://doi.org/10.1007/s40430-024-04676-6
16. Virdi, R.L., Chatha, S.S., Singh, H.: Experiment evaluation of grinding properties under Al_2O_3 nanofluids in minimum quantity lubrication. Mater. Res. Express **6**(9), 096574 (2019). https://doi.org/10.1088/2053-1591/ab301f
17. Qu, S., Yao, P., Gong, Y., Chu, D., Yang, Y., Li, C., Wang, Z., Zhang, X., Hou, Y.: Environmentally friendly grinding of C/SiCs using carbon nanofluid minimum quantity lubrication technology. J. Clean. Prod. **366**, 132898 (2022). https://doi.org/10.1016/j.jclepro.2022.132898
18. Li, W., Zeng, Z., Le, S., Zhu, K., Huang, X., Hegab, H., Ibrahim, A.M.M.: Investigation of a green nanofluid added with graphene and Al_2O_3 nano-additives for grinding hard-to-cut materials. Tribol. Int.. Int. **195**, 109580 (2024). https://doi.org/10.1016/j.triboint.2024.109580
19. Mao, C., Tang, X., Zou, H., Huang, X., Zhou, Z.: Investigation of grinding characteristic using nanofluid minimum quantity lubrication. Int. J. Precis. Eng. Manuf. **13**(10), 1745–1752 (2012). https://doi.org/10.1007/s12541-012-0229-6
20. Sharma, A., Chaudhari, A., Yusufzai, M.Z.K., Vashista, M.: Effectiveness of using liquid nitrogen cryogen in grinding to enhance the grinding performance of hard steel. Proc. Instit. Mech. Eng. Part B: J. Eng. Manuf. **238**(1–2), 315–327 (2024). https://doi.org/10.1177/09544054221147622
21. Paul, S., Chattopadhyay, A.B.: Determination and control of grinding zone temperature under cryogenic cooling. Int. J. Mach. Tools ManufManuf **36**(4), 491–501 (1996). https://doi.org/10.1016/0890-6955(95)00053-4
22. Li, C.H., Lu, B.H., Ding, Y.C., Cai, G.Q.: Innovative technology investigation into cryogenic cooling green grinding using liquid nitrogen jet. Paper presented at the 2009 International Conference on Management and Service Science (MASS), Beijing, China (2009). https://doi.org/10.1109/ICMSS.2009.5302527
23. Elanchezhian, J., Pradeep Kumar, M., Manimaran, G.: Grinding titanium Ti-6Al-4V alloy with electroplated cubic boron nitride wheel under cryogenic cooling. J. Mech. Sci. Technol. **29**(11), 4885–4890 (2015). https://doi.org/10.1007/s12206-015-1036-7
24. Zhang, J., Li, C., Zhang, Y., Yang, M., Jia, D., Liu, G., Hou, Y., Li, R., Zhang, N., Wu, Q., Cao, H.: Experimental assessment of an environmentally friendly grinding process using nanofluid minimum quantity lubrication with cryogenic air. J. Clean. Prod. **193**, 236–248 (2018). https://doi.org/10.1016/j.jclepro.2018.05.009
25. Cui, X., Li, C., Zhang, Y., Said, Z., Debnath, S., Sharma, S., Ali, H.M., Yang, M., Gao, T., Li, R.: Grindability of titanium alloy using cryogenic nanolubricant minimum quantity lubrication. J. Manuf. Process. **80**, 273–286 (2022). https://doi.org/10.1016/j.jmapro.2022.06.003
26. Tawakoli, T., Azarhoushang, B.: Influence of ultrasonic vibrations on dry grinding of soft steel. Int. J. Mach. Tools ManufManuf **48**(14), 1585–1591 (2008). https://doi.org/10.1016/j.ijmachtools.2008.05.010
27. Nik, M.G., Movahhedy, M.R., Akbari, J.: Ultrasonic-assisted grinding of Ti-6Al4-V alloy. Procedia CIRP **1**, 353–358 (2012). https://doi.org/10.1016/j.procir.2012.04.063
28. Wen, Y., Tang, J., Zhou, W., Zhu, C.: Study on contact performance of ultrasonic-assisted grinding surface. Ultrasonics **91**, 193–200 (2019). https://doi.org/10.1016/j.ultras.2018.08.009

29. Molaie, M.M., Akbari, J., Movahhedy, M.R.: Ultrasonic assisted grinding process with minimum quantity lubrication using oil-based nanofluids. J. Clean. Prod. **129**, 212–222 (2016). https://doi.org/10.1016/j.jclepro.2016.04.080
30. Chen, H., Tang, J., Shao, W., Zhao, B.: An investigation on surface functional parameters in ultrasonic-assisted grinding of soft steel. Int. J. Adv. Manuf. Technol. **97**(5–8), 2697–2702 (2018). https://doi.org/10.1007/s00170-018-2164-x
31. Bhaduri, D., Soo, S.L., Aspinwall, D.K., Novovic, D., Bohr, S., Harden, P., Webster, J.A.: Ultrasonic assisted creep feed grinding of gamma titanium aluminide using conventional and superabrasive wheels. CIRP Ann. **66**(1), 341–344 (2017). https://doi.org/10.1016/j.cirp.2017.04.085
32. Maroju, N.K., Jin, X.: Effects of vibration assistance on surface residual stress in grinding of Ti-6Al-4V alloy. Procedia Manuf. **10**, 171–182 (2017). https://doi.org/10.1016/j.promfg.2017.07.045

Sunil Kumar Singh is pursuing his PhD in the Department of Production and Industrial Engineering at National Institute of Technology, Jamshedpur. He received his M. Tech degree from Kalyani Government Engineering College, West Bengal, in 2013 and B. Tech from R.V.S College of Engineering and Technology, Jamshedpur, in 2010. With nearly 10 years of teaching experience, his research interests focus on sustainable manufacturing and optimization techniques.

Dr. Ashok Kumar Jha is an Associate Professor in the Production and Industrial Engineering department at NIT Jamshedpur. He earned his B. Tech. from Visvesvaraya National Institute of Technology, Nagpur, in 1983, his M. Tech. from the Indian Institute of Technology, Kharagpur, in 1987, and his Ph. D. from Jadavpur University in 2018. Dr. Jha has published several journal articles and conference proceedings in highly reputed journals. He has served more than 30 years in an academic institute. His research interest is robotics and manufacturing.

Dr. Amitava Ray is the Professor and Principal of Jalpaiguri Government Engineering College in West Bengal. He received his Bachelor of Engineering (B. E.) in Production Engineering from Jadavpur University in 1994, followed by a Master of Engineering (M. E.) in Production Engineering with a specialization in Manufacturing Engineering from Mumbai University in 2001. He completed his Ph. D. in Engineering at Jadavpur University in 2010. Dr. Ray has published over 100 research papers in both national and international journals. He has six years of experience working in industry and has spent 25 years in academic institutions. His research supervision includes topics in green manufacturing, green energy, non-traditional machining processes, and supply chain management.

Chapter 15
An Analysis of the Precision of Dental Teeth Manufactured Through Milling Machines and 3D Printing Techniques

Surendra Singh Thakur, Sharad Kumar Pradhan, S. S. Mahapatra, and Emmanuel Philip Ododo

Abstract This study examines the accuracy in terms of surface finish of the dental teeth manufactured by using two different techniques: milling machines and DLP based 3D printing machine. The dental tooth was scanned by using intraoral scanner and modifying the shape of teeth through the 3D dental design software (CAD). The precision of each method was compared to determine which technique produces more accurate shape of dental teeth. Ra, Rt, Rz, roughness criteria were measured. The study found that both techniques have their advantages and disadvantages, but overall, dental teeth made through milling machines have higher value Ra = 1.237, Rq = 1.492 and Rz = 6.08 as compared to those made through 3D printing machine Ra = 0.872, Rq = 1.251 and Rz = 6.08. These findings could potentially impact the dental industry and help dentists make more informed decisions when choosing which technique to use for their patients.

Keywords Computer-aided design (CAD) · Digital light processing techniques (DLP) · Roughness average (Ra) · Light-emitting diodes (LEDs) · Stereo lithography (SLA)

Preface

This chapter explores the comparative evaluation of two advanced manufacturing techniques-milling and DLP-based 3D printing techniques in the fabrication of

S. S. Thakur (✉)
3D Engineering Automation LLP, Pune, Maharashtra, India
e-mail: thakur_surendra2003@rediffmail.com

S. K. Pradhan
THDC-Institute of Hydropower Engineering and Technology, Tehri, Uttrakhand, India

S. S. Mahapatra
National Institute of Technology, Rourkela, Odisha, India

E. P. Ododo
University of Uyo, Uyo, Nigeria

© The Author(s), under exclusive license to Springer Nature Switzerland AG 2025
K. C. Rath et al. (eds.), *Engineering, Technology and Management*, Information Systems Engineering and Management 54, https://doi.org/10.1007/978-3-031-93654-8_15

dental restorations. By analyzing key parameters such as surface roughness and dimensional accuracy, the study highlights the unique advantages and limitations of each approach. Utilizing state-of-the-art tools like intraoral scanners and CAD software, the research underscores the importance of digital workflows in transforming modern dental practices. The findings offer valuable guidance for dental professionals, enabling them to make informed choices when selecting technologies that ensure precision and efficiency in prosthetic production.

15.1 Introduction

Digital technology has significantly impacted dentistry, and 3D printing is a prime example. This innovative technology offers several benefits, including faster production times, lower costs, and the ability to create complex structures [1]. As a result, it's transforming dentistry, particularly in the creation of dental prosthetics like crowns and artificial teeth. 3D printers use digital files to build 3D objects layer by layer, unlike traditional subtractive manufacturing that removes material to create the desired shape [2]. This technology has the potential to streamline the production of dental models, which are currently made using physical impressions. By eliminating the need for physical models, 3D printing could significantly reduce production time and allow for the creation of multiple, identical copies.

One common application of 3D-printed models in dentistry is the fabrication of thermoplastic orthodontic appliances [3]. Ideally, these appliances could be directly printed instead of using a physical model as an intermediary step. However, for clinical use, the accuracy of 3D-printed models is critical. Currently, 3D printing may not match the precision of computer numeric control (CNC) processing, a subtractive manufacturing method. In some cases, 3D-printed products may require post-processing to achieve a smooth surface. Therefore, choosing the right type of 3D printer with suitable features is essential for orthodontic applications. The introduction of 3D dental design software (CAD) has revolutionized the fabrication of dental prosthetics in dental labs. This technology offers significant advantages over traditional casting methods, including reduced human error and improved accuracy in the surface finish of teeth [4]. Studies have shown that 3D-CAD software produces more precise dental restorations compared to conventional techniques [5]. The 3D-CAD process involves a sequential workflow. First, an oral scanner captures a digital impression of the patient's teeth. This digital model can then be manipulated within the software to design the desired restoration or modify existing tooth structures. Finally, the design is translated into a physical object using either milling machines or 3D printing techniques [2]. A key benefit of 3D-CAD software is the ability to fabricate intricate dental shapes in significantly less time compared to traditional methods. Traditional techniques often result in greater material waste and limitations in accuracy due to machine wear and tear. Additionally, casting methods can introduce cracks and brittleness in the final restoration [6]. These issues with surface finish and material integrity can be significantly reduced through advanced techniques like 3D printing.

3D printing is an additive manufacturing process that builds dental prosthetics layer-by-layer using a digital file. This technology has gained significant traction in recent years. Studies have shown that 3D printing produces restorations with superior marginal fit compared to milling machines [7]. The integration of 3D printing with intraoral scanning, which captures intricate details of the oral cavity and translates them into digital models, further enhances patient outcomes. While 3D printing holds immense promise, further research is necessary to optimize its potential in various dental applications. Beyond dentistry, 3D printing finds application in surgical implants, surgical guides, and various fields like automobile manufacturing and medicine [8]. This report aims to compare the accuracy of dental restorations produced using milling machines and 3D printing techniques.

15.1.1 Overview of CAD/CAM Technologies for Dentistry

Dental restorations may now be made precisely and effectively thanks to CAD/CAM (Computer-Aided Design/Computer-Aided Manufacturing) technology, which have completely changed the field. Dentists may now create and produce personalized dental implants, crowns, bridges, and dentures because to these technologies. Dental practitioners can attain more precision, better fit, and superior aesthetics with CAD/CAM, which will eventually improve patient results and satisfaction.

(a) **Desktop scanner**: Dental scanners utilize light technologies to capture detailed 3D representations of a patient's jaw and teeth [9]. These scanners offer a faster alternative to traditional mechanical methods. Two primary illumination techniques are employed: laser and structured light. Laser scanners rely on the principle of triangulation. A laser beam and a camera are positioned at specific angles relative to each other. As the laser sweeps across the teeth, the camera captures the position of the reflected light. Software then utilizes trigonometry to calculate the 3D structure from these images. While optical scanners boast faster scanning times, their accuracy can be impacted by several factors. The surface characteristics of the teeth, such as color, curvature, and roughness, can influence the scan data. Additionally, ambient lighting and camera focus need to be carefully controlled to ensure optimal results. Due to these sensitivities, optical scanners may not achieve the same level of precision as mechanical scanners.

(b) **Intraoral scanner**: Traditional dentistry relied on desktop scanners that captured the geometry of teeth by scanning physical plaster models. However, this process could lead to a loss of detail, particularly regarding margins, due to multiple conversions between the physical and digital forms. In 1987, the CEREC® system emerged as the first commercially available chairside CAD/CAM system for dental restorations [10]. This technology utilized blue light-emitting diodes (LEDs) to project patterns onto the teeth, enabling a more precise capture of the virtual dental model. The blue LEDs minimized distortions, even at the edges of the scan area [11]. While desktop scanners offered a digital workflow, they were limited in capturing the nuances of teeth spacing due to physical constraints within the clinical setting. Intraoral scanners addressed this limitation by directly scanning the teeth within the mouth, providing a more accurate digital representation. However, intraoral scanners present their own challenge: multi-patch registration errors. Traditional methods relied on mechanical parameters for coarse positioning, leading to more accurate results. Digital models acquired with intraoral scanners can accumulate discrepancies, particularly in the reconstruction of full arches, especially for posterior teeth.

15.1.2 Dental Manufacturing

Dental manufacturing in CAD/CAM dentistry is the automated use of computer-controlled machinery to create dental restorations like crowns, bridges, and implants. This procedure shortens manufacturing times and improves uniformity and quality by enabling the quick and accurate manufacture of personalized dental goods.

15.1.3 Milling Operation

After being generated by computer-aided design (CAD) software, the construction data is transformed into milling instructions (toolpaths) through CAM processing [12]. These instructions are then uploaded to the milling machine for fabrication. Milling machines come in various configurations, categorized by their number of controllable axes. The most common type is the 3-axis machine [12], followed by 4-axis and 5-axis machines. A 3-axis milling machine can move along the three primary spatial axes: X, Y, and Z. This allows for precise control of the cutting tool path using just these three coordinates, minimizing the complexity of calculations. Notably, all 3-axis machines used in dentistry can rotate the workpiece by 180 degrees, enabling efficient milling of both the inner and outer surfaces of crowns or capping [13]. This simplicity translates to faster milling times and easier operation compared to their multi-axis counterparts. Consequently, 3-axis milling machines are generally more affordable than machines with additional axes.

15.1.4 Dental 3D Printing

Dental 3D printing has emerged as a transformative technology in modern dentistry, streamlining workflows, enhancing accuracy, and personalizing patient care [14].

This paper delves into the principles, working procedures, and diverse applications of dental 3D printing machines. Dental 3D printers operate on the principle of additive manufacturing. Unlike traditional subtractive methods that remove material to create a desired shape, these printers build objects layer by layer using digital data [15]. This data is typically generated from intraoral scans or computer-aided design (CAD) software.

15.1.5 Working Procedure

In order to make a digital impression of the patient's mouth, CAD/CAM dentistry uses software to design the restoration. The finished result is then milled or 3D printed from a block of ceramic or composite material. The working procedure of a dental 3D printer can be broadly summarized into four stages:

(i) **Preparation**: The digital model is uploaded to a slicing software that converts it into thin, horizontal slices. Printing parameters, such as layer thickness and exposure time, are defined in this stage.
(ii) **Printing**: The printer utilizes a vat of light-curable resin. A light source of digital light processing, selectively cures the resin based on the sliced data, solidifying each layer. The build platform lowers incrementally after each layer, allowing the object to be built from the bottom up.
(iii) **Post-Processing**: Once printing is complete, the object is removed from the build platform and undergoes a cleaning process to remove excess resin. Depending on the material, post-curing with ultraviolet light might be required to achieve optimal strength.
(iv) **Finishing**: The final stage involves support removal, sanding, and polishing to achieve the desired surface finish and functionality.

15.1.6 Applications

Dental 3D printing offers a vast array of applications, impacting nearly every aspect of dental care. Here are some prominent examples.

(i) **Crown and Bridge Production**: Highly accurate crowns, bridges, and inlays can be 3D printed, offering a faster and more predictable alternative to traditional methods.
(ii) **Surgical Guides**: 3D printed surgical guides aid in precise implant placement during dental implant surgery, minimizing complications and improving predictability.
(iii) **Removable Dentures**: Customizable dentures with improved fit and comfort can be fabricated using 3D printing technology.

(iv) **Orthodontic Models**: Clear aligner therapy and other orthodontic treatments benefit from 3D printed models for accurate planning and treatment visualization.
(v) **Night Guards**: Personalized night guards for teeth grinding and bruxism can be efficiently 3D printed for improved patient comfort.

15.1.7 Objective of the Work

The primary objective of this research is to analyze and compare the precision of dental teeth manufactured using two advanced techniques: milling machines and digital light processing 3D printing. By evaluating key surface roughness parameters (Ra, Rq, Rz), this study aims to determine the relative accuracy and quality of dental prosthetics produced by each method. The work seeks to provide insights into the benefits and limitations of both techniques, ultimately assisting dental professionals in selecting the most suitable manufacturing method based on precision and patient-specific requirements.

15.2 3D Printing

3D printing has revolutionized various industries beyond its initial application in rapid prototyping. In medicine and dentistry, for instance, 3D models reconstructed from CT scans aid in complex surgical procedures [16]. These models, encompassing bone structures, implants, and foreign bodies, improve diagnosis and preoperative planning [17]. This study explores the use of a compact intraoral scanner to capture a patient's dental anatomy and generate 3D virtual data of the tooth surface through mesh reconstruction. Successful data acquisition requires slow and steady movement of the scanner to ensure continuous focus and real-time registration [18]. The resulting 3D STL model is then used for virtual orthodontic treatment planning using in-house software. Following virtual planning, physical models are fabricated using two techniques: CNC 5-axis milling and 3D printing [19]. The CNC milling machine, with its three-axis milling spindle and two-axis rotary table, utilizes dry milling for cost-effectiveness and to prevent moisture absorption by the material. The Next Dent™ 5100 dental 3D printer, on the other hand, employs a micro jet head that deposits 16-micron layers of photopolymer material onto the build tray [20]. This material is instantly cured by UV light, eliminating the need for post-curing and allowing for immediate product testing. This efficient approach to model creation facilitates orthodontic treatment planning and evaluation.

15.2.1 Case Study

The paper describes an orthodontic case involving an internal premolar rotation. An intra-oral scanner, a non-contact structured-light device with high technologies (MHT), was used to digitally capture the patient's 3D oral geometry. Real-time registration and surface reconstruction algorithms processed the data to create a precise orthodontic setup model. The mesh underwent post-processing steps like topology checking and curvature-based hole-filling to ensure a water-tight STL model. This model was then utilized in custom-developed orthodontic planning software to design the correction process. The software generated eight incremental variations, which directly translated into the creation of eight clear aligner sets for the patient's treatment. Table 15.1 details the planned movement for each affected tooth throughout the correction process.

15.2.1.1 Selection of Material for 3D Printing Process

The success of a 3D printed product hinges on choosing the right material. Several factors come into play during material selection, including the final product's desired properties, the 3D printing process itself, cost, and environmental and safety considerations.

(i) **Properties**: This refers to the characteristics you want in the final product, such as strength, flexibility, heat resistance, and more.
(iii) **Printing Process**: Different 3D printers have specific requirements for resin materials, like viscosity (thickness) and curing time (how long it takes to harden).
(iv) **Cost**: Material prices can vary significantly, so it's important to factor this into your decision.
(v) **Environmental and Safety**: Choose a non-toxic material that's safe to handle during printing to protect yourself and the environment.

For example, stereolithography (SLA) 3D printing commonly uses Tough Gry-15, Dental Photopolymer Resin (Next Dent Model 2.0, Next Dent C&B MFH, Next Dent Cast, Next Dent Model Ortho. These resins offer a variety of properties to suit different project needs.

Table 15.1 RMS value of tooth model using milling machine

Sr. no.	Ra value of milling tooth		
	Ra	Rq	Rz
1	1.237	1.492	6.088
2	1.033	1.322	6.392

15.2.1.2 Digital Flow from Patient Diagnosis to Treatment

In modern dentistry, advanced technologies have revolutionized the path from initial examination to receiving treatment. This digital workflow begins with a comprehensive examination where the dentist utilizes intraoral scanners to capture detailed digital images of the patient's teeth [21] (refer Fig. 15.1). These images are then uploaded to a computer system for analysis and creation of a precise 3D dental model. Software specifically designed for dental applications comes into play next. This software allows the dentist to meticulously plan the treatment process. It provides a virtual environment where the dentist can design the desired outcome, adjust as needed, and finalize the plan before proceeding with the actual treatment. Once the design is finalized, the data is transferred to either a 3D printer or a milling machine. These machines utilize computer-aided manufacturing (CAM) technology to precisely fabricate dental restorations from a variety of materials such as ceramics, resins, or metals, depending on the specific needs of the patient.Finally, the restorations are carefully placed onto the patient's teeth, ensuring a perfect fit and an aesthetically pleasing result. This level of precision is achievable due to the digital planning and CAM technology employed throughout the process.

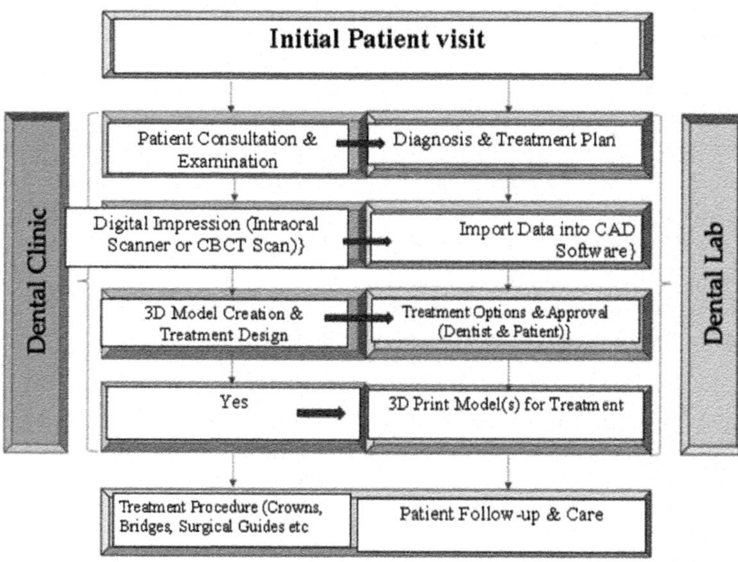

Fig. 15.1 Digital workflow of dental diagnosis and treatment using dental design software (CAD) and 3D printing technology

15.2.1.3 Comparison and Discussion

This research investigated the application of 5-axis machining and 3D printing in creating dental setup models. Digital models were obtained via intra-oral scanning and used to fabricate eight models through both techniques. These models were then compared to the original digital plan to assess the accuracy and efficiency of each method. A 5-axis milling machine specifically designed for dental applications was used. This machine offered automated toolpath generation for efficient production of various dental prosthetics [22]. For 3D printing, a machine known for its use in medical models and surgical stents was employed. Analysis of the finished models revealed key differences between the two fabrication methods. 5-axis machining produced models with smoother surfaces, particularly on the crown portions. However, there were recurring issues with over-cutting of crowns and deviations on the inner gum wall. This suggests a directional error inherent to the 5-axis machining process that doesn't affect repeatability. Conversely, 3D printed models exhibited lower surface quality due to the layer-by-layer printing process. Errors were concentrated on the crown ridges, particularly in horizontal directions, indicating a strong correlation between error and printing direction. The research then discusses the inherent differences between CNC machining (including 5-axis) and 3D printing. CNC machining is a subtractive process that removes material from a block, while 3D printing is an additive process that builds the model layer by layer. This difference translates to distinct advantages for each method. CNC machining is limited to single-model production, while 3D printing allows for efficient mass production. Additionally, calculating tool-paths for complex dental geometries can be challenging in CNC machining, whereas 3D printing only requires slicing the model for printing. In terms of model quality, CNC machining offers superior accuracy and smoother surfaces. 3D printing's limitations include potential for stair-stepping on surfaces and model deformation. Cost-wise, 3D printing can be more expensive for dental applications due to the materials used.

The paper concludes that while the dental industry demands high customization, 3D printing currently offers the best solution for both customization and mass production needs. However, for specific areas requiring exceptional accuracy and surface quality, CNC machining remains irreplaceable. The future holds promise for 3D printing to potentially replace CNC machining entirely, provided advancements continue in 3D printing technology and materials.

15.3 Literature Review

3D printing has become a game-changer in many industries, including dentistry. In the past, dentists used expensive and time-consuming methods to create custom parts. But 3D printing is faster, cheaper, and more accurate [23]. New technology like 3D scanners and design software allows dentists to create highly precise models of teeth and jaws. This lets them print custom crowns, teeth, and even surgical guides. 3D printing is better than old-fashioned techniques because it can create complex shapes that traditional methods. This is especially useful for tissue engineering, where 3D printing can create scaffolds to help new tissue grow. Studies have shown that 3D-printed teeth are just as good as, or even better than, traditionally made teeth [24]. This makes 3D printing a valuable tool for dentists and their patients. This research will compare two methods for making temporary crowns: 3D printing and a traditional technique. The researchers will use a computer-designed model to create the crowns [25]. Dentists create custom structures to help tissues heal better. These structures can be designed in unique shapes and sizes, and they're strong enough to support the healing process. 3D printing works by building things up layer by layer, kind of like icing a cake. This method is being used in many different fields, including dentistry. In dentistry, 3D printing can be used to make implants, dentures, and even models of teeth. This can help dentists plan surgeries, train for procedures, and create more accurate restorations. 3D printing is even more precise than older methods used to make teeth. For example, one study showed that 3D printing dentures with a special ceramic material (zirconia) resulted in dentures that were stronger and took less time to make. This study will compare two ways of making temporary crowns for teeth: 3D printing and a more traditional method. Both methods will start with a computer-generated design of the crown.

15.3.1 Applications of Additive Manufacturing in Dentistry

The emergence of 3D printing has significantly transformed dentistry, enabling dental professionals to create personalized dental solutions. This cutting-edge technology has revolutionized the field by enhancing efficiency, precision, and even opening doors for further medical exploration.

(i) **Implants made easy**: 3D printing plays a critical role in crafting patient-specific 3D models for replacing missing teeth. This method surpasses traditional techniques by offering superior surgical accuracy, improved comfort, and cost-effectiveness. Additionally, 3D-printed models boast exceptional surface finishes.

(ii) **Repairing and replacing teeth**: Dentists offer patients a more comfortable experience with digital impressions and can now swiftly assess a patient's mouth and generate a digital 3D model of their teeth and gums. Using dental computer-aided design (CAD) software, they can virtually design the required tooth, and 3D printing brings that design to life, facilitating a smoother workflow.

(iii) **Surgical guides with precision**: This innovative technology allows for the creation of highly accurate surgical guides. By scanning a patient's teeth and converting the data into a CAD format, 3D printing produces a physical guide for optimal surgical precision.

(iv) **Orthodontic aligners on demand**: 3D printing streamlines orthodontic treatment for misaligned teeth and jaws. Dentists can now scan a patient's teeth, design a customized orthodontic appliance using CAD software, and directly print it, expediting the treatment process.
(v) **Custom molding for implants**: 3D printing offers the unique ability to create personalized molds for intricate implants. These molds enable the efficient production of custom implants using various materials within a short time frame.
(vi) **Tailored surgical tools**: The technology extends beyond dental implants. Dentists can leverage 3D printing to design and create customized surgical instruments suited to their specific needs, fostering greater efficiency during procedures.

15.4 Experimentation Work

The goal of CAD/CAM dentistry experimentation work is to attain the best possible outcomes by testing and improving the design and production process. This includes evaluating different materials, software settings, and machining parameters to ensure accuracy, precision, and aesthetic appeal of the final dental restoration.

15.4.1 Material and Methods

This study explored the design and 3D printing process for creating a permanent tooth reference model. A 3D intraoral scanner with 67 microns of precision was used to capture a digital scan of a patient's mouth. The scan data was converted into a standard tessellation language (STL) file, a common format used in 3D printing for generating instructions for creating physical models (refer Fig. 15.2). 3D modeling software was employed to design reference markers on the scanned dental STL files. These modified files were then used for 3D printing the tooth model with a digital light processing technique (details in Fig. 15.3) using figure 4 standalone 3D printer (Figs. 15.4 and 15.5). For comparison, a separate tooth milling model was manufactured using an X-Mill 500 milling machine (Korea) based on the reference STL file. Zirconia was the material used for this milled tooth. Burs with diameters ranging from 1mm to 2.5 mm were used for milling (Figs. 15.6 and 15.7). To ensure consistency, a single set of burs was used to mill a single zirconia block.

Following fabrication, both the 3D printed and milled models were examined for surface roughness. A surface profile meter was employed to measure the roughness of both 2D and 3D surfaces. To confirm the 2D data, measurements were also taken with a 2D contact surface profile meter. The 2D roughness profiles were captured perpendicular to the feed direction during milling. Two surface measurements were obtained per specimen, and roughness values were determined from the center of

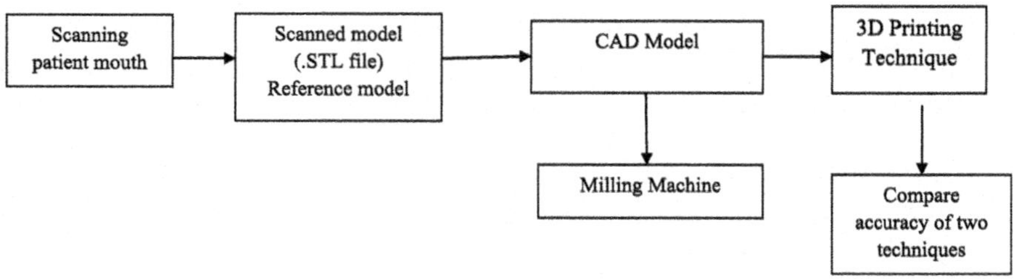

Fig. 15.2 Flow diagram of implementing project work

Fig. 15.3 Standalone 3D printing machine

the milled areas. Average roughness (Ra), maximum peak-to-valley height (Rq), and mean peak-to-valley height (Rz) were used as criteria to assess the 2D roughness [26]

15.4.2 Results and Discussion

The accuracy of two manufacturing methods was assessed using a profile meter. The results are detailed in Tables 15.1 and 15.2. Similarly, Table 15.3 represent the trueness value of the tooth created by both technique and Table 15.4 presents

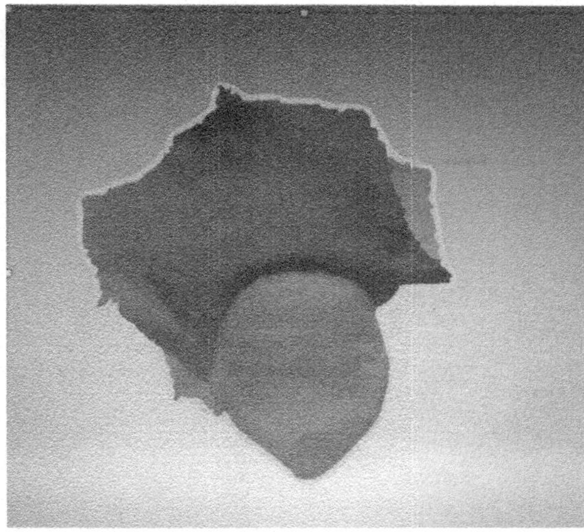

Fig. 15.4 STL model of dental tooth manufactured by 3D printing machine

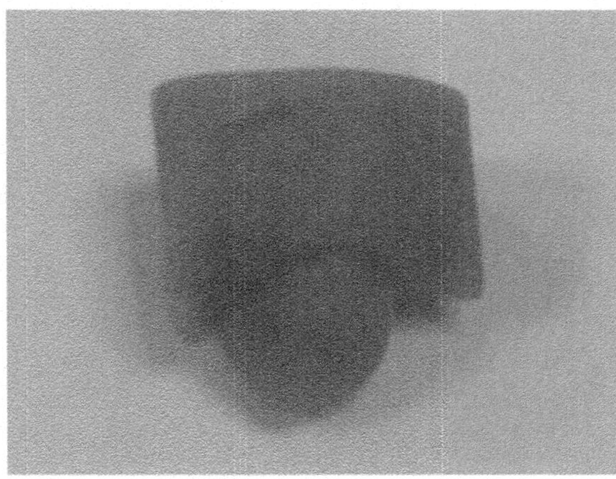

Fig. 15.5 Dental tooth manufactured by 3D printing machine

a comparison of the manufactured teeth. Notably, the Ra values for the models produced by the additive manufacturing 3D printer were consistently lower than those manufactured using a milling machine. This trend was observed throughout the measurement of both types of models using designated measuring points.

The field of dentistry is witnessing significant progress in the use of computer-aided design (CAD) and manufacturing (CAM) technologies, alongside 3D printing, for creating dental prostheses. These methods offer a promising alternative to traditional casting techniques, which were often plagued by issues like deformation, shrinkage, and expansion of the final restoration [27]. While the initial introduction of CAD/CAM systems in dentistry yielded solutions with limited clinical applicability, advancements in scanning technology, design software, and machining

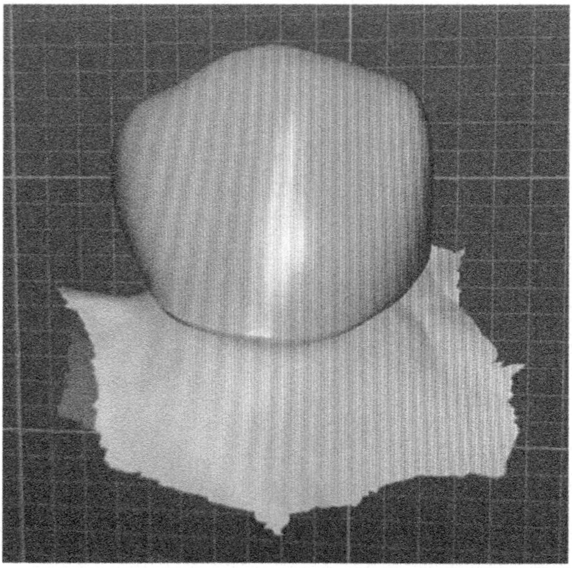

Fig. 15.6 CAD model of dental tooth manufactured by milling machine

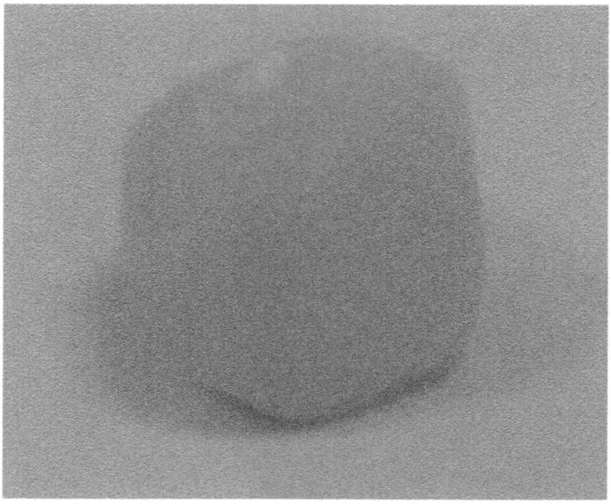

Fig. 15.7 Dental tooth manufactured by milling machine

Table 15.2 RMS value of tooth model using 3D printing machine

Sr. no	Ra value of 3D printing machine tooth		
	Ra	Rq	Rz
1	0.872	1.251	7.592
2	1.268	1.567	7.198

Table 15.3 Trueness value of tooth model using milling machine and 3D printing

Trueness	Minimum value	Maximum value	Average mismatch (mm)
Milling machine	−12	11.80	0.020
3D printing	−12	11.90	0.010
P value			

Table 15.4 Comparison of cost between 3D pining and milling

Sr no.	3D printing technique	Milling machine
1	Low	High

processes have led to a significant improvement in their effectiveness [28]. Today, CAD/CAM-based prosthetics are widely accepted for their clinical suitability.

Similar advancements are being observed in 3D printing technologies. Both milling and 3D printing are finding extensive applications in dental prosthesis production. However, existing comparative studies primarily focus on the fit of prosthetics created using these techniques. The rise of intraoral scanners has paved the way for model-less prosthesis fabrication using CAD systems. However, studies evaluating the replica technique for assessing 3D-printed zirconia teeth (using a milling machine) highlight the need for further improvements in standardization, reproducibility, and efficiency. The working model plays a crucial role in ensuring the longevity and stability of dental prostheses. This study compared the accuracy and trueness of two manufacturing methods:

(i) 5-axis milling machine: This method yielded a higher surface roughness value ($Ra = 1.237$) for the fabricated teeth.
(ii) DLP-based 3D printer with 50 μm layer thickness: This method produced teeth with a lower surface roughness value ($Ra = 0.872$).

These findings suggest that 3D printing offers a statistically significant advantage over milling in terms of surface smoothness and trueness. The data also indicates a greater difference in surface roughness with the milling method compared to 3D printing [29].

(i) **Manufacturing Cost**: Manufacturing cost of tooth created from 3D printing is low as compared to conventional machine.

15.5 Conclusion

This study concludes that milling machines produce dental teeth with a higher surface finish accuracy compared to DLP-based 3D printing machines, with Ra, Rq, and Rz values of 1.237, 1.492, and 6.08, respectively. These findings have significant implications for the dental industry, enabling dentists to make informed decisions when selecting a technique for their patients, ultimately leading to improved treatment outcomes and patient satisfaction.

15.5.1 Current Challenges and Future Directions

The dental industry is undergoing a significant shift towards digital workflows, with 3D printing emerging as a favorable alternative to traditional substractive manufacturing techniques used in computer-aided manufacturing (CAM). This technology offers several advantages, including the creation of highly accurate dental models with minimal material waste, using materials like metals, plastics, and ceramics. However, there are some limitations to address. Certain materials used for surgical guides might not be compatible with sterilization procedures, hindering their application. Additionally, the precision of 3D-printed models hinges on the quality of intraoral scans. These scans may have limitations for full arch captures or uneven surfaces. As the adoption of 3D printing and intraoral scanners continues to grow, the dental field is poised for a more digital future. This shift offers potential benefits such as reduced physical storage needs for dental records. However, it also necessitates careful consideration of ethical concerns surrounding patient data privacy and confidentiality, as digital information becomes more readily accessible for research purposes.

References

1. Tian, Y., et al.: A review of 3D printing in dentistry: Technologies, affecting factors, and applications. Scanning **1**, 9950131 (2021)
2. Jandyal, A., et al.: 3D printing–A review of processes, materials and applications in industry 4.0. Sustain. Oper. Comput. **3**, 33–42 (2022)
3. Gupta, D.K., Tuli, A., Jain, A.: 3D printed material application in orthodontics. Mater. Today Proc. **28**, 1635–1642 (2020)
4. Cicciù, M., et al.: 3D digital impression systems compared with traditional techniques in dentistry: a recent data systematic review. Materials **13**(8), 1982 (2020)
5. Alammar, A., Att, W., Beuer, F.: The accuracy of 3d-printed fixed dental restorations (2024)
6. Padrós, R., et al.: Mechanical properties of cocr dental-prosthesis restorations made by three manufacturing processes influence of the microstructure and topography. Metals **10**(6), 788 (2020)
7. Camargo, B., et al.: 3D printing and milling accuracy influence full-contour zirconia crown adaptation. Dent. Mater. **38**(12), 1963–1976 (2022)
8. Pavan Kalyan, B.G., Kumar, L.: 3D printing applications in tissue engineering, medical devices, and drug delivery. Aaps Pharmscitech **23**(4), 92 (2022)
9. Lee, J.D., et al.: Facial scanners in dentistry: an overview. Prosthesis **4**(4), 664–678 (2022)
10. Shi, H.Y., et al.: Overview of several typical ceramic materials for restorative dentistry. BioMed Res. Int. 2022(1), 8451445 (2022)
11. Hou, X., et al.: An overview of three-dimensional imaging devices in dentistry. J. Esthetic Restorative Dent. **34**(8), 1179–1196 (2022)

12. Alsawaf, A.F.B.: Computer aided design/Computer aided manufacturing and 3D printing for fabrication of complete dentures. Diss. University of Baghdad (2023)
13. Yau, H.T., Yang, T.J., Lin, Y.K.: Comparison of 3-D printing and 5-axis milling for the production of dental e-models from intra-oral scanning. Comput. Aided Des. Appl. **13**(1), 32–38 (2016)
14. Alauddin, M.S., Baharuddin, A.S., Mohd Ghazali, M.I.: The modern and digital transformation of oral health care: a mini review. Healthcare **9**(2) (2021)
15. Jiménez, M., et al.: Additive manufacturing technologies: an overview about 3D printing methods and future prospects. Complexity **2019**(1), 9656938 (2019)
16. Zoabi, A., et al.: 3D printing and virtual surgical planning in oral and maxillofacial surgery. J. Clin. Med. **11**(9), 2385 (2022)
17. Wang, J., et al.: Recent advances in digital technology in implant dentistry. J. Dent. Res., 00220345241253794 (2024)
18. Cen, Y., et al.: Application of three-dimensional reconstruction technology in dentistry: a narrative review. BMC Oral Health **23**(1), 630 (2023)
19. Budzik, G., et al.: Analysis of the accuracy of reconstructed two teeth models manufactured using the 3DP and FDM technologies. Strojniški vestnik-J. Mech. Eng. **62**(1), 11–20 (2016)
20. Sheng, R.: 3D Printing: A Revolutionary Process for Industry Applications. Woodhead Publishing (2022)
21. Angelone, F., et al.: Diagnostic applications of intraoral scanners: a systematic review. J. Imaging **9**(7), 134 (2023)
22. Kot-Thompson, D.: 5-axis milling for micro-machining of implantable flow sensor (2022)
23. Nulty, A.B.: 3D printing part 1: A history and literature review of 3D printing in dentistry (2021)
24. Mahrous, A., et al.: A comparison of pre-clinical instructional technologies: natural teeth, 3D models, 3D printing, and augmented reality. J. Dent. Educ. **85**(11), 1795–1801 (2021)
25. Deswal, H., et al.: Use of additive manufacturing in surgical tools/guides for dental implants. In: Additive Manufacturing with Medical Applications, pp. 41–71. CRC Press (2022)
26. Chavali, R., Nejat, A.H., Lawson, N.C.: Machinability of CAD-CAM materials. J. Prosthet. Dent. **118**(2), 194–199 (2017)
27. Rezaie, F., et al.: 3D printing of dental prostheses: current and emerging applications. J. Compos. Sci. **7**(2), 80 (2023)
28. Valenti, C., et al.: Mechanical properties of 3D printed prosthetic materials compared with milled and conventional processing: a systematic review and meta-analysis of in vitro studies. J. Prosthet. Dentist. **132**(2), 381–391 (2024)
29. Panda, S.K., et al.: Revolutionizing product development: The growing importance of 3D printing technology. Mater. Today Proc. (2023)

Dr. Surendra Singh Thakur, has 20 years of academic and industrial experience and he is currently working as a Senior Project Engineer at 3D Engineering Automation LLP, Bhopal. Earlier he has worked as Associate professor in various engineering institute. He did his B.Tech (Mechanical Engineering) from National Institute of Technology, Surat (Gujrat) and did Ph.D. from MATS University, Raipur. His research contributions include Non Conventional Machining, Advance Material, Additive Manufacturing, Internal Combustion Engine.

Dr. Sharad Kumar Pradhan is working as Director, THDC-Institute of Hydropower Engineering and Technology, Tehri, Govt of Uttrakhand on deputation from NITTTR Bhopal, where he was Professor, Mechanical Engineering, Dean, Corporate & International Relations and Associate Director, NITTTR Bhopal-Siemens Center of Excellence at National Institute of Technical Teachers' Training and Research (NITTTR), Bhopal under the aegis of Government of India. He has developed Outcome Based Curriculum of Mechanical Engineering and allied disciplines for four states of the country and contributed in four MOOCs (Instructional Planning and Delivery NITTT-AICTE, Advanced Instructional Methods, Accreditation of PG Engineering Programme, Accreditation of UG Engineering Programme, 3D Printing and Design). His research contributions include Finite element model updating, Structural dynamic design, Optimization of machining parameters, Simulation of Manufacturing Processes, Robotics and Additive Manufacturing. He has a number of research papers, one book, four patents and a DST funded project to his credit.

Dr. S. S. Mahapatra is affiliated to Department of Mechanical Engineering, National Institute of Technology, Rourkela. He is is currently providing services as Professor. He has authored and co-authored multiple peer-reviewed scientific papers and presented works at many national and International conferences. His contributions have acclaimed recognition from honourable subject experts around the world. He is actively associated with different societies and academies. His academic career is decorated with several reputed awards and funding. His research interests include Industrial Engineering Group: [Production Engineering Division].

Emmanuel Philip Ododo is an Assistant Lecturer at the University of Uyo, Nigeria, specializing in computer Science. He has contributed significantly to academia with several publications in reputable journals. His scholarly contributions extend to book chapters also. Emmanuel has presented papers at various national and international conferences, showcasing his research. His work reflects a commitment to advancing knowledge, fostering innovation, and inspiring academic growth.

The manufacturer's authorised representative in the EU is Springer Nature Customer Service Centre GmbH, Europaplatz 3, 69115 Heidelberg, Germany. If you have any concerns regarding our products, please contact ProductSafety@springernature.com

Printed and bound by CPI Group (UK) Ltd, Croydon, CR0 4YY
30/09/2026
02225500-0005